# Advances in Carbon Nanotubes: Synthesis, Properties, and Cutting-Edge Applications

# Advances in Carbon Nanotubes: Synthesis, Properties, and Cutting-Edge Applications


Guest Editors

**Guohai Chen**
**Dai-Ming Tang**



MDPI

Basel • Beijing • Wuhan • Barcelona • Belgrade • Novi Sad • Cluj • Manchester

*Guest Editors*
Guohai Chen
National Institute of Advanced Industrial Science and Technology (AIST)
Ibaraki
Japan

Dai-Ming Tang
Research Center for Materials Nanoarchitectonics (MANA)
National Institute for Materials Science (NIMS)
Tsukuba
Japan

*Editorial Office*
MDPI AG
Grosspeteranlage 5
4052 Basel, Switzerland

This is a reprint of the Special Issue, published open access by the journal *Nanomaterials* (ISSN 2079-4991), freely accessible at: https://www.mdpi.com/journal/nanomaterials/special_issues/52U2MW2O08.

For citation purposes, cite each article independently as indicated on the article page online and as indicated below:

| Lastname, A.A.; Lastname, B.B. Article Title. *Journal Name* **Year**, *Volume Number*, Page Range. |
|---|

**ISBN 978-3-7258-7707-2 (Hbk)**
**ISBN 978-3-7258-7708-9 (PDF)**
**https://doi.org/10.3390/books978-3-7258-7708-9**

Cover image courtesy of Dai-Ming Tang

# Contents

# About the Editors

**Guohai Chen**

Guohai Chen is a Chief Senior Researcher at the Nanocarbon Material Research Institute, National Institute of Advanced Industrial Science and Technology (AIST), Japan. His research focuses on the synthesis, characterization, and application of nanomaterials, particularly carbon nanotubes. His current work emphasizes machine learning-assisted strategies for achieving high crystallinity and high yield CNT synthesis, as well as the development of CNT-based electronic, thermal, and electrical materials.

**Dai-Ming Tang**

Dai-Ming Tang is a Principal Researcher at the National Institute for Materials Science (NIMS), Japan. Their research fields include nanomechanics, nanoelectronics, carbon nanotubes, and in situ electron microscopy.

*nanomaterials* MDPI

*Editorial*

# Advances in Carbon Nanotubes: Synthesis, Properties, and Cutting-Edge Applications

**Guohai Chen [1,*] and Dai-Ming Tang [2,3,*]**

1 Nanocarbon Material Research Institute, National Institute of Advanced Industrial Science and Technology (AIST), Tsukuba Central 5, 1-1-1 Higashi, Tsukuba 305-8565, Japan
2 Research Center for Materials Nanoarchitectonics (MANA), National Institute for Materials Science (NIMS), Tsukuba 305-0044, Japan
3 Institute of Pure and Applied Sciences, University of Tsukuba, Tsukuba 305-8571, Japan
* Correspondence: guohai-chen@aist.go.jp (G.C.); tang.daiming@nims.go.jp (D.-M.T.)

Carbon nanotubes (CNTs) have remained at the forefront of nanoscience for more than three decades, owing to their unique cylindrical structures, exceptional physical, chemical, and mechanical properties, and broad potential in electronics, energy, composites, and biomedical applications. Significant advances have been achieved in scalable synthesis, structural control, and multifunctional integration. More recently, the rapid development of artificial intelligence (AI) and machine learning (ML) methodologies has opened new opportunities to optimize CNT growth, tailor structures, and predict performance with unprecedented precision.

This Special Issue of *Nanomaterials*, "Advances in Carbon Nanotubes: Synthesis, Properties, and Cutting-Edge Applications," assembles ten contributions from several research groups worldwide [1–10]. The collection comprises five research articles and five review articles that span fundamental synthesis studies, advanced characterizations, and diverse application demonstrations, reflecting both the maturity of the field and the emergence of new directions.

## 1. Synthesis and Structural Engineering

A central topic in CNT research remains the precise control of synthesis and its structures. Several contributions focus on refining CNT growth strategies and tailoring their structures for targeted applications. Shimizu et al. [1] introduced a chromatographic approach to effectively assess CNT–organic interactions, enabling improved CNT dispersibility in suspensions. Their study provides a versatile platform for understanding interfacial chemistry, which is critical for tailoring CNT compatibility with polymers, solvents, and functional molecules in advanced applications. Matsumoto et al. [2] reported the direct growth of CNTs on aluminum nitride substrates, demonstrating a marked improvement in composite thermal conductivity with reduced filler ratios. This approach represents a significant advancement in the design of thermally conductive materials, with potential implications for a wide range of applications. Nwanno et al. [3] fabricated copper-filled vertically aligned CNTs (VACNTs) directly on copper foil substrates, achieving enhanced field emission performance. Their findings demonstrate the viability of thin copper substrates in creating dense and highly conductive copper-filled VACNT arrays for advanced electronic and nanoelectronics applications.

 https://doi.org/10.3390/nano15201595

## 2. Properties from Nano to Macro

Translating the intrinsic properties of CNTs into macroscopic performance remains another core challenge. Xiang et al. [4] comprehensively review the electrical properties and measurement techniques of CNTs, spanning from individual nanotubes to macroscopic assemblies. They highlight the difficulties in transferring the electrical properties from nanoscale to bulk materials and proposed strategies for future research and development directions to boost the electrical conductivity of CNT assemblies. Rezaee et al. [5] fabricated single-walled CNT (SWCNT) thin films using the floating catalyst chemical vapor deposition method and systematically explored their electrical characteristics after acid treatment through Hall effect measurements. Their study offers new insights into the correlation between structural, electrical, and optical properties, contributing to a deeper understanding of structure–property relationships.

## 3. Application-Oriented Developments

The versatility of CNTs is further illustrated through application-focused contributions. Ishihara et al. [6] developed a purge-free, actuator-driven formaldehyde gas sensor based on SWCNT chemiresistors. By leveraging periodic actuation, the sensor achieved reliable detection of formaldehyde gas over a wide concentration range (0.05–15 ppm) with excellent selectivity over other volatile organic compounds and stability, underscoring the utility of CNTs in environmental monitoring. Almansoori et al. [7] present a comprehensive review of CNT/graphene-reinforced ceramics, detailing how these hybrid materials can achieve superior toughness, hardness, and thermal stability compared to conventional ceramics, while also highlighting challenges in processing, scalability, and interfacial compatibility that must be addressed for practical applications. Alfei et al. [8] review antimicrobial CNT composites, emphasizing their potential in medical, packaging, and environmental applications. They explore how CNTs can inhibit microbial growth while also addressing concerns of cytotoxicity, environmental impact, and regulatory acceptance, thereby offering balanced insights into performance and safety considerations. Snowdon et al. [9] review strategies to enhance product durability of CNT-based materials, focusing on synergistic molecular assembly, intrinsic and engineered self-repair, and advanced characterization techniques. By identifying current challenges and future research frontiers, the review underscores that the creation of truly durable materials depends on an integrated understanding of how to build, repair, and precisely measure CNT-based systems.

## 4. Emerging Trends and AI Integration

A particularly exciting frontier is the integration of AI and ML into CNT research. Chen et al. [10] review the role of ML as a "catalyst" for accelerating CNT research, covering areas such as synthesis optimization, characterization, property prediction, and application development. The review highlights how data-driven approaches can close the loop between experiment and theory, potentially enabling autonomous laboratories for discovery and design of CNT-related research. Such approaches are likely to become increasingly important for accelerating innovation in nanomaterials science.

## 5. Outlook

Taken together, the contributions in this Special Issue capture the dynamic progress of CNT research. They reveal how advances in synthesis and structural engineering, the translation of properties from nano to macro scales, and the expansion of applications into sensing, composites, and biomedicine are converging to unlock the full potential of CNTs. At the same time, emerging AI-driven methodologies are reshaping how the field approaches discovery and optimization.

We hope this collection not only highlights the representative advances in CNT research but also inspires future interdisciplinary innovation. In particular, the integration of AI with CNT science has the potential to redefine material design, accelerate translation into real-world technologies, and drive the next breakthroughs in nanomaterials.

Looking forward, key priorities include bridging the gap between nanoscale performance and scalable industrial implementation, developing sustainable and cost-effective synthesis routes, integrating AI-guided design into mainstream workflows, and addressing safety, environmental, and regulatory considerations to support broader commercialization. With these combined efforts, CNTs will continue to inspire and enable transformative technologies in the years ahead.

**Author Contributions:** Conceptualization, G.C. and D.-M.T.; investigation, G.C. and D.-M.T.; writing—original draft preparation, G.C.; writing—review and editing, G.C. and D.-M.T.; funding acquisition, G.C. and D.-M.T. All authors have read and agreed to the published version of the manuscript.

**Funding:** This work was supported by JSPS KAKENHI Grant Number JP23K04552. D.T. discloses support from JSPS Kakenhi (grants JP25820336, JP20K05281, JP23H01796), JST-FOREST Program (grant JPMJFR223T, Japan), WPI-MANA 'Challenging Research Program (CRP)', NIMS 'Support system for curiosity-driven research', and "Advanced Research Infrastructure for Materials and Nanotechnology in Japan (ARIM)" of the Ministry of Education, Culture, Sports, Science and Technology (MEXT). Proposal Number JPMXP1224NM5238.

**Data Availability Statement:** Not applicable.

**Acknowledgments:** We extend our sincere gratitude to all the authors for their excellent contributions and to the reviewers for their valuable time and constructive feedback, which ensured the high quality of this Special Issue. We also acknowledge the *Nanomaterials* editorial team for their professional and efficient management and guidance.

**Conflicts of Interest:** The authors declare no conflicts of interest.

## References

1. Shimizu, T.; Kishi, R.; Hirano, A.; Kokubo, K.; Hata, K. Chromatographic Assessment of Organic Compounds Using Carbon Nanotubes: The Relationship between Affinity and Dispersibility. *Nanomaterials* **2024**, *14*, 824. [CrossRef] [PubMed]
2. Matsumoto, N.; Futaba, D.N.; Yamada, T.; Kokubo, K. Enhancing the Thermal Conductivity of CNT/AlN/Silicone Rubber Composites by Using CNTs Directly Grown on AlN to Achieve a Reduced Filler Filling Ratio. *Nanomaterials* **2024**, *14*, 528. [CrossRef] [PubMed]
3. Nwanno, C.E.; Thapa, A.; Watt, J.; Bendayan, D.S.; Li, W.Z. Field Emission Properties of Cu-Filled Vertically Aligned Carbon Nanotubes Grown Directly on Thin Cu Foils. *Nanomaterials* **2024**, *14*, 988. [CrossRef] [PubMed]
4. Xiang, Y.X.; Zhang, L.L.; Liu, C. Electrical Properties of Carbon Nanotubes: From Individual to Assemblies. *Nanomaterials* **2025**, *15*, 1165. [CrossRef] [PubMed]
5. Rezaee, M.D.; Dahal, B.; Watt, J.; Abrar, M.; Hodges, D.R.; Li, W.Z. Structural, Electrical, and Optical Properties of Single-Walled Carbon Nanotubes Synthesized through Floating Catalyst Chemical Vapor Deposition. *Nanomaterials* **2024**, *14*, 965. [CrossRef] [PubMed]
6. Ishihara, S.; Chahal, M.K.; Labuta, J.; Tanaka, T.; Kataura, H.; Hill, J.P.; Nakanishi, T. Actuator-Driven, Purge-Free Formaldehyde Gas Sensor Based on Single-Walled Carbon Nanotubes. *Nanomaterials* **2025**, *15*, 962. [CrossRef] [PubMed]
7. Almansoori, A.; Balázsi, K.; Balázsi, C. Advances, Challenges, and Applications of Graphene and Carbon Nanotube-Reinforced Engineering Ceramics. *Nanomaterials* **2024**, *14*, 1881. [CrossRef] [PubMed]
8. Alfei, S.; Schito, G.C. Antimicrobial Nanotubes: From Synthesis and Promising Antimicrobial Upshots to Unanticipated Toxicities, Strategies to Limit Them, and Regulatory Issues. *Nanomaterials* **2025**, *15*, 633. [CrossRef] [PubMed]

9. Snowdon, M.R.; Rathod, S.; Liang, R.F.L.; Freire-Gormaly, M. Architecting Durability: Synergies in Assembly, Self-Repair, and Advanced Characterization of Carbon Nanotube Materials. *Nanomaterials* **2025**, *15*, 1352. [CrossRef] [PubMed]
10. Chen, G.; Tang, D.-M. Machine Learning as a "Catalyst" for Advancements in Carbon Nanotube Research. *Nanomaterials* **2024**, *14*, 1688. [CrossRef] [PubMed]

nanomaterials

MDPI

*Article*

# Chromatographic Assessment of Organic Compounds Using Carbon Nanotubes: The Relationship between Affinity and Dispersibility

**Taiyo Shimizu [1], Ryoichi Kishi [1,*], Atsushi Hirano [2], Ken Kokubo [1] and Kenji Hata [1]**

[1] Nano Carbon Device Research Center, National Institute of Advanced Industrial Science and Technology (AIST), AIST Tsukuba Central 5, 1-1-1 Higashi, Tsukuba 305-8565, Ibaraki, Japan; shimizu-t@aist.go.jp (T.S.)

[2] Nanomaterials Research Institute, National Institute of Advanced Industrial Science and Technology (AIST), AIST Tsukuba Central 5, 1-1-1 Higashi, Tsukuba 305-8565, Ibaraki, Japan

* Correspondence: r-kishi@aist.go.jp

**Abstract:** The affinity between carbon nanotubes (CNTs) and organic compounds is of substantial importance since it strongly relates to the dispersibility of CNTs in those compounds. Several affinity evaluation methods have been developed so far, and the concept of the Hansen solubility parameter is a representative method widely used in the field of nanocarbon materials. Here, we demonstrate that CNT-loaded silica columns can effectively assess the affinity of organic compounds for CNT surface by exploiting the chromatographic retention time as a criterion. Obtained trends of the affinity of organic compounds for CNT were compared to those based on Hansen solubility parameter distance values. Most organic compounds showed similar trends, but one exceptional compound was observed. Simple CNT dispersion tests were conducted with these organic compounds to demonstrate the advantage of the chromatographic assessment. Further, we conducted comparison experiments using a pyrene-functionalized column and other CNT-loaded columns to elucidate the characteristics of each CNT column. The chromatographic approaches using CNT columns would be beneficial for realizing CNT suspensions with improved CNT dispersibility.

**Keywords:** carbon nanotube; chromatography; retention time; Hansen solubility parameter; dispersibility

## 1. Introduction

The affinity between carbon nanotubes (CNTs) and organic molecules is of crucial importance because CNTs are prone to agglomerate in any media without the aid of dispersant (e.g., surfactant) [1]. Insufficiently dispersed CNTs do not display their superior properties such as high electrical conductivity, thermal conductivity, and so on. To fabricate composite materials exploiting the full potentials of CNTs, for instance, requires efficient exfoliation of CNTs [2,3]. Basically, media possessing high affinity to CNTs can disperse them more stably than media with low affinity. Determining the affinity between them can lead to a rational design for experimental systems.

Several methods for determining affinities between CNTs and organic compounds have been developed so far. One of the representative methods for determining the affinity is based on solubility parameters. The well-known Hildebrand solubility parameter has been widely used for estimating the solubility of polymers in specific solvents. The Hildebrand solubility parameters have been modified to Hansen solubility parameters (HSPs), composed of three parameters ($\delta_D$: dispersion term; $\delta_P$: dipolar term; $\delta_H$: hydrogen bonding term) divided from the Hildebrand ones [4]. In addition to the prediction of polymer solubility, Hansen solubility parameters have been widely investigated in terms of dispersion stability of nanomaterials not limited to spherical inorganic nanoparticles [5], but including nanocarbons such as CNTs [6], graphene [7], graphene oxide [8], nano-onion [9], and so on. In particular, CNTs are principally not dispersible in most solvents, and thus the estimation of solubility parameters for them possesses practical importance to survey the dispersibility to diverse kinds of solvents. Determining the HSPs of CNTs, however, requires a lot of experimental data on CNT dispersions using diverse kinds of solvents with different HSPs, which takes a lot of effort. In addition, the methodology to determine the degree of dispersibility, which is required to calculate HSPs, relies on the discretion of individual researchers and lacks consensus [10,11]. Although HSPs have advantages in predicting the affinity (dispersibility) not only for single solvents but also for solutions composed of multiple solvents, the methodology could be problematic when dealing with diverse kinds of CNTs.

Another potential method to determine the affinity between CNT and organic compounds is exploiting CNTs as a stationary phase for column chromatography. In the initial stage of the research, column chromatography using CNTs as a stationary phase has been investigated to utilize the high specific surface area of CNTs [12]. CNTs intrinsically possess high surface areas and thus the interaction between CNT surface and organic molecules is expected to effectively affect the retention of these molecules. Immobilization of CNTs has been made by methods such as direct growth on column beads [13], covalent attachment via chemical modifications [14–16], and so on. Organic compounds such as small polycyclic aromatic hydrocarbons [15] can be separated due to the strong interactions such as $\pi$-$\pi$ interactions between those compounds and CNT surfaces. After that, several studies have investigated the applicability of chromatography using noncovalently loaded CNT columns to quantitatively determine the affinities of organic compounds [17,18]. Retention of organic molecules on columns originates from the interaction between the CNT surface and molecules, and thus strong retention implies their strong interaction, namely strong affinity. These techniques have the potential to realize the facile and rapid determination of the affinity of organic compounds, which could aid in the selection of appropriate dispersing media for CNTs. However, the number of research studies concerning this characterization method is limited and thus the actual influence on dispersibility in selected solvents is still not clear.

In this work, we have fabricated liquid chromatography columns containing single-walled CNTs (SWCNTs) with a relatively large diameter (so-called supergrowth SWCNT) deposited on the amine-functionalized silica beads according to the literature [17,18] and characterized the affinity between CNTs and organic molecules. While previous research investigated the affinity of SWCNTs (HiPco CNTs) for the molecular design of CNT solubilizers [17], and to understand the interaction between protein and CNTs [18], our study has focused on the relationship between the chromatography-based affinity of organic compounds and its effect on the dispersibility of CNTs. Further, we have characterized HSPs as another evaluation methodology to quantify the affinity of CNTs and compared it to that quantified by the CNT column chromatography to extract the difference between these techniques. The comparison experiment using a pyrene-functionalized column and other CNT (oxidized SWCNT and multi-walled CNT (MWCNT)) columns was also conducted to elucidate the characteristics of these CNT columns.

## 2. Materials and Methods

### 2.1. Materials

In most parts of this manuscript, single-walled CNTs (ZEONANO® SG101, commercially available CNTs fabricated by the supergrowth procedure [19], produced by Zeon corporation, Tokyo, Japan) were used as a loaded material for column applications. For comparison, oxidized supergrowth SWCNT (fabricated by heating to 500 °C with a ramp rate at 1 °C/min under an air flow of 2 L/min) and MWCNT Knanos-400T (Kumho Petrochemical Co., Ltd., Seoul, Republic of Korea) were also used. Amine-functionalized silica beads CHEMCOBOND 3.5NH2 were purchased from Chemco Plus Scientific Co., Ltd., Osaka, Japan. *N*-methylpyrrolidone (NMP), methanol, toluene, *p*-xylene (p-Xy), 1,2,4-trimethylbenzene (1,2,4-TMB), p-chlorotoluene (p-CT), o-chlorotoluene (o-CT), p-bromotoluene (p-BT), N,N-dimethylacetamide, methyl ethyl ketone, acetophenone, chloroform, diethylene glycol, hexane, phenyl benzoate, naphthalene, 1-methylnaphthalene, and 2-methylnaphthalene were purchased from FUJIFILM Wako Pure Chemical, Osaka, Japan. All organic reagents were used without further purification. A pyrene-functionalized silica column (COSMOSIL 5PYE: pyrenylethyl groups are bonded to the surface) was purchased from Nakalai Tesque, Kyoto, Japan.

### 2.2. Fabrication Procedure of CNT Columns

Figure 1 shows the photographs of each step of the fabrication procedure for SWCNT columns used in this study, designed by referring to the literature [17,18]. Details of the procedure were as follows: first, SWCNTs were dispersed in NMP by a pre-treatment using an ultrasonic processor Nanoruptor NR-350 (Cosmo Bio Co., Ltd., Tokyo, Japan) for 1 min in a high-power mode and following intensive dispersion using an ultrasonic homogenizer Sonifier 250D Advanced (Branson, Brookfield, CO, USA) with 30% amplitude for 1 h under stirring. After the ultrasonication, obtained CNT dispersions were centrifuged by an ultracentrifuge Micro-ultracentrifuge GXII (Hitachi-koki, Tokyo, Japan) at 57,000 rpm (210,000× g) for 10 min, followed by extraction of supernatant (upper 3.5 mL from ca. 5 mL of dispersion and sediment) to be used for CNT loading. The CNT concentration of the obtained supernatant was adjusted to 0.025 mg/mL by adding an appropriate amount of NMP via referring to the calibration curves determined from UV-vis spectra obtained by using a UV-vis spectrometer UV-3600 (Shimadzu, Kyoto, Japan). Then, 20 mL of the diluted supernatant was mixed with amine-functionalized silica beads (0.50 g), and the mixture was rotated for 1 h using a rotator RT-5 (TAITEC, Saitama, Japan). Obtained CNT-loaded silica beads were packed into a rod as follows: 20 mL of the suspension was centrifuged at 3000 rpm (1650× g) for 3 min using a centrifuge H-36α (KOKUSAN, Saitama, Japan), and then 16 mL of the supernatant was replaced with chloroform/diethylene glycol (4:1 v/v). These centrifugation and supernatant exchange processes were repeated three times. Finally, 4 mL of the slurry was collected and then packed in a 2.1 mm i.d. × 125 mm empty column (Chemco Plus Scientific Co., Ltd.) by flowing chloroform/hexane (1:1 v/v) at 20 MPa. In this paper, the term "SWCNT column" indicates the column loading pristine supergrowth SWCNTs.

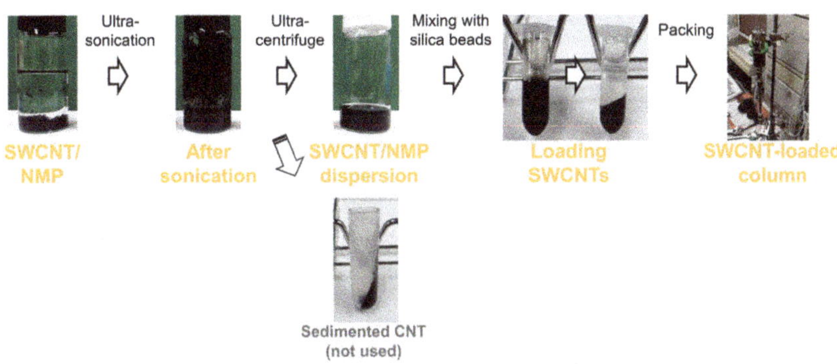


**Figure 1.** Photographic images of the preparation procedure of SWCNT columns.

The fabrication of columns containing other kinds of CNTs was carried out using the same procedure as described above. For MWCNTs, 50 mL of 0.05 mg/mL CNT supernatant was mixed with the silica beads.

### 2.3. Chromatographic Analysis

Fabricated CNT columns have been characterized by using a gel permeation chromatography system HLC-8320GPC EcoSEC (Tosoh Bioscience, Tokyo, Japan) operating as a liquid chromatography system. The eluent was 50/50 ($v/v$) water/methanol solution at 40 °C, and the flow rate was fixed to 0.2 mL/min. Analytes dissolved in methanol were injected and detected by UV light at the wavelength of 220 nm. Blank measurement was conducted without dissolving analytes. The values of retention times were adopted from the peak top of the chromatograms and adjusted retention times were calculated by subtracting holdup time determined by blank measurement from retention time.

### 2.4. Evaluation of Dispersibility of CNTs

Evaluation of dispersibility of CNT was conducted by dispersing CNTs by stirring and measuring the height of CNT agglomerates after dispersion: CNTs (20 mg) were dried under vacuum for 3 h at 180 °C, and then 40 mL of dispersing solvents were added to them, and the resulting mixtures were stirred by a magnetic stirrer for 24 h at 1000 rpm. The degree of dispersion was determined by dividing the height of the CNT agglomerations observed after 2 h from the dispersion process by the height of the dispersing solvent.

### 2.5. Other Characterizations

Optical microscope images were obtained with a digital microscope VHX-7000 (Keyence, Osaka, Japan). HSPs of organic compounds for the calculation of HSP distance values were adopted from a database in software for Hansen solubility parameter HSPiP (version 5.3.05).

## 3. Results and Discussion

Figure 2 shows the chromatograms of several kinds of aromatic compounds for a SWCNT column. A vertical axis of Figure 2 indicates the degree of detected absorption of UV light, closely related to the amount of analyte. Obviously, the retention times of all compounds differ from each other, indicating that each organic molecule interacts with the CNT surface to varying degrees. Toluene was detected at almost the same retention time as the blank, indicating almost no interaction with the CNT surface. Halogen-substituted compounds resulted in longer retention times than toluene, and Br substitution (for *p*-BT) had a stronger effect on the affinity than Cl substitution (*p*-CT and *o*-CT). These results are presumably derived from halogen–π interactions, which are reported in a paper investigating the interaction between halogen-substituted benzene and polycyclic aromatic hydrocarbons or fullerenes [20]. Further, while constitutional isomers (*p*-CT and *o*-CT) were detected at almost identical retention times, retention times of a di-substituted compound (*p*-Xy) and tri-substituted compound (1,2,4-TMB) were quite different. A similar tendency has been observed for porous graphitic carbons [21], and analysis showed that an increasing number of methyl substituents resulted in more negative changes in molar enthalpy on adsorption. These results indicate that the affinity to the CNT surface can be significantly varied by the number of substitution groups even when the polarity of the molecule is virtually not changed. In contrast, the chromatograms shown in the inset of Figure 2 exhibit almost the same retention time for all compounds. These results indicate that while the CNT surface can interact with aromatic compounds and the degree of interaction between them differs from each other, the amine-functionalized silica surface without CNTs possesses virtually no interaction with all investigated compounds. In this work, we immobilized supergrowth SWCNTs in the columns, which exhibit relatively low G/D ratios (low degrees of crystallinity). This structural feature will contribute to the diversity of the chemical environment on the CNT surface, such as the existence of structural defects, surface functional groups, etc., and presumably lead to different results from other SWCNTs. The difference in organic compound–CNT affinity derived from the chemical environments will be briefly discussed in the latter part of this article.

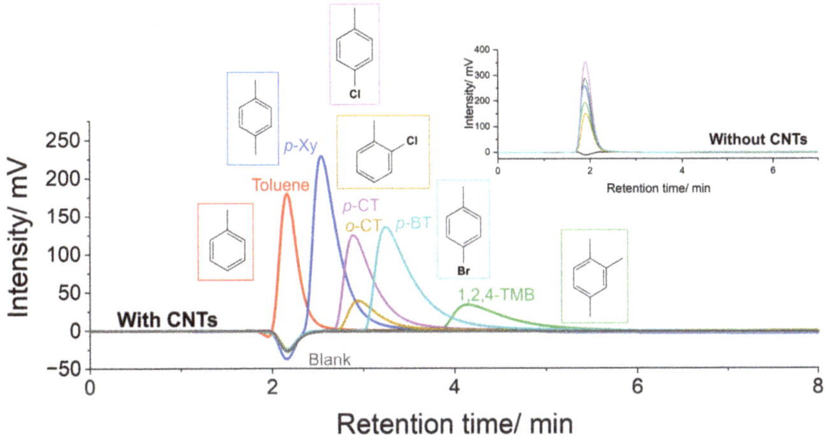


**Figure 2.** Liquid column chromatograms of several kinds of aromatic compounds using a SWCNT column. (inset) Liquid column chromatograms of several kinds of aromatic compounds using an amine-modified silica column without SWCNTs.

To compare the characterization of affinity with another analytical technique, we adopted the HSPs described in the Introduction. Figure 3a shows the concept of the HSP distance. As shown in Figure 3a, the HSP distance is essentially equal to the difference in the position in Hansen space, except for the factor of 4 on the dispersion term [4]. HSPs of used SWCNTs were adopted from the literature (dispersive term: $\delta_D$ = 19.4, polar term: $\delta_P$ = 6.0, hydrogen-bonding term: $\delta_H$ = 4.5) [22]. Basically, compounds with similar HSPs show high compatibility and, in many cases, high solubility or dispersibility among themselves. Thus, the above-defined HSP distance indicates the degree of incompatibility, and increasing HSP distance is prone to result in lower solubility (dispersibility).

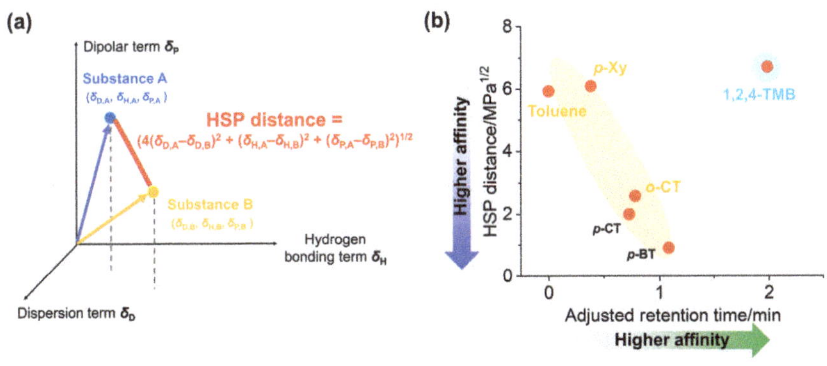


**Figure 3.** (**a**) Concept of the HSP distance in Hansen space. (**b**) Relationship between adjusted retention times obtained from SWCNT column and HSP distance values (from SWCNTs to organic compounds) of several kinds of aromatic compounds.

Figure 3b shows the relationship between HSP distance values calculated from HSPs in a database and adjusted retention times obtained from SWCNT column chromatography for several aromatic compounds shown in Figure 2. Adjusted retention times were calculated by subtracting holdup time from retention times. It can be observed that most compounds shown in Figure 3b showed the trend in which decreasing HSP distance values lead to increasing adjusted retention times (highlighted in orange). As expected, HSP distance and (adjusted) retention times are closely related, and the compounds with similar HSPs to that of CNTs are estimated to exhibit long retention times. However, it can be clearly seen that 1,2,4-TMB displays a specifically longer retention time than expected from its high HSP distance value (highlighted in blue). These results indicate that (adjusted) retention times and HSP distance values do not necessarily exhibit similar tendencies for all compounds.

In general, solvents exhibiting a high affinity for materials are expected to show high dispersibility for them. To demonstrate the efficacy of the chromatographic analysis in searching CNT dispersants, we performed simple dispersion tests of CNTs using toluene, *p*-Xy, *o*-CT, and 1,2,4-TMB; the first three solvents exhibited a relationship between adjusted retention times and HSP distance values, while the last displays longer adjusted retention time than expected (Figure 3b). The degree of exfoliation was determined under the same dispersion conditions; specifically, we compared the percentage of the height of CNT agglomerates sufficiently after (more than 40 min) the dispersion process (stirring), where the sedimentation of CNT agglomerates reached the stable states. The properties of these solvents and the results of dispersion tests are listed in Table 1. It should be noted that while the degree of dispersibility of CNTs by stirring can be improved by using dispersant with high viscosity (typically > 50 mPa s) irrespective of molecular structure [23], these four kinds of solvents exhibit similar and relatively low viscosity (0.59 mPa s to 0.9 mPa s, shown in Table 1), and thus the effect derived from viscosity on the dispersibility could be negligible. The appearance of the obtained CNT dispersions by stirring is shown in Figure 4. The height of the CNT is relatively low for toluene and *p*-Xy (39% and 41%, respectively), indicating low CNT dispersibility, but it is higher for 1,2,4-TMB (59%). Note that the apparently high dispersibility in *o*-CT (80%) can be attributed to the high density of this solvent (Table 1), which provides buoyancy to CNTs, in addition to its low HSP distance value. The dispersibility tendency (1,2,4-TMB > toluene ≈ *p*-Xy) is consistent with the order of retention times but it cannot be explained by the order of HSP distance values (toluene ≈ *p*-Xy ≈ 1,2,4-TMB). The inconsistency in the HSP data is ascribable to the fact that HSPs are typically calculated from the partial contributions of functional groups [24], and thus HSP differences in polarity between compounds with a similar chemical structure (toluene, *p*-Xy, and 1,2,4-TMB) are basically small. Thus, the (adjusted) retention times can be useful for the estimation of CNT dispersibility using solvents with a similar chemical structure. Note that in the case of more intensive dispersion using ultrasonication resulted in different situations, as shown in Figure S1 and Table S1 in Supplementary Materials. In this dispersing condition, the highest CNT dispersibility can be obtained when 1,2,4-TMB is used as a solvent. Although further research is needed, the CNT column approach demonstrated in this paper has a high potential for estimating the dispersibility of CNTs, especially in the case of using solvents with similar chemical structures and low polarity.

**Table 1.** Properties of dispersing solvents used in Figure 4 and their dispersibility for CNTs.

| | Toluene | *p*-Xylene (*p*-Xy) | 1,2,4-Trimethylbenzene (1,2,4-TMB) | *o*-Chlorotoluene (*o*-CT) |
|---|---|---|---|---|
| Viscosity/mPa s (20 °C) | 0.59 | 0.6475 | 0.9 | 0.884 (30 °C) |
| Density/g cm$^{-3}$ | 0.867 | 0.861 | 0.876 | 1.087 |
| CNT height percentage after stirring [1]/% | 39 | 41 | 59 | 80 |

[1] CNT concentration: 0.5 mg/mL.

Further, to highlight the characteristic properties of SWCNT columns used in this study, we performed comparison experiments using a pyrene-functionalized column and other CNT columns. Figure 5a displays the comparison of retention times of various organic compounds (mainly including one-benzene-ring aromatic compounds) for a SWCNT column and a pyrene-functionalized column. The pyrene-functionalized column is commercially available and displays strong $\pi$-$\pi$ interactions. The pyrene-functionalized column exhibited long retention times (i.e., strong affinity) for aromatic compounds. In particular, this column showed a clear correlation with that of the SWCNT column for one-benzene-ring aromatic compounds (highlighted by orange color) (Figure 5a). However, organic compounds possessing two benzene rings (phenyl benzoate, naphthalene, and 1- and 2-methylnaphthalene) exhibited behavior different from those of the other compounds. These two-benzene-ring aromatic compounds showed relatively higher affinity to the SWCNT column compared to the one-benzene-ring aromatic compounds. This difference presumably originated from the difference in the degree of the extension of $\pi$-conjugated systems. While pyrene is a relatively small polycyclic aromatic hydrocarbon, CNTs correspond to rolled graphene sheets with a large contact area; note that the contact area governs the affinity (aromaphilicity) of aromatic substances for CNT surfaces [25]. Thus, the interaction of two-benzene-ring aromatic compounds with pyrene is relatively weak compared to that with CNT surfaces. These results indicate that the pyrene-functionalized column is not applicable for estimating the affinity of compounds, especially those multiple benzene rings, for CNTs.

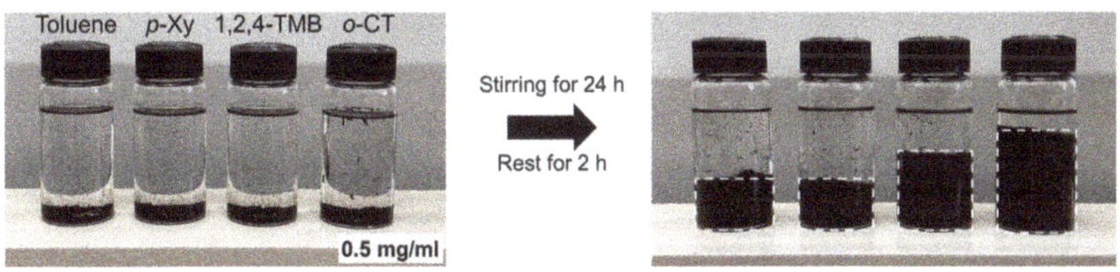


**Figure 4.** Appearance of CNTs in 4 kinds of solvents before and after the dispersion via vigorous stirring.

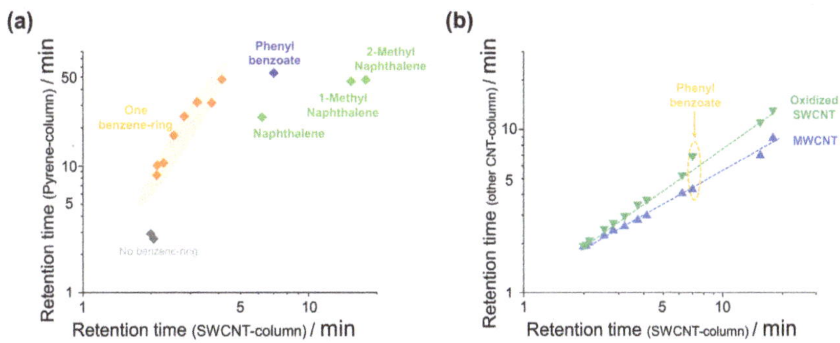


**Figure 5.** (**a**) Comparison of retention times of SWCNT column and pyrene-functionalized column. (**b**) Comparison of retention times of the SWCNT column to those of columns containing other kinds of CNTs.

Moreover, we conducted comparison experiments on other two kinds of CNTs: oxidized SWCNTs and MWCNTs. Oxidized SWCNTs were fabricated by mild air oxidation, which can typically introduce carboxyl and carbonyl groups to the CNT surface [26]. The MWCNTs used to possess a relatively long length (ca. 100 μm according to the supplier). Columns with these CNTs were fabricated via the same procedure as that employed in the pristine SWCNT column preparation, except for adjusting the loading amount of MWCNTs. Liquid chromatography analysis on these columns was conducted under identical conditions to that used for the pristine SWCNT columns. The obtained retention times for several organic compounds on the columns were plotted as a function of those on the pristine SWCNT columns (Figure 5b). Apparently, the retention times on both columns, in particular the MWCNT column, exhibited a trend similar to that of the pristine SWCNT column. The MWCNTs used in this study show relatively low crystallinity and long length [27], and both structural features can also be seen in the characteristics of the supergrowth SWNCTs. The results suggest that CNT curvature has an insignificant effect on the retention of these aromatic compounds. The oxidized SWCNT column also exhibited a trend similar to that of the pristine one, but only phenyl benzoate (emphasized by an arrow in Figure 5b) displayed a slightly longer retention time than expected, which may be a sign of its interaction with polar functional groups introduced by the oxidation process.

## 4. Conclusions

We have successfully fabricated SWCNT columns containing supergrowth SWCNTs and characterized their affinity with organic compounds by using them as a stationary phase for liquid chromatography. The tendency of affinity between organic compounds and SWCNTs was estimated by the adjusted retention times and compared to those estimated from Hansen solubility parameter (HSP) distance values. It was found that most of the investigated organic compounds exhibited a similar tendency to that obtained from HSP distance values, but 1,2,4-TMB deviated from this correlation. Simple dispersing experiments elucidated that affinities determined from retention times of SWCNT columns are effective in estimating the dispersibility of CNTs in organic compounds. The comparison experiments, using a pyrene-functionalized column, an oxidized SWCNT column, and an MWCNT column, demonstrated that the pristine SWCNT column exhibited relatively longer retention times for compounds possessing two benzene rings, indicating the affinity of CNT for large aromatic compounds. Although the differences in affinity for aromatic compounds are minimal among the pristine SWCNT column and the other CNT columns, the oxidized CNT column exhibited a slight influence on the affinity of phenyl benzoate, implying the possibility of achieving desired affinity for specific compounds by tuning the CNT surface environment through surface modification techniques. The chromatographic approaches developed in this work exhibited the practicality of realizing CNT suspensions with good dispersibility. We believe that our methodology could be extended to other nanocarbon materials with the capability to be loaded to the columns (potentially, e.g., graphene oxide) for assessing their dispersibility in organic compounds.

**Supplementary Materials:** The following supporting information can be downloaded at: https://www.mdpi.com/article/10.3390/nano14100824/s1, Figure S1: appearance of CNT dispersions for 4 kinds of solvent dispersed via ultrasonication; Table S1: the dispersibility of CNTs in 4 kinds of solvents dispersed via ultrasonication; Figure S2: scanning electron microscope (SEM) images of SWCNTs on Si substrates deposited by spin coating the ultrasonicated CNT dispersions of 1,2,4-trimethylbenzene (1,2,4-TMB) and *o*-chlorotoluene (*o*-CT).

**Author Contributions:** Conceptualization, R.K. and K.K.; methodology, R.K., A.H. and K.K.; investigation, T.S., R.K. and A.H.; resources, T.S., R.K. and A.H.; visualization, T.S. and R.K.; writing—original draft, T.S.; writing—review and editing, T.S., R.K., A.H., K.K. and K.H.; project administration, K.H. All authors have read and agreed to the published version of the manuscript.

**Funding:** This work was partially supported by JST, PRESTO Grant Number JPMJPR23N8, Japan.

**Data Availability Statement:** Data are contained within the article and Supplementary Materials.

**Acknowledgments:** The authors are grateful for technical assistance from A. Nishikawa and M. Wada. Helpful discussions with K. Kobashi were also appreciated.

**Conflicts of Interest:** The authors declare no conflicts of interest.

## References

1. Ausman, K.D.; Piner, R.; Lourie, O.; Ruoff, R.S.; Korobov, M. Organic Solvent Dispersions of Single-Walled Carbon Nanotubes: Toward Solutions of Pristine Nanotubes. *J. Phys. Chem. B* **2000**, *104*, 8911–8915. [CrossRef]
2. Choi, J.; Lee, C.; Park, S.; Embleton, T.J.; Ko, K.; Jo, M.; Saqib, K.S.; Yun, J.; Jo, M.; Son, Y.; et al. Analysis of Electrochemical Performance with Dispersion Degree of CNTs in Electrode According to Ultrasonication Process and Slurry Viscosity for Lithium-Ion Battery. *Nanomaterials* **2022**, *12*, 4271. [CrossRef] [PubMed]
3. Shimizu, T.; Kishi, R.; Kobashi, K.; Morimoto, T.; Okazaki, T.; Yamada, T.; Hata, K. Improved thermal stability of silicone rubber nanocomposites with low filler content, achieved by well-dispersed carbon nanotubes. *Compos. Commun.* **2020**, *22*, 100482. [CrossRef]
4. Hansen, C.M. 50 Years with solubility parameters—Past and future. *Prog. Org. Coat.* **2004**, *51*, 77–84. [CrossRef]

5. Wieneke, J.U.; Kommoß, B.; Gaer, O.; Prykhodko, I.; Ulbricht, M. Systematic Investigation of Dispersions of Unmodified Inorganic Nanoparticles in Organic Solvents with Focus on the Hansen Solubility Parameters. *Ind. Eng. Chem. Res.* **2011**, *51*, 327–334. [CrossRef]
6. Cheng, Q.; Debnath, S.; O'neill, L.; Hedderman, T.G.; Gregan, E.; Byrne, H.J. Systematic Study of the Dispersion of SWNTs in Organic Solvents. *J. Phys. Chem. C* **2010**, *114*, 4857–4863. [CrossRef]
7. Hernandez, Y.; Lotya, M.; Rickard, D.; Bergin, S.D.; Coleman, J.N. Measurement of Multicomponent Solubility Parameters for Graphene Facilitates Solvent Discovery. *Langmuir* **2009**, *26*, 3208–3213. [CrossRef] [PubMed]
8. Konios, D.; Stylianakis, M.M.; Stratakis, E.; Kymakis, E. Dispersion behaviour of graphene oxide and reduced graphene oxide. *J. Colloid Interface Sci.* **2014**, *430*, 108–112. [CrossRef] [PubMed]
9. Zuaznabar-Gardona, J.C.; Fragoso, A. Determination of the Hansen solubility parameters of carbon nano-onions and prediction of their dispersibility in organic solvents. *J. Mol. Liq.* **2019**, *294*, 111646. [CrossRef]
10. Ma, J.; Larsen, R.M. Comparative Study on Dispersion and Interfacial Properties of Single Walled Carbon Nanotube/Polymer Composites Using Hansen Solubility Parameters. *ACS Appl. Mater. Interfaces* **2013**, *5*, 1287–1293. [CrossRef]
11. Bergin, S.D.; Sun, Z.; Rickard, D.; Streich, P.V.; Hamilton, J.P.; Coleman, J.N. Multicomponent Solubility Parameters for Single-Walled Carbon Nanotube−Solvent Mixtures. *ACS Nano* **2009**, *3*, 2340–2350. [CrossRef] [PubMed]
12. Valcárcel, M.; Cárdenas, S.; Simonet, B.M. Role of Carbon Nanotubes in Analytical Science. *Anal. Chem.* **2007**, *79*, 4788–4797. [CrossRef] [PubMed]
13. Saridara, C.; Mitra, S. Chromatography on Self-Assembled Carbon Nanotubes. *Anal. Chem.* **2005**, *77*, 7094–7097. [CrossRef]
14. Li, Y.; Chen, Y.; Xiang, R.; Ciuparu, D.; Pfefferle, L.D.; Horváth, C.; Wilkins, J.A. Incorporation of Single-Wall Carbon Nanotubes into an Organic Polymer Monolithic Stationary Phase for μ-HPLC and Capillary Electrochromatography. *Anal. Chem.* **2005**, *77*, 1398–1406. [CrossRef]
15. Yuan, L.-M.; Ren, C.-X.; Li, L.; Ai, P.; Yan, Z.-H.; Zi, M.; Li, Z.-Y. Single-Walled Carbon Nanotubes Used as Stationary Phase in GC. *Anal. Chem.* **2006**, *78*, 6384–6390. [CrossRef]
16. Chang, Y.X.; Zhou, L.L.; Li, G.X.; Li, L.; Yuan, L.M. Single?Wall Carbon Nanotubes Used as Stationary Phase in HPLC. *J. Liq. Chromatogr. Relat. Technol.* **2007**, *30*, 2953–2958. [CrossRef]
17. Yoo, J.; Ozawa, H.; Fujigaya, T.; Nakashima, N. Evaluation of affinity of molecules for carbon nanotubes. *Nanoscale* **2011**, *3*, 2517–2522. [CrossRef] [PubMed]
18. Iwashita, K.; Shiraki, K.; Ishii, R.; Tanaka, T.; Hirano, A. Liquid Chromatographic Analysis of the Interaction between Amino Acids and Aromatic Surfaces Using Single-Wall Carbon Nanotubes. *Langmuir* **2015**, *31*, 8923–8929. [CrossRef] [PubMed]
19. Hata, K.; Futaba, D.N.; Mizuno, K.; Namai, T.; Yumura, M.; Iijima, S. Water-Assisted Highly Efficient Synthesis of Impurity-Free Single-Walled Carbon Nanotubes. *Science* **2004**, *306*, 1362–1364. [CrossRef]
20. Kanao, E.; Morinaga, T.; Kubo, T.; Naito, T.; Matsumoto, T.; Sano, T.; Maki, H.; Yan, M.; Otsuka, K. Separation of halogenated benzenes enabled by investigation of halogen–π interactions with carbon materials. *Chem. Sci.* **2019**, *11*, 409–418. [CrossRef]
21. Zhang, Y.; McGuffin, V.L. Thermodynamic and Kinetic Characterization of Porous Graphitic Carbon in Reversed-Phase Liquid Chromatography. *J. Liq. Chromatogr. Relat. Technol.* **2007**, *30*, 1551–1575. [CrossRef]
22. Takeyama, Y.; Ueno, M. Elastomer Composition, Method for Producing Same, Crosslinked Substance, and Molded Body. WO 2021/172555 A1, 26 February 2021.
23. Kobashi, K.; Sekiguchi, A.; Yamada, T.; Muroga, S.; Okazaki, T. Dispersions of High-Quality Carbon Nanotubes with Narrow Aggregate Size Distributions by Viscous Liquid for Conducting Polymer Composites. *ACS Appl. Nano Mater.* **2020**, *3*, 1391–1399. [CrossRef]
24. Stefanis, E.; Panayiotou, C. Prediction of Hansen Solubility Parameters with a New Group-Contribution Method. *Int. J. Thermophys.* **2008**, *29*, 568–585. [CrossRef]
25. Hirano, A.; Kameda, T. *Aromaphilicity Index* of Amino Acids: Molecular Dynamics Simulations of the Protein Binding Affinity for Carbon Nanomaterials. *ACS Appl. Nano Mater.* **2021**, *4*, 2486–2495. [CrossRef]
26. Osswald, S.; Havel, M.; Gogotsi, Y. Monitoring oxidation of multiwalled carbon nanotubes by Raman spectroscopy. *J. Raman Spectrosc.* **2007**, *38*, 728–736. [CrossRef]
27. Lin, D.; Futaba, D.N.; Kobashi, K.; Zhang, M.; Muroga, S.; Chen, G.; Tsuji, T.; Hata, K. A Microwave-Assisted, Solvent-Free Approach for the Versatile Functionalization of Carbon Nanotubes. *ACS Nano* **2023**, *17*, 3976–3983. [CrossRef]

*nanomaterials* MDPI

*Article*

# Enhancing the Thermal Conductivity of CNT/AlN/Silicone Rubber Composites by Using CNTs Directly Grown on AlN to Achieve a Reduced Filler Filling Ratio

**Naoyuki Matsumoto *, Don N. Futaba, Takeo Yamada and Ken Kokubo**

Nano Carbon Device Research Center, National Institute of Advanced Industrial Science and Technology (AIST), Tsukuba Central 5, 1-1-1 Higashi, Tsukuba 305-8565, Ibaraki, Japan
* Correspondence: matsumoto-naoyuki@aist.go.jp; Tel.: +81-29-861-4551

**Abstract:** Achieving the thermal conductivity required for efficient heat management in semiconductors and other devices requires the integration of thermally conductive ceramic fillers at concentrations of 60 vol% or higher. However, an increased filler content often negatively affects the mechanical properties of the composite matrix, limiting its practical applicability. To address this issue, in this paper, we present a new strategy to reduce the required ceramic filler content: the use of a thermally conductive ceramic composite filler with carbon nanotubes (CNTs) grown on aluminum nitride (AlN). We combined catalyst coating technology with vacuum filtration to ensure that the catalyst was uniformly applied to micrometer-sized AlN particles, followed by the efficient and uniform synthesis of CNTs using a water-assisted process in a vertical furnace. By carefully controlling the number of vacuum filtration cycles and the growth time of the CNTs, we achieved precise control over the number and length of the CNT layers, thereby adjusting the properties of the composite to the intended specifications. When AlN/CNT hybrid fillers are incorporated into silicone rubber, while maintaining the mechanical properties of rubber, the thermal diffusivity achieved at reduced filler levels exceeds that of composites using AlN-only or simultaneous AlN and CNTs formulations. This demonstrates the critical influence of CNTs on AlN surfaces. Our study represents a significant advancement in the design of thermally conductive materials, with potential implications for a wide range of applications.

**Keywords:** carbon nanotubes; thermal conductivity; aluminum nitride; chemical vapor deposition; anti-hydrolysis; mechanical property

## 1. Introduction

The rise in the demand for compact electronic devices and high-power LED lighting systems has underscored the critical importance of efficient thermal management alongside advances in battery and device technology [1–4]. As a critical aspect of thermal management, the effective dissipation of heat generated by these devices to the surrounding environment has received considerable attention. The corresponding efforts include the use of various techniques, methods, and materials aimed at minimizing thermal resistance and utilizing materials with high thermal conductivity [4–7]. Among these approaches and tools, thermal interface materials (TIMs) are increasingly emerging as critical components that bridge the gap between heat sources (e.g., devices) and heat sinks (typically metal surfaces) [3,4,8]. Previous research has explored various avenues, including the integration of high-thermal-conductivity metals and alloys [5,9]; carbon-based materials such as graphene, graphene oxide (GO), and carbon nanotubes (CNTs) [10–13]; and ceramic fillers such as aluminum nitride (AlN), boron nitride (BN), silicon carbide (SiC), aluminum oxide ($Al_2O_3$), and zinc oxide (ZnO) [14–21] into silicone rubber or epoxy resin matrices. In addition, efforts have extended to the development of composite materials with malleable thermally conductive fillers capable of conforming to intricate device and heat sink geometries, thereby improving processability and heat dissipation efficiency [22,23].

AlN exhibits remarkable properties, including corrosion resistance, wear resistance, and exceptional thermal stability. Compared to materials such as $Al_2O_3$ (25–40 W/(m K)) [24] and silicon nitride ($Si_3N_4$, 180 W/(m K)) [25], AlN has a significantly higher isotropic thermal conductivity range (170–230 W/(m K), with a theoretical upper limit of 320 W/(m K)) [26]. This level is comparable with the isotopically high thermal conductivity of h-BN (in-plane: 200–280 W/(m K)) [27], a material characterized by its anisotropic thermal behavior. In addition, AlN has remarkable electrical resistivity (>1014 Ω cm) and a coefficient of thermal expansion similar to that of silicon semiconductors ($4.3 \times 10^{-6}$ $K^{-1}$ at room temperature) and is resistant to the halogen gas plasma often used in semiconductor-manufacturing processes [28]. These properties make AlN a promising ceramic filler candidate for improving the thermal conductivity of resins and rubbers [14,29,30].

Despite the remarkable properties of AlN, its susceptibility to water-induced instability is a serious problem. When exposed to moisture, AlN oxidizes, denitrates, and hydrolyzes, forming aluminum hydroxide and corrosive ammonia. The result is a decrease in thermal conductivity, as the formation of non-thermal species takes precedence. Improving water resistance becomes an indispensable pursuit as we move toward the practical use of high-thermal-conductivity ceramic fillers such as AlN [31–36]. Generally, in order to effectively manage heat in semiconductors and similar devices, it is imperative to incorporate thermally conductive ceramic fillers at concentrations exceeding 60% by volume, thereby attaining the necessary thermal conductivity (Figure 1a). While increasing filler content increases thermal conductivity, it also affects the mechanical properties of TIM composites. Critical properties, such as tensile strength and elongation, are susceptible to decreasing [37,38]. In addition, increasing filler content affects the inherent flexibility of matrix resins or rubbers and subsequently compromises thermal dissipation efficiency by degrading adhesion to the contact surfaces of devices and heat sinks. Therefore, a critical challenge in the quest for superior thermal transfer properties remains strengthening the stability of AlN coupled with strategies to reduce the filler content in the rubber or resin matrix while maintaining high thermal conductivity.

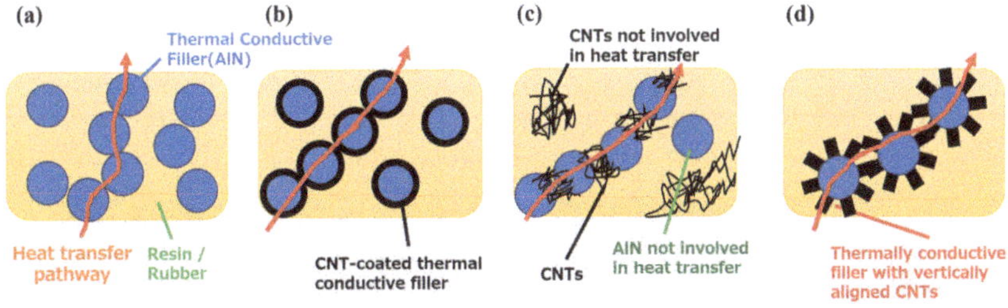


**Figure 1.** Concepts behind filler addition methods for CNT and AlN rubber/resin composites. (**a**) AlN, (**b**) CNT-coated AlN filler, (**c**) simultaneous addition of CNT + AlN, and (**d**) CNT/AlN composite filler. Red arrows: heat transfer pathway.

Numerous efforts have focused on reducing the need for high AlN filler concentrations, including strategies such as particle coarsening and amalgamation with other fillers [39,40]. On the other hand, these methods, often resulting in research on improving AlN stability, particularly in terms of thwarting hydrolysis, have spawned diverse approaches. These include chemical treatments, such as surface conditioning and the use of surfactants, that modify surface chemistry [33]. In addition, shielding layers of materials such as GO, yttrium oxide ($Y_2O_3$), and silicon oxycarbonitride ceramics have been used to encapsulate AlN surfaces and insulate them from hydrolysis [34–36]. However, there is a potential problem: the inadvertent deposition of materials with inferior thermal conductivity compared to AlN could obstruct thermal conduction pathways, resulting in a decrease in overall thermal conductivity. In addition, the presence of organic substances, including surfactants, can induce coalescence in processes such as AlN powder molding and sintering [32]. It may also affect the stability and mechanical properties of rubber or resin compounds that interact with matrix components. Furthermore, coatings containing conductive substances, such as GO, could potentially affect the insulating properties of AlN [32]. While reports utilizing these strategies demonstrate increased water resistance, they do not report on the critical issue of reducing the filler content within the composite matrix.

CNTs, analogous to GO in their function of enhancing AlN stability (water resistance), have recently received considerable attention as potential fillers for high thermal conductivity composites, particularly TIMs. CNTs have a one-dimensional (1D) configuration, a high aspect ratio, and remarkable thermal conductivity (3000 W/(m K) for individual multiwall carbon nanotubes (MWCNTs) and 200 W/(m K) for bulk multi-walled CNTs at room temperature). With these properties, MWCNTs represent a promising high-thermal-conductivity filler, similar to AlN [41–44]. While pure silicone rubber typically has low thermal conductivity (0.1–0.3 W/(m K)), both its corrosion resistance and thermal conductivity significantly improve when it is filled with high-thermal-conductivity ceramic powders [24]. This improvement in properties expands its range of applications. Its blending with CNTs has demonstrated increased thermal conductivity and reduced the required filler content in composites, including those with ceramic fillers such as AlN and CNTs [20,41]. However, the simultaneous integration of ceramics, such as AlN and CNTs, can be complicated. In addition, the difficulty of AlN stabilization (water resistance) persists, and the presence of non-contributing CNTs within thermal pathways can potentially trigger a decline in mechanical properties due to the need for increased filler content (Figure 1c).

We have previously reported the successful synthesis of vertically aligned CNTs on flat AlN substrates using water-assisted chemical vapor deposition (CVD), i.e., the "super-growth method" [45]. The commonly used method of the vapor-phase deposition of an Fe catalyst is not suitable for the deposition of spherical particles, as shown in Figure 1. In this paper, we present a new strategy for reducing the filler content of AlN and improving the water resistance of AlN: CNT/AlN fillers, in which CNTs are grown on AlN particles used as a thermal conductive filler, as shown in Figure 1d. Specifically, we have focused on the application of Fe catalyst coating methods to both AlN filler particles and CNT synthesis processes. In this pursuit, we used wet-based catalyst coating techniques, in particular dip coating and vacuum filtration coating, to coat AlN particles with an Fe catalyst. We then proceeded to vertically grow CNTs directly on the AlN particles, a step aimed at improving the water resistance of AlN. In addition, we demonstrate the exceptional properties of this CNT/AlN composite as a thermally conductive filler in silicone rubber matrices. By creating efficient thermal conduction pathways within the matrix, as shown in Figure 1d, we increased thermal conductivity while ensuring reduced filler content. For this purpose, we directly mixed CNT/AlN particles with silicone rubber. In the process of conducting this study, we successfully exercised precise control over the structural properties (length and number of layers) of the synthesized CNTs. This was made possible by meticulously adjusting the catalyst coating conditions applied to the AlN particles and regulating the CNT synthesis duration through a straightforward methodology. Moreover, by integrating the synthesized CNT/AlN filler with silicone rubber, we achieved about twice the thermal conductivity of AlN alone at about half the filler addition rate while maintaining the mechanical properties of rubber. This was evident when compared with previous reports involving other strategies for producing AlN-based composites either in isolation or in conjunction with CNTs.

## 2. Materials and Methods

### *2.1. Coating of Fe Catalyst onto AlN Particles*

AlN particles selected for this study had a particle size distribution characterized by $D_{50}$ = 80 μm and $D_{90}$ = 122 μm (HFS-80, Tokuyama Corporation, Tokyo, Japan). Prior to use, these AlN particles were ultrasonically cleaned using acetone and ethanol for 5 min, followed by vacuum drying at 100 °C. The dried AlN particles were then immersed in a 30 mM $(CH_3COO)_2Fe$/ethanol solution. A vacuum pump and filter were used for the vacuum-filtration-coating procedure. This setup facilitated the controlled filtration of the iron acetate solution onto the surfaces of the AlN particles, resulting in the deposition of iron acetate (Fe catalyst) onto the AlN substrates. To achieve the desired amount of iron acetate deposition, both the deposition and vacuum filtration steps were repeated iteratively, ranging from 1 to 5 cycles. In the dip-coating process, the prepared dried AlN particles were enclosed in a stainless-steel mesh container. This container had an aperture size of 0.20 mm, a wire diameter of 0.53 mm, and an open porosity of 62.6%. The AlN-loaded container was then immersed in a 30 mM $(CH_3COO)_2Fe$/ethanol solution. The container was pulled out of the solution at a controlled rate of 0.5 mm/s, and this immersion/pulling sequence was repeated three times, as shown in the catalyst-coating process in Figure 2.

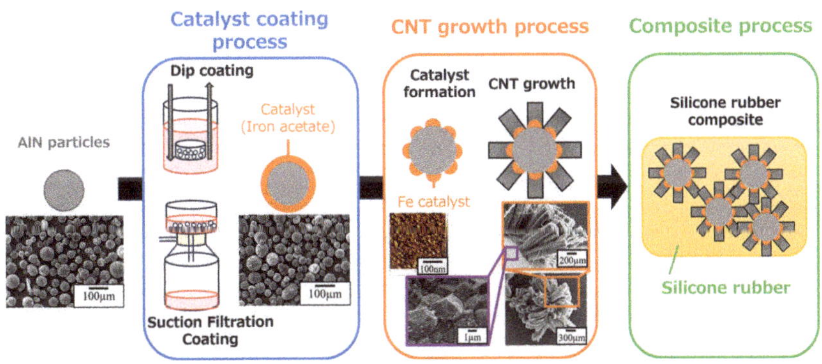


**Figure 2.** Scheme for the preparation of CNT/AlN composite filler and CNT/AlN/Silicone rubber composite. CNT growth process inset shows SEM and AFM images of Fe particles on an AlN substrate after partial exfoliation of CNTs grown over the entire particle after CNT synthesis, as well as SEM images showing vertical CNT growth from the AlN substrate.

### 2.2. Synthesis of Vertically Aligned CNTs on Fe/AlN Particles

AlN particles coated with iron acetate (Fe catalyst) via both vacuum filtration and dip-coating methods were introduced into a mesh dish with an aperture size of 0.35 mm, a wire diameter of 0.21 mm, and an open porosity of 29.9%. This mesh dish was then placed in a horizontal 1-inch quartz tube furnace. The temperature of the furnace was ramped up to 750 °C at a controlled rate of 50 °C/min. The furnace was then held at this temperature for an additional 10 min while maintaining an atmosphere of $H_2$:$N_2$ at a ratio of 9:1. This particular environment was created to facilitate the formation of Fe catalyst nanoparticles. Next, the synthesis of CNTs was performed using water-assisted chemical vapor deposition (CVD). A mixture consisting of a carbon feedstock of $C_2H_2$ (0.5%), a growth enhancer of $H_2O$ at a concentration of about 250 ppm, and a $N_2$ carrier gas was introduced into the furnace. This introduction was carried out over a period of 0–20 min while maintaining the temperature at 750 °C. This process, referred to in Figure 2 as the CNT growth process, allowed the growth of vertically aligned CNTs [46].

### 2.3. Preparation of CNT/AlN/Silicone Rubber Nanocomposites

CNT/AlN/silicone rubber nanocomposites were prepared in accordance with a standardized protocol [47]. First, CNT/AlN particles were dispersed directly into a silicone precursor (KE-106 silicone rubber base compound, Shin-Etsu Chemical Co., Ltd., Tokyo, Japan) via overnight stirring at 50 rpm. A curing agent (CAT-RG, Shin-Etsu Chemical Co., Ltd., Tokyo, Japan) was then added to the CNT/silicone precursor mixture at a weight ratio of 10% with respect to the silicone precursor. The resulting mixture was homogenized and degassed using a planetary centrifugal vacuum mixer (ARV-310, THINKY Co., Ltd., Laguna Hills, CA, USA). To fabricate the CNT/AlN/silicone rubber nanocomposites, the mixture was hardened in a heated mold under a pressure of approximately 10 MPa at 150 °C for 30 min (Figure 2, Composite Process). This process resulted in the formation of the CNT/AlN/silicone rubber nanocomposites.

### *2.4. Characterizations*

The color difference between Fe/AlN particles subjected to suction filtration and dip coating was evaluated using a colorimeter/whiteness meter (NW-12, Nippon Denshoku Industries CO., Ltd., Tokyo, Japan) with a sample size ($n$) of 10. The Fe and Al proportions were quantified using a fluorescence X-ray instrument (EDXL 300, Rigaku Corporation, Tokyo, Japan), with $n = 5$. Both digital microscopy (VHX-5000, Keyence, Osaka, Japan) and scanning electron microscopy (SEM; VE-9800, Keyence) were used to investigate the general morphology of CNTs on AlN particles. The determination of the CNT length measured from the AlN surface to the CNT tip via the SEM images was performed using using the SEM's built-in high-resolution dimensional measurement function. A minimum of 20 samples were taken for each sample. Quantification of the CNT layers was performed using transmission electron microscopy (TEM; EM-002B, TOPCON, Tokyo, Japan). For TEM observations, CNT-ethanol dispersions were deposited on Cu TEM grids, and the number of layers was measured over more than 50 CNTs. The thermal diffusivity (through plane) of the CNT/AlN/rubber composite was evaluated at room temperature under ambient conditions using a flash analyzer (LFA-464, NETZSCH, Selb, Germany) equipped with a Xenon lamp. To visualize the thermal conduction pathways (CNT/AlN) within the silicone rubber, the rubber was immersed in chloroform. Dissolution of the silicone rubber was carried out via stirring at high speed (500 rpm) overnight using a high-speed shaker (ASCM-1, High-Speed Shaker, AS ONE Corporation, Osaka, Japan). After dissolution, the resulting sample (CNT/AlN) was observed using SEM.

## 3. Results and Discussion

### *3.1. Coating of Fe Catalyst on AlN Particles*

First, we investigated how the uniformity of the Fe catalyst on AlN particles changes with different catalyst-coating methods. In particular, we found that Fe acetate as a catalyst source could be more uniformly distributed on the AlN particle surface via suction filtration compared to dip coating. The uniform distribution of Fe acetate was confirmed through color difference measurements. When Fe (II) acetate was applied to AlN, the sample surface changed to a reddish-brown color. This color change was quantified in the L*a*b* color space, a color model commonly used to represent object colors. In the Lab color space, both a* and b* values become positive: L* represents lightness, a* represents the green-to-red direction, and b* represents the blue-to-yellow direction [48]. As the values increase, the color becomes more vivid, while values near the center (where a* and b* are both zero) indicate more muted colors. In essence, a thicker (higher amount of) Fe acetate (the Fe catalyst) coating on AlN results in increased a* and b* values, resulting in decreased brightness (L*). Digital microscope images of the AlN particles before being coated with the Fe catalyst and of AlN particles coated with Fe acetate (Fe catalyst) via suction filtration or dip coating (100 °C, 1 h, vacuum-dried) are shown in Figure 3a–c. In addition, Table 1 shows the mean and standard deviation of color differences for the AlN particles and Fe/AlN particles measured at 10 points for each condition. From the visual inspection of Figure 3 and the comparison of color differences in Table 1, it is evident that the use of suction filtration to apply a 30 mM Fe acetate solution resulted in a more uniform distribution of Fe acetate compared to dip coating. When comparing the color differences, the dip coating ($a^* = 3.5 \pm 1.9$, $b^* = 18.9 \pm 3.4$, $L^* = 82.8 \pm 4.2$) showed color differences similar to the suction filtration coating (applied five times) ($a^* = 3.9 \pm 1.1$, $b^* = 15.6 \pm 1.9$, $L^* = 77.4 \pm 2.8$) but with approximately half the standard deviation. This quantitative assessment by color difference confirms the uniform distribution of Fe acetate (Fe catalyst) not only visually but also quantitatively. Similar results were obtained from the semi-quantitative values of Al and Fe, as well as the standard deviation of the Al/Fe ratio obtained from X-ray fluorescence (XRF) measurements, as shown in Table S1. In addition, the comparison of coating repetition (three vs. five times) in Table 1 shows that repeated application of the same 30 mM solution via suction filtration led to an increase in color differences: a*—1.6 to 3.9, b*—10.3 to 15.6, and L*—79.3 to 77.4. This shift indicates a darker red-brown color attributed to Fe acetate (Fe catalyst). This result illustrates how the amount of Fe acetate on AlN particles (Fe catalyst film thickness) can be controlled by adjusting the number of suction filtration rounds, demonstrating the potential of this simple technique for controlling the amount of Fe acetate through filtration repetitions.

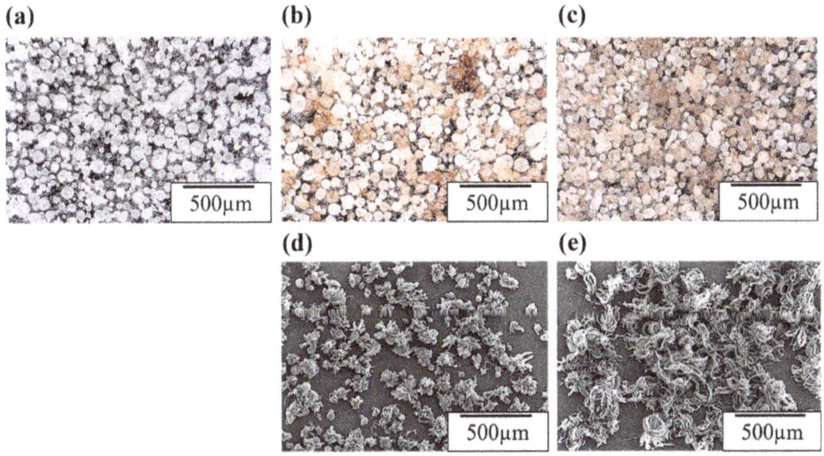


**Figure 3.** Overall view of AlN particles coated with Fe catalyst source (iron acetate) using different coating methods: (**a**) AlN particles (before coating), (**b**) dip coating, and (**c**) suction filtration coating. SEM images of CNT/AlN composite filler after the synthesis of CNTs directly on Fe/AlN particles prepared using different coating methods: (**d**) dip coating; (**e**) suction filtration coating.

**Table 1.** Change in color difference of AlN beads due to iron acetate application.

| | | a* | | b* | | L* | |
|---|---|---|---|---|---|---|---|
| | | Value | STD | Value | STD | Value | STD |
| AlN Beads | | 0.03 | 0.01 | 4.0 | 0.3 | 95.0 | 2.2 |
| Vacuum filtration | 3 times | 1.6 | 0.85 | 10.3 | 1.6 | 79.3 | 2.4 |
| | 5 times | 3.9 | 1.1 | 15.6 | 1.9 | 77.4 | 2.8 |
| Dip coating | | 3.5 | 1.9 | 18.9 | 3.4 | 82.8 | 4.2 |

Next, we demonstrate the controlled and uniform synthesis of CNTs on AlN particles coated with an Fe catalyst using either the suction filtration or dip-coating methods. Figure 3d,e show SEM images of CNTs synthesized via the CVD method on Fe acetate/AlN substrates prepared via suction filtration or dip coating, respectively. When the Fe catalyst was applied using suction filtration, the CNTs grew uniformly, while dip coating resulted in non-uniform CNT growth, with some particles showing no CNT growth at all. As shown in the AFM and SEM insets of the growth process in Figure 2, the CNTs grew vertically from the Fe catalyst particles on the AlN substrate. The CNT lengths were determined from the SEM images (corresponding to the distance from the AlN particle surface to the CNT tip) in Figure 3d,e, confirming that the CNT lengths for suction filtration and dip coating were 350 ± 52 μm and 368 ± 97 μm, respectively. The smaller error in CNT length for suction filtration further supports the uniform growth of CNTs. Although the resultant

CNT growth was highly dependent on the synthesis conditions, it was also strongly dependent on the thickness and uniformity of the Fe catalyst coating as well as the support layer (such as $Al_2O_3$) [49]. Essentially, under synthesis conditions where the carbon source gas (such as ethylene) is uniformly fed to the Fe catalyst on AlN particles, optimal CNT growth conditions vary with Fe catalyst thickness. Consequently, the non-uniform CNT growth observed in the dip-coated specimens (as shown in Figure 3d) suggests that the non-uniform thickness of the Fe acetate catalyst layer is a contributing factor. This notion is consistent with the results regarding the uniformity of the Fe catalyst coating as indicated by the color difference data shown in Figure 3d,e and Table 1.

### *3.2. Control of CNT Growth on AlN Particles*

We further investigated the influence of the catalyst-coating methods, namely, suction filtration and dip coating, on the growth of CNTs on Fe/AlN particles. The number of CNT layers was determined from TEM images ($n = 50$, as shown in the inset of Figure 4a) of CNTs synthesized on Fe/AlN particles coated using both methods. The relationship between the Fe-to-Al ratio on AlN and the number of CNT layers was established and shown in Figure 4a. The Fe-to-Al ratio was calculated based on the Fe and Al content (%) obtained via the X-ray fluorescence analysis of Fe/AlN particles, and the average value of ten measurements was considered to be the Fe/Al ratio. In addition, the adjustment of the Fe acetate (Fe catalyst) content in the dip-coating process was controlled by modifying the number of dip cycles, similar to the suction filtration process: an increase in the number of dips resulted in a higher Fe acetate content. Regardless of the catalyst-coating method employed, an increase in the Fe/Al ratio corresponded to an increase in the number of CNT layers synthesized. However, when comparing the error bars on each plot showing the layer number distribution, it becomes evident that the layer number distribution was narrower for suction filtration synthesis compared to that for dip coating. We clarified the relationship between the thickness (deposition amount) of the Fe catalyst and the number of CNT layers. An increase in the thickness of the Fe catalyst resulted in an increase in the number of CNT layers (diameter). This trend is consistent with previous research indicating that the observed increase in number of layers of the CNTs with the increased Fe thickness likely results from the increase in the average catalyst nanoparticle size, which has been reported previously [49].

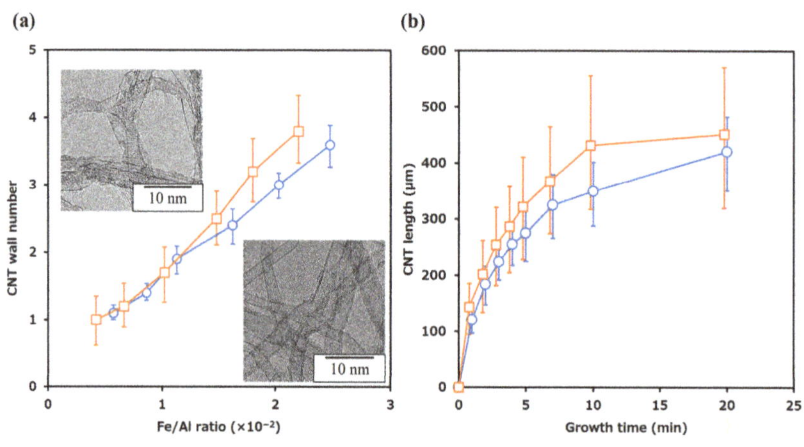


**Figure 4.** Structural changes of CNTs synthesized on AlN with varying amounts of Fe catalyst and synthesis time: (**a**) catalyst amount and number of CNT layers; (**b**) synthesis time and CNT length. Squares (orange lines): dip coating; circles (blue lines): vacuum filtration coating.

Therefore, the wider distribution of layer counts observed in dip coating can be attributed to the lower uniformity of catalyst thickness, while suction filtration, which allows for uniform catalyst application, showed a narrower layer count distribution. Figure 4b illustrates the relationship between CNT length (from the AlN particle surface to the CNT tip) and synthesis time. The CNT length was measured via SEM images using the measurement software provided with the scanning electron microscope, and measurements were taken for 30 CNT/AlN particles to calculate the average length and deviation. Figure 4b shows that regardless of the coating method, the CNT length increased with longer synthesis times. These growth processes are consistent with previous findings and confirm that the CNT length on AlN particles can be controlled by adjusting the synthesis time [50].

With these results, we have demonstrated the feasibility of controlling the number of layers (diameter) and length of CNTs grown on AlN substrates using the simple adjustment methods of Fe catalyst coating, either by varying the number of dips in dip coating or the synthesis time. Furthermore, we found that the ability of suction filtration to uniformly coat the Fe catalyst leads to uniform CNT growth, which is different from dip coating. These results are important, as the number of CNT layers and CNT length are two factors that determine electrical and thermal properties. We realize that a number of other factors, such as straightness and the absence of crystalline defects, were within the scope of our synthetic control. We found that as the number of layers increased, the thermal conductivity of the CNTs improved. Further, we found that CNTs with 2–3 layers had the highest electrical conductivity [51]. On the other hand, CNT length affects not only electrical and thermal properties but also mechanical properties. Longer CNTs lower the percolation threshold and improve electrical conductivity and tensile strength when used to fabricate CNT films [52,53]. The process presented in this paper, which facilitates the control of CNT structures (regarding number of layers and length) in a straightforward manner, is well suited for synthesizing CNTs with tailored properties and creating composites with application-specific material properties.

### 3.3. Thermal Properties of CNT/AlN/Slicon Rubber Composites

Using the suction filtration method to uniformly coat the Fe catalyst and enable uniform growth of CNTs on AlN particles (CNT/AlN composite fillers), we fabricated composites by combining these fillers with silicone rubber (CNT/AlN/silicone rubber composites). We have demonstrated that these composites exhibit superior thermal properties at a lower filler content than that of the AlN alone and of AlN with CNTs (AlN + CNT). Specifically, we fabricated CNT/AlN composite fillers and silicon rubber composites and examined the relationship between thermal conductivity and filler content. Figure 5a shows the thermal conductivity of composites consisting of AlN only and with the simultaneous addition of AlN and CNTs and of CNT/AlN composite fillers synthesized via the suction filtration method for Fe catalyst deposition on AlN added to silicon rubber (CNT/AlN/silicon rubber composites). The preparation conditions of the CNT/AlN composite fillers used in Figure 5a were adjusted so that they were structurally similar to CNTs (ZEONANO® SG101, Tokyo, Japan) when added simultaneously with AlN. The filtration cycle in suction filtration was set to three (Fe/Al ratio: 0.0066) using Fe/AlN particles with an average layer number of 1.2 and an average length of 350 μm (10 min synthesis time). With the incorporation of CNT/AlN composite fillers, the onset of thermal conductivity increased for composites containing AlN only or both AlN and CNTs and occurred at relatively high filler proportions (50 vol% and 30 vol%, respectively). In contrast, the CNT/AlN composite exhibited an exceptionally early-onset filling level at ~20 vol%. In addition, the filler content required to achieve ~20 W/mK occurs at 30 vol%, ~45 vol%, and >60 vol% for CNT/AlN, AlN + CNT, and AlN, respectively. Furthermore, the maximum observed thermal conductivity occurred for the CNT/AlN composite and at a relatively low filler content. This result demonstrates the importance of incorporating the CNTs and AlN into the polymer matrix together rather than as separate components.

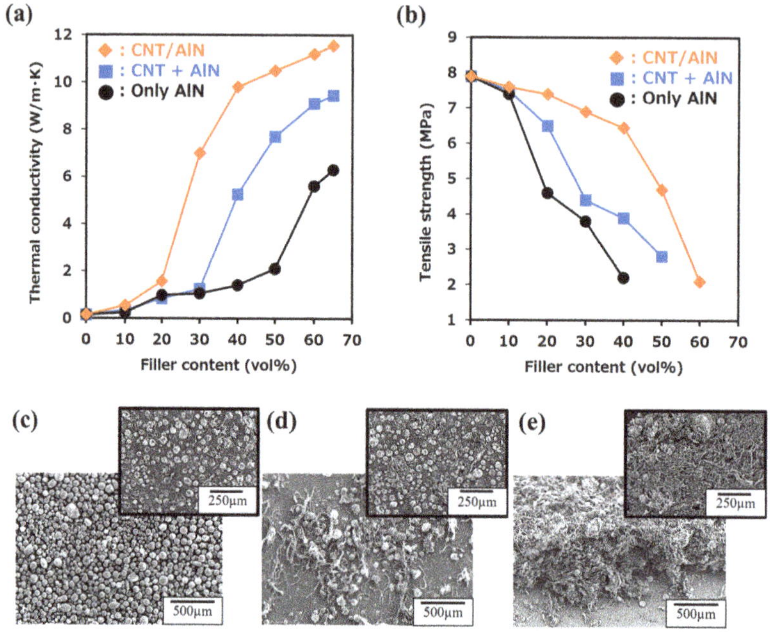


**Figure 5.** Variation in (**a**) thermal conductivity and (**b**) tensile strength with filler content in silicone rubber composites with AlN filler, AlN + CNT filler addition, and CNT/AlN composite filler. SEM images of CNT and AlN filler network after removal of silicone rubber: (**c**) AlN filler, (**d**) AlN + CNT filler addition, and (**e**) CNT/AlN composite filler (inset: SEM images of the surface of each composite). The filler content of AlN, AlN + CNT, and CNT/AlN in each composite was 65, 60, and 50 vol%, respectively.

The use of CNT/AlN composite fillers facilitates the formation of thermal conduction paths via the bonding of CNTs to AlN particles, resulting in a lower minimum filler content (percolation threshold) required to exhibit thermal conductivity, as shown in the conceptual diagram in Figure 1. In addition, by using CNT/AlN composite fillers, the non-contributing CNTs and AlN in terms of thermal conductivity are minimized by the composite formation of CNTs and AlN compared to adding the same quantities of CNTs and AlN simultaneously. Thus, a further reduction in filler content can be achieved compared to the simultaneous addition of AlN and CNTs. Furthermore, Figure 5a shows that the thermal conductivity of CNT/AlN composites was 10.5 W/m K (filler content: 50 vol%), while composites containing only AlN or CNTs or both AlN and CNT exhibited thermal conductivities of 6.3 W/m K (filler content: 65 vol%) and 9.1 W/m K (filler content: 60 vol%), respectively, indicating high thermal conductivity regardless of filler content. Figure 5b also shows the tensile strength of these composites. When AlN alone or AlN + CNT alone were combined with silicone rubber, the tensile strength increased with the addition of AlN alone or CNT alone, but in both cases, the tensile strength decreased significantly at additions above 20 vol%. On the other hand, in the case of CNT/AlN filler addition, the tensile strength remained almost unchanged up to 40 vol%: the tensile strength decreased above 40 vol%. These results indicate that the CNT/AlN composite filler not only reduces the amount of filler added to about half of that of AlN alone while maintaining the mechanical properties (tensile strength) but also improves thermal conductivity. In addition, the simultaneous incorporation of AlN and CNTs, which can be easily integrated into composites, improves thermal conductivity and mechanical properties compared to AlN alone. Therefore, the choice between CNT and AlN or the simultaneous introduction of CNTs and AlN should be based on the thermal conductivity, mechanical properties, and cost considerations relevant to the target application.

To elucidate the mechanism of enabling high thermal conductivity at a low filler content while maintaining mechanical strength, we tried to observe the filler network (AlN and CNT) in each composite, as shown in Figure 5c–e. Although we tried to freeze-slice these composites using a cryomicrotome, it was difficult to prevent the CNTs from becoming detached during sectioning. Therefore, surface observation (SEM) and a technique involving the dissolution of the base material (silicone rubber) with chloroform were used to observe each composite. In the case of the addition of AlN alone, shown in Figure 5c, contact between AlN particles was observed. When AlN and CNTs were added simultaneously (Figure 5d), contact between AlN particles and the bridging of CNTs with AlN particles were observed, along with the presence of individual AlN and CNTs that did not depend on heat transfer. Figure 5e confirms the three-dimensional interconnection of CNTs grown on AlN with other CNTs within the composite structure. From these observations, as shown schematically in Figure 1, it was concluded that the simultaneous addition of AlN and CNTs results in a lower filler content and higher thermal conductivity due to the contact between AlN and CNTs and the bridging between them compared to the addition of AlN. In contrast, when a CNT/AlN composite filler is used, the direct growth of CNTs on AlN suppresses the thermal conductivity of the non-participating CNTs and AlN (Figure 1d). In addition, the formation of a three-dimensional thermal conduction network between the fillers (Figure 5e) enables efficient heat transfer, resulting in the development of thermal conductivity at lower filler additions compared to that for the simultaneous addition of CNTs and AlN, as shown in Figure 5a. This phenomenon highlights the important role of 3D CNT networks in promoting thermal conduction within composites and, in particular, suggests that networks formed by CNTs synthesized directly on AlN are an important factor in improving thermal properties even at low filler concentrations.

On the other hand, the CNTs were uniformly dispersed and properly aligned within the matrix, as shown in Figure 5e; they interacted with the matrix to increase material resistance to mechanical stress. First, the three-dimensional network of CNTs on AlN effectively disperses mechanical (tensile) stress from external sources, reducing stress concentrations [53]. As a result, the overall strength of the material is increased, and the occurrence of cracks and fractures is prevented. In addition, when CNTs are uniformly dispersed and aligned within the matrix, the interactions between individual CNTs are optimized, maximizing the tensile strength of the material [53]. Furthermore, the right density of CNTs promotes good adhesion to the matrix, ensuring that tensile loads are evenly distributed. As a result of these interactions, the material can withstand higher tensile stresses, resulting in improved durability. This is provided by the uniform distribution and correct density of the CNTs within the material. Therefore, CNT/matrix composites show potential as materials with high tensile strength and excellent durability network.

## 4. Conclusions

This study provides evidence of the utility of vertically oriented CNT/AlN composite fillers on thermally conductive ceramic fillers, namely, AlN particles. In addition, uniform synthesis of CNTs on AlN surfaces was achieved by using the wet catalyst coating method of suction filtration. By adjusting the filtration cycle of the iron catalyst source (iron acetate), it was possible to control the number of layers and the length of the CNTs, thus controlling the growth time of the CNTs. In addition, because this process is performed at room temperature and atmospheric pressure, the fabrication time of CNT/AlN composite fillers, from AlN catalyst deposition to CNT growth, is remarkably short (within 1 h). This fast and convenient process is a significant advantage. The composite of the fabricated CNT/AlN fillers with silicone rubber showed that even with reduced filler content, superior thermal conductivity can be achieved compared to that of conventional AlN alone or the simultaneous addition of AlN and CNT. CNT synthesis on the AlN surface also contributes to the reduction in odor (ammonia odor) resulting from the reaction between AlN and moisture, as shown in Figure S1. These results suggest that CNTs encapsulating AlN and Fe particles may also help prevent direct contact between the Fe catalyst and the silicone rubber, thereby mitigating rubber degradation. Therefore, when these CNT/AlN composite fillers are incorporated into resins or rubbers as thermal conductivity enhancers for TIM materials, they are expected to provide high thermal conductivity without compromising the mechanical properties of the matrix material, even at reduced filler content. However, it should be noted that the CNT/AlN/silicon rubber composite presented in this study exhibits conductivity on the order of several mS/m due to the synthesis of CNTs on AlN surfaces. For applications requiring thermal materials with insulating properties, such as semiconductors, ensuring insulation is often imperative. Recent studies have reported the wrapping of CNTs with insulating BN [54], and the adoption of these insulation technologies for CNT/AlN composites is expected to be pursued in the future. In this study, we have not only introduced the concept of novel fillers vertically oriented CNTs on the surface of thermally conductive materials but also demonstrated their ability to efficiently create three-dimensional heat conduction paths, thereby enhancing material properties. This concept is expected to have a ripple effect on materials such as CNT composites, which exhibit excellent electrical properties. In addition to the demonstrated improvements in thermal conductivity and mechanical properties, the developed CNT/AlN/silicone rubber composites hold promise for various practical applications, including TIMs, electronic packaging, and thermal management systems.

**Supplementary Materials:** The following supporting information can be downloaded at https://www.mdpi.com/article/10.3390/nano14060528/s1. Table S1: Amount of Fe catalyst on AlN and its standard deviation in X-ray fluorescence (XRF) measurement; Figure S1: Variation in ammonia concentration in AlN or AlN/CNT composite filler as a function of time of atmospheric exposure (atmospheric exposure conditions: Temperature: 24.3 °C, Humidity: 75%).

**Author Contributions:** Methodology, N.M. and D.N.F.; Formal analysis, N.M.; Investigation, N.M.; Writing—Original Draft, N.M.; Visualization, N.M.; Project administration, N.M. and D.N.F.; Writing—Review & Editing, N.M., D.N.F., T.Y. and K.K.; Supervision, T.Y. and K.K. All authors have read and agreed to the published version of the manuscript.

**Funding:** This research received no external funding.

**Data Availability Statement:** Data are contained within the article.

**Conflicts of Interest:** The authors declare no conflicts of interest.

## References

1. Wu, W.; Wang, S.; Wu, W.; Chen, K.; Hong, S.; Lai, Y. A critical review of battery thermal performance and liquid based battery thermal management. *Energy Convers. Manag.* **2019**, *182*, 262–281. [CrossRef]
2. Xia, G.; Cao, L.; Bi, G. A review on battery thermal management in electric vehicle application. *J. Power Sources* **2017**, *367*, 90–105. [CrossRef]
3. Feng, C.P.; Yang, L.Y.; Yang, J.; Bai, L.; Bao, R.Y.; Liu, Z.Y.; Yang, M.B.; Lan, H.B.; Yang, W. Recent advances in polymer-based thermal interface materials for thermal management: A mini-review. *Compos. Commun.* **2020**, *22*, 100528. [CrossRef]
4. Xing, W.; Xu, Y.; Song, C.; Deng, T. Recent Advances in Thermal Interface Materials for Thermal Management of High-Power Electronics. *Nanomaterials* **2022**, *12*, 3365. [CrossRef] [PubMed]
5. Qu, X.H.; Zhang, L.; Wu, M.; Ren, S.B. Review of metal matrix composites with high thermal conductivity for thermal management applications. *Prog. Nat. Sci. Mater. Int.* **2011**, *21*, 189–197. [CrossRef]
6. Ruan, K.; Shi, X.; Guo, Y.; Gu, J. Interfacial thermal resistance in thermally conductive polymer composites: A review. *Compos. Commun.* **2020**, *22*, 100518. [CrossRef]
7. Zhao, C.; Li, Y.; Liu, Y.; Xie, H.; Yu, W. A critical review of the preparation strategies of thermally conductive and electrically insulating polymeric materials and their applications in heat dissipation of electronic devices. *Adv. Compos. Hybrid Mater.* **2023**, *6*, 27. [CrossRef]
8. Cui, Y.; Li, M.; Hu, Y. Emerging interface materials for electronics thermal management: Experiments, modeling, and new opportunities. *J. Mater. Chem. C* **2020**, *8*, 10568–10586. [CrossRef]
9. NematpourKeshteli, A.; Iasiello, M.; Langella, G.; Bianco, N. Enhancing PCMs thermal conductivity: A comparison among porous metal foams, nanoparticles and finned surfaces in triplex tube heat exchangers. *Appl. Therm. Eng.* **2022**, *212*, 118623. [CrossRef]

10. Renteria, J.D.; Nika, D.L.; Balandin, A.A. Graphene Thermal Properties: Applications in Thermal Management and Energy Storage. *Appl. Sci.* **2014**, *4*, 525–547. [CrossRef]
11. Li, J.; Zhao, X.; Zhang, Z.; Xian, Y.; Lin, Y.; Ji, X.; Lu, Y.; Zhang, L. Construction of interconnected $Al_2O_3$ doped rGO network in natural rubber nanocomposites to achieve significant thermal conductivity and mechanical strength enhancement. *Compos. Sci. Technol.* **2020**, *186*, 107930. [CrossRef]
12. Ouyang, Y.; Li, X.; Ding, F.; Bai, L.; Yuan, F. Simultaneously enhance thermal conductive property and mechanical properties of silicon rubber composites by introducing ultrafine $Al_2O_3$ nanospheres prepared via thermal plasma. *Compos. Sci. Technol.* **2020**, *190*, 108019. [CrossRef]
13. Gojny, F.H.; Wichmann, M.H.; Fiedler, B.; Kinloch, I.A.; Bauhofer, W.; Windle, A.H.; Schulte, K. Evaluation and Identification of Electrical and Thermal Conduction Mechanisms in Carbon Nanotube/Epoxy Composites. *Polymer* **2006**, *47*, 2036–2045. [CrossRef]
14. Pinto, R.M.; Gund, V.; Calaza, N.K.; Vinayakumar, K. Piezoelectric aluminum nitride thin films: A review of wet and dry etching techniques. *Microelectron. Eng.* **2022**, *257*, 111753. [CrossRef]
15. Song, H.; Liu, J.; Liu, B.; Wu, J.; Cheng, H.M.; Kang, F. Two-Dimensional Materials for Thermal Management Applications. *Joule* **2018**, *2*, 442–463. [CrossRef]
16. Nishimura, T.; Hirosaki, N. Electric current assisted sintering of AlN ceramics: Thermal conductivity and transparency. *Adv. Appl. Ceram.* **2014**, *113*, 89–93. [CrossRef]
17. Maier, H.J.; Gawlytta, R.; Fromm, A.; Klose, C. Increasing thermal conductivity in aluminium-copper compound castings: Modelling and experiments. *Mater. Sci. Technol.* **2023**, *39*, 1903–1913. [CrossRef]
18. Kim, B.J.; Bae, K.M.; An, K.H.; Park, S.J. Effects of surface nitrification on thermal conductivity of modified aluminum oxide nanofibers-reinforced epoxy matrix nanocomposites. *Bull. Korean Chem. Soc.* **2012**, *33*, 3258–3264. [CrossRef]
19. AbdulKarim-Talaq, M.; Hassan, K.T.; Hameed, D.A. Improvement of thermal conductivity of novel asymmetric dimeric coumarin liquid crystal by doping with boron nitride and aluminium oxide nanoparticles. *Mater. Chem. Phys.* **2023**, *297*, 127367. [CrossRef]
20. Dong, M.; Zhang, J.; Liu, L.; Hou, G.; Yu, Y.; Yuan, C.; Wang, X. New gutta percha composite with high thermal conductivity and low shear viscosity contributed by the bridging fillers containing ZnO and CNTs. *Compos. Part B Eng.* **2019**, *173*, 106903. [CrossRef]
21. Lee, E.S.; Lee, S.M.; Shanefield, D.J.; Cannon, W.R. Enhanced Thermal Conductivity of Polymer Matrix Composite via High Solids Loading of Aluminum Nitride in Epoxy Resin. *J. Am. Ceram. Soc.* **2008**, *91*, 1169–1174. [CrossRef]
22. Chiu, H.T.; Sukachonmakul, T.; Kuo, M.T.; Wang, Y.H.; Wattanakul, K. Surface modification of aluminum nitride by polysilazane and its polymer-derived amorphous silicon oxycarbide ceramic for the enhancement of thermal conductivity in silicone rubber composite. *Appl. Surf. Sci.* **2014**, *292*, 928–936. [CrossRef]
23. Prasher, R. Thermal Interface Materials: Historical Perspective, Status, and Future Directions. *Proc. IEEE Inst. Electr. Electron. Eng.* **2006**, *94*, 1571–1586. [CrossRef]
24. Gao, B.Z.; Xua, J.Z.; Peng, J.J.; Kanga, F.Y.; Dua, H.D.; Lia, J.; Chiang, S.W.; Xu, C.J.; Hu, N.; Ning, X.S. Experimental and theoretical studies of effective thermal conductivity of composites made of silicone rubber and $Al_2O_3$ particles. *Thermochim. Acta* **2014**, *614*, 1–8. [CrossRef]
25. Zhou, Y.; Hyuga, H.; Kusano, D.; Yoshizawa, Y.; Hirao, K. A Tough Silicon Nitride Ceramic with High Thermal Conductivity. *Adv. Mater.* **2011**, *23*, 4563–4567. [CrossRef]
26. Kim, S.Y.; Choi, D.K.; Yeo, D.H.; Shin, H.S.; Yoon, H.G. Influence of ball milling contamination on properties of sintered AlN substrates. *J. Eur. Ceram. Soc.* **2020**, *40*, 5349–5356. [CrossRef]
27. Zhou, H.; Zhu, J.; Liu, Z.; Yan, Z.; Fan, X.; Lin, J.; Wang, G.; Yan, Q.; Yu, T.; Ajayan, P.M.; et al. High thermal conductivity of suspended few-layer hexagonal boron nitride sheets. *Nano Res.* **2014**, *7*, 1232–1240. [CrossRef]
28. Bartel, C.J.; Muhich, C.L.; Weimer, A.W.; Musgrave, C.B. Aluminum Nitride Hydrolysis Enabled by Hydroxyl-Mediated Surface Proton Hopping. *ACS Appl. Mater. Interfaces* **2016**, *8*, 18550–18559. [CrossRef] [PubMed]
29. Ohashi, M.; Kawakami, S.; Yokagawa, Y. Spherical Aluminum Nitride Fillers for Heat-Conducting Plastic Packages. *J. Am. Ceram. Soc.* **2005**, *88*, 2615–2618. [CrossRef]
30. Yu, H.; Li, L.; Kido, T.; Xi, G.; Xu, G.; Guo, F. Thermal and insulating properties of epoxy/aluminum nitride composites used for thermal interface material. *J. Appl. Polym. Sci.* **2012**, *124*, 669–677. [CrossRef]
31. Kocjan, A.; Dakskobler, A.; Krnel, K.; Kosmac, T. The course of the hydrolysis and the reaction kinetics of AlN powder in diluted aqueous suspensions. *J. Eur. Ceram. Soc.* **2011**, *31*, 815–823. [CrossRef]
32. Zhang, C.Y.; Yu, Y.L.; Yang, H.; Wu, Y.Y.; Zhong, M.F.; Lin, S.M.; Zhang, Z.J.; Xu, W.; Wu, L.G. Mechanism for the hydrolysis resistance of aluminum nitride powder modified by boric acid. *Ceram. Int.* **2022**, *48*, 32696–32702. [CrossRef]
33. Binner, J.; Zhang, Y. Surface chemistry and hydrolysis of a hydrophobic-treated aluminium nitride powder. *Ceram. Int.* **2005**, *31*, 469–474. [CrossRef]
34. O'Toole, R.J.; Hill, C.; Buur, P.J.; Bartel, C.J.; Gump, C.J.; Musgrave, C.B.; Weimer, A.W. Hydrolysis protection and sintering of aluminum nitride powders with yttria nanofilms. *J. Am. Ceram. Soc.* **2022**, *105*, 3123–3127. [CrossRef]
35. Hu, L.H.; Wang, Y.K.; Wang, S.C. Aluminum nitride surface functionalized by polymer derived silicon oxycarbonitride ceramic for anti-hydrolysis. *J. Alloys Compd.* **2019**, *772*, 828–833. [CrossRef]

36. Chen, C.; Xue, Y.; Li, Z.; Wen, Y.; Li, X.; Wu, F.; Li, X.; Shi, D.; Xue, Z.; Xie, X. Construction of 3D boron nitride nanosheets/silver networks in epoxy-based composites with high thermal conductivity via in-situ sintering of silver nanoparticles. *Chem. Eng. J.* **2019**, *369*, 1150–1160. [CrossRef]
37. Lee, G.W.; Park, M.; Kim, J.; Lee, J.I.; Yoon, H.G. Enhanced thermal conductivity of polymer composites filled with hybrid filler. *Compos. Part A Appl. Sci.* **2006**, *37*, 727–734. [CrossRef]
38. Kim, K.; Kim, J. Magnetic aligned AlN/epoxy composite for thermal conductivity enhancement at low filler content. *Compos. Part B Eng.* **2016**, *93*, 67–74. [CrossRef]
39. Choi, S.; Kim, J. Thermal conductivity of epoxy composites with a binary-particle system of aluminum oxide and aluminum nitride fillers. *Compos. Part B Eng.* **2013**, *51*, 140–147. [CrossRef]
40. Gaska, K.; Rybak, A.; Kapusta, C.; Sekula, R.; Siwek, A. Enhanced thermal conductivity of epoxy–matrix composites with hybrid fillers. *Polym. Adv. Technol.* **2015**, *26*, 26–31. [CrossRef]
41. Zhou, T.; Wang, X.; Liu, X.; Xiong, D. Improved thermal conductivity of epoxy composites using a hybrid multi-walled carbon nanotube/micro-SiC filler. *Carbon* **2010**, *48*, 1171–1176. [CrossRef]
42. Yang, D.J.; Zhang, Q.; Chen, G.; Yoon, S.F.; Ahn, J.; Wang, S.G.; Zhou, Q.; Wang, Q.; Li, J.Q. Thermal conductivity of multiwalled carbon nanotubes. *Phys. Rev. B* **2002**, *66*, 165440. [CrossRef]
43. Moisala, A.; Li, Q.; Kinloch, I.A.; Windle, A.H. Thermal and electrical conductivity of single- and multi-walled carbon nanotube-epoxy composites. *Compos. Sci. Technol.* **2006**, *66*, 1285–1288. [CrossRef]
44. Yang, K.; Gu, M.; Guo, Y.; Pan, X.; Mu, G. Effects of carbon nanotube functionalization on the mechanical and thermal properties of epoxy composites. *Carbon* **2009**, *47*, 1723–1737. [CrossRef]
45. Tsuji, T.; Matsumoto, N.; Takai, H.; Sakurai, S.; Futaba, D.N. Millimetre-scale growth of single-wall carbon nanotube forests using an aluminium nitride catalyst underlayer. *MRS Adv.* **2019**, *4*, 177–183. [CrossRef]
46. Hata, K.; Futaba, D.N.; Mizuno, K.; Namai, T.; Yumura, M.; Iijima, S. Water-Assisted Highly Efficient Synthesis of Impurity-Free Single-Walled Carbon Nanotubes. *Science* **2004**, *306*, 1362–1364. [CrossRef]
47. Shimizu, T.; Kishi, R.; Kobashi, K.; Morimoto, T.; Okazaki, T.; Yamada, T.; Hata, K. Improved thermal stability of silicone rubber nanocomposites with low filler content, achieved by well-dispersed carbon nanotubes. *Compos. Commun.* **2020**, *22*, 100482. [CrossRef]
48. Mclaren, K. XIII—The development of the CIE 1976 (L*a*b*) uniform colour space and colour-difference formula. *J. Soc. Dye. Colour.* **1976**, *92*, 338–341. [CrossRef]
49. Yamada, T.; Namai, T.; Hata, K.; Futaba, D.N.; Mizuno, K.; Fan, J.; Yudasaka, M.; Yumura, M.; Iijima, S. Size-selective growth of double-walled carbon nanotube forests from engineered iron catalysts. *Nat. Nanotech.* **2006**, *1*, 131–136. [CrossRef] [PubMed]
50. Futaba, D.N.; Hata, K.; Yamada, T.; Mizuno, K.; Yumura, M.; Iijima, S. Kinetics of Water-Assisted Single-Walled Carbon Nanotube Synthesis Revealed by a Time-Evolution Analysis. *Phys. Rev. Lett.* **2005**, *95*, 056104. [CrossRef]
51. Chen, G.; Futaba, D.N.; Kimura, H.; Sakurai, S.; Yumura, M.; Hata, K. Absence of an Ideal Single-Walled Carbon Nanotube Forest Structure for Thermal and Electrical Conductivities. *ACS Nano* **2013**, *7*, 10218–10224. [CrossRef] [PubMed]
52. Sakurai, S.; Kamada, F.; Futaba, D.N.; Yumura, M.; Hata, K. Influence of lengths of millimeter-scale single-walled carbon nanotube on electrical and mechanical properties of buckypaper. *Nanoscale Res. Lett.* **2013**, *8*, 546. [CrossRef] [PubMed]
53. Ata, S.; Kobashi, K.; Yumura, M.; Hata, K. Mechanically Durable and Highly Conductive Elastomeric Composites from Long Single-Walled Carbon Nanotubes Mimicking the Chain Structure of Polymers. *Nano Lett.* **2012**, *12*, 2710–2716. [CrossRef] [PubMed]
54. Xiang, R.; Inoue, T.; Zheng, Y.; Kumamoto, A.; Qian, Y.; Sato, Y.; Liu, M.; Tang, D.; Gokhale, D.; Guo, J.; et al. One-dimensional van der Waals heterostructures. *Science* **2020**, *367*, 537–542. [CrossRef]

nanomaterials

Article

# Field Emission Properties of Cu-Filled Vertically Aligned Carbon Nanotubes Grown Directly on Thin Cu Foils

Chinaza E. Nwanno [1], Arun Thapa [1], John Watt [2], Daniel Simkins Bendayan [1] and Wenzhi Li [1,*]

[1] Department of Physics, Florida International University, Miami, FL 33199, USA; cnwan007@fiu.edu (C.E.N.); dsimk005@fiu.edu (D.S.B.)
[2] Center for Integrated Nanotechnologies, Los Alamos National Laboratory, Los Alamos, NM 87545, USA
* Correspondence: wenzhi.li@fiu.edu; Tel.: +1-(305)-348-7257

**Abstract:** Copper-filled vertically aligned carbon nanotubes (Cu@VACNTs) were grown directly on Cu foil substrates of 0.1 mm thicknesses at different temperatures via plasma-enhanced chemical vapor deposition (PECVD). By circumventing the need for additional catalyst layers or intensive substrate treatments, our in-situ technique offers a simplified and potentially scalable route for fabricating Cu@VACNTs with enhanced electrical and thermal properties on thin Cu foils. Comprehensive analysis using field emission scanning microscopy (FESEM), transmission electron microscopy (TEM), energy-dispersive X-ray spectroscopy (EDS) mappings, and X-ray diffraction (XRD) revealed uniform Cu filling within the VACNTs across a range of synthesis temperatures (650 °C, 700 °C, and 760 °C). Field emission (FE) measurements of the sample synthesized at 700 °C (S700) showed low turn-on and threshold fields of 2.33 V/µm and 3.29 V/µm, respectively. The findings demonstrate the viability of thin Cu substrates in creating dense and highly conductive Cu-filled VACNT arrays for advanced electronic and nanoelectronics applications.

**Keywords:** VACNTs; copper; PECVD; field emission

## 1. Introduction

Since their discovery in 1991 by Ijima [1], CNTs have been extensively researched for possible applications in myriads of fields. They exhibit exceptional chemical stability, high aspect ratio, low work function, large field enhancement factor, superior mechanical strength, and excellent electrical and thermal conductivities, making them promising materials for field emission applications [2–7]. VACNTs, a special kind of CNT, consist of individual CNTs aligned perpendicular to the substrate [8–10]. Although they share similar alignment with the three-dimensional (3D) vertically aligned graphene nanosheets (VAGNAs) comprising interconnected graphene sheets arranged vertically relative to the substrate and forming a porous, 3D network [11–13], both materials have marked differences. For instance, the individual CNTs forming the VACNTs possess hollow cylindrical (tube-like) structures characterized by a high aspect ratio, whereas the VAGNAs retain the planar shape (sheet-like) structure of the component graphene sheets while providing a large surface area. The hollow nature of the CNTs provides a unique advantage over the 3D VAGNAs for applications requiring material encapsulation, allowing the CNTs to be filled with other foreign materials. Studies have suggested that due to the hollow cavities of the CNTs, filling them with foreign materials will drastically improve their intrinsic properties and enhance their performances in field emission devices and several other nanoscale applications, including nanosensors [14], nanomagnets [15], nanoswitches [16], nanothermometers [17], batteries [18], supercapacitors [19], etc.

The resulting filled CNTs (X@CNTs) display the characteristic features of both the host CNTs and the foreign filler materials, together with elongated cavities acting as templates to generate 1D nanostructures such as nanowires [20]. For instance, studies have shown that CNTs filled with ferromagnetic materials (Fe, Co, and Ni) display excellent magnetoresistance and, thus, can be used as nanomagnets in magnetic data storage devices [21–24]. The CNTs ensure durability and stability by protecting the ferromagnetic filler from possible oxidation. In field emission devices, the electrical and thermal conductivities of CNTs are greatly improved by encapsulating highly conductive nanowires inside the cores of the CNT arrays. This greatly increases the field enhancement factor and reduces the turn-on and threshold fields. In 2021, our group reported very low turn-on and threshold fields of 1.57 V/μm and 2.43 V/μm, respectively, and a high field enhancement factor (ß = 3061) for Cu-filled VACNTs (Cu@VACNTs) grown on bulk Cu disks [25]. These excellent field emission properties resulted from synergetic contributions from the VACNTs and the highly conductive encapsulated Cu nanowires.

Depending on whether filling occurs during or after CNT growth, the techniques are classified into two broad categories: ex-situ and in-situ. The ex-situ techniques, which include the sonication-assisted wet chemical method [26], supercritical fluid chemical deposition (SFCD) [27], the solution infusion method [28], the chemical fluid deposition method (CFD) [29], and the electrochemical deposition method [30], involve encapsulating foreign materials into the hollow cavities of CNTs post CNT synthesis. On the other hand, in-situ filling occurs simultaneously with CNT growth. The arc-discharge [31] and chemical vapor deposition (CVD) [32] methods are the most used in-situ methods. Meanwhile, unlike the ex-situ techniques, which involve multiple steps that require opening the closed ends of CNTs, the in-situ methods present a single-step synthesis route for filled CNTs (X@CNTs) with well-preserved capsules and closed ends [33].

Several foreign materials, including metals [34–36] and metal oxides [37–39], organic molecules [40–42], carotenes [43–45], fullerenes [46–48], fluorescent NPs [49], etc., have been used to fill CNTs for several applications. Among these materials, Cu-filled CNTs (Cu@CNTs) have gained huge recognition owing to their excellent electrical and thermal conductivities, the low cost of Cu, and weak interaction with carbon. Furthermore, studies have demonstrated that the inherent high conductivity of CNTs can be increased by encapsulating Cu inside the core of the CNTs [50–52]. Despite these impressive results, the filling methods employed are fraught with several issues, such as poor yield of filled CNTs, multiple intricate steps, improper alignment of the resulting Cu@CNTs, and the need for Cu salts.

Furthermore, to achieve Cu@CNTs via the in-situ technique, the CNTs need to be grown directly on the Cu substrates without any extra catalyst layers. However, reports have shown that it is challenging to synthesize CNTs on Cu because of its poor catalytic activity and extremely low carbon solubility [53]. Cu has occupied 3d orbitals, which prohibit the formation of covalent bonds with hydrocarbon molecules. The small binding energy of Cu with carbon also suppresses CNT graphitization during growth [54]. Atthipalli et al. [55] grew multi-walled CNTs (MWCNTs) on bulk Cu wafers but by first depositing Ni film as a catalyst layer. Similarly, Yin et al. [56] reported bamboo-like CNTs on an oxygen-free Cu substrate by first sputtering Ni particles as catalysts on the Cu substrate. Rao et al. [57] also reported CNT growth on TEM Cu grids coated with Ni. Sepahvand et al. [58] obtained dense VACNT arrays by depositing Ni and chromium (Cr) as both catalyst and barrier layers, respectively. Meanwhile, it is important to point out that prior to the catalyst deposition, an intermediate buffer layer is first deposited onto the substrate material to prevent the diffusion of the catalyst particles into the Cu substrates. Aside from this being a tortuous procedure, the presence of the intermediate layer increases the contact resistance between the CNTs and the Cu substrates. In addition, it is impossible to achieve in-situ filling of the CNTs with Cu following the procedure described above because the CNTs grow on the deposited catalyst particles and are more likely to be filled with them. There are a few successful reports on the direct synthesis of MWCNTs on Cu substrates by first activating the catalytic properties of the Cu substrates by either acid or sulfur treatment for many hours [53,59,60]. However, this process is long and expensive. Our group has tried to address these intricacies by developing a facile in-situ method for filling VACNTs with Cu [25]. Although the technique obviates the deposition of extra layers and acid treatment, the Cu@VACNTs were grown directly on thick Cu disks (thickness = 0.6 mm), which are not good for practical micro- and nanoelectronics applications.

Herein, we report the direct synthesis of Cu@VACNTs on thin Cu foils (thickness = 0.1 mm) at different temperatures. Generally, compared to thick substrates, it is more difficult to grow CNTs on their thinner counterparts because they lack the structural integrity to withstand the high mechanical and thermal stress during the CNT growth process. They crack, melt, or undergo other structural deformations, which could in turn impact the uniformity and quality of the resulting CNTs.

## 2. Experimental Section

### 2.1. Cu@VACNTs Synthesis

Arrays of copper-filled VACNTs (Cu@VACNTs) were synthesized on 0.1 mm thick Cu foil substrates using the direct current (DC) PECVD method described in our previous report [61]. The Cu foil substrates were ultrasonicated with acetone and isopropyl alcohol for 10 min each to eliminate organic and inorganic contaminants. After cleaning, the substrates were allowed to dry in the open air before being transferred to the PECVD chamber for CNT growth.

The growth chamber was pumped down to a base pressure of 0.01 Torr. Subsequently, ammonia ($NH_3$) gas was introduced at a flow rate of 110 sccm, maintaining a constant chamber pressure of 7 Torr to create the reduced environment necessary for CNT growth. Under these conditions, the Cu substrates were heated to growth temperatures of 650 °C, 700 °C, and 760 °C at a rate of 50 °C/min. Uniform catalytic active sites were formed on the substrate surface due to the etching effect of $NH_3$ on the substrate at elevated temperatures. These catalytic sites were necessary for the nucleation of CNTs. Upon reaching the desired growth temperature, the DC plasma was turned on, and the power was maintained at 70 W. This was immediately followed by introducing acetylene ($C_2H_2$) as the carbon precursor gas at a constant flow rate of 30 sccm. The system was turned off after 30 min of growth and was allowed to cool down at the base pressure. The samples of Cu@VACNTs synthesized at 650 °C, 700 °C, and 760 °C were named as S650, S700, and S760, respectively.

### 2.2. Material Characterization

A field emission scanning electron microscope (FESEM) was used to characterize the surface structure of the as-synthesized Cu@VACNTs at an accelerating voltage of 15 kV. The XRD patterns of the different samples were obtained using the Siemens Diffractometer D5000 (Munich, Germany) with Cu K$\alpha$ radiation ($\lambda$ = 1.54 Å). The nanostructures of the as-synthesized Cu@VACNTs were examined using an image-aberration-corrected ThermoFisher Titan 80–300 (Waltham, MA, USA) fitted with an EDAX Octane Elite T solid-state X-ray spectrometer operated at 300 kV. The Cu@VACNT samples were prepared for TEM by gently scrapping them off the Cu foil substrate with a surgical blade and flushed off onto the TEM grid by dropping alcohol on the blade. Low-magnification images were taken to reveal the continuity of the Cu filling inside the CNTs. High-magnification images of the CNT shell and the filled Cu core were taken to show their structures and interfaces. TEM-EDS mapping was also carried out to confirm the chemical constituent of the guest material inside the cores of the CNTs.

### 2.3. FE Measurement

As shown in the Electronic Supplementary Information (ESI), Figure S1, a diode configuration was used to measure the field emission properties of the sample S700 Cu@VACNTs in a vacuum chamber with a base pressure of approximately $1 \times 10^{-6}$ Torr. Using a silver paste, the cathode was prepared by gluing the as-synthesized Cu@VACNTs onto the stainless-steel plate. The anode was a solid cylindrical stainless-steel rod with a 0.803 $cm^2$ diameter. The separation distance between both electrodes was maintained at 615 μm with the aid of ceramic and plastic spacers. Fifty-volt increments of voltage bias were applied using a DC power source (Matsusada AU-15P20, Otsu, Japan), and the emission current was measured with a Keithley 236 unit. At least five cycles of current density vs. electric field (J vs. E) characteristics were recorded with multiple samples to ensure the reproducibility of data.

## 3. Results and Discussion

### *3.1. VACNT Structure*

The FESEM images in Figure 1a–c reveal the effect of different growth temperatures on the morphology and density of the VACNT arrays. In Figure 1a, sparsely grown VACNTs of non-uniform height (between 0.4 and 10 μm), diameter (between 300 and 670 nm), and inter-tube spacings can be observed, some of which appear in bundles. The poor growth, non-uniformity, and bundled nature of these VACNTs can be attributed to the fact that the growth temperature (650 °C) was insufficient to break the surface of the Cu substrate into uniform and distinct nanosized islands for CNT growth. In addition, 650 °C may not be high enough to sufficiently dissociate the $C_2H_2$ into carbon atoms, leading to insufficient carbon species for further dissolution into the available nano islands. At a much higher synthesis temperature of 700 °C, the VACNTs appear as freestanding uniform VACNTs with average diameter and height of 940 nm and 14 μm, respectively (Figure 1b). At 760 °C (Figure 1c), the average diameter of the VACNTs was measured as 1.2 μm. Interestingly, the CNTs were slightly shorter than the S700 samples, with an average height of 8.5 μm. This might be attributed to the excessive carbon deposition at a higher temperature, which caused early growth termination. Also, most CNTs appear as bundles due to Ostwald ripening, which causes individual nanoparticles to coalesce and form bigger ones. As a result, numerous individual VACNTs grow from these bigger but single catalyst sites to form bundles [62].

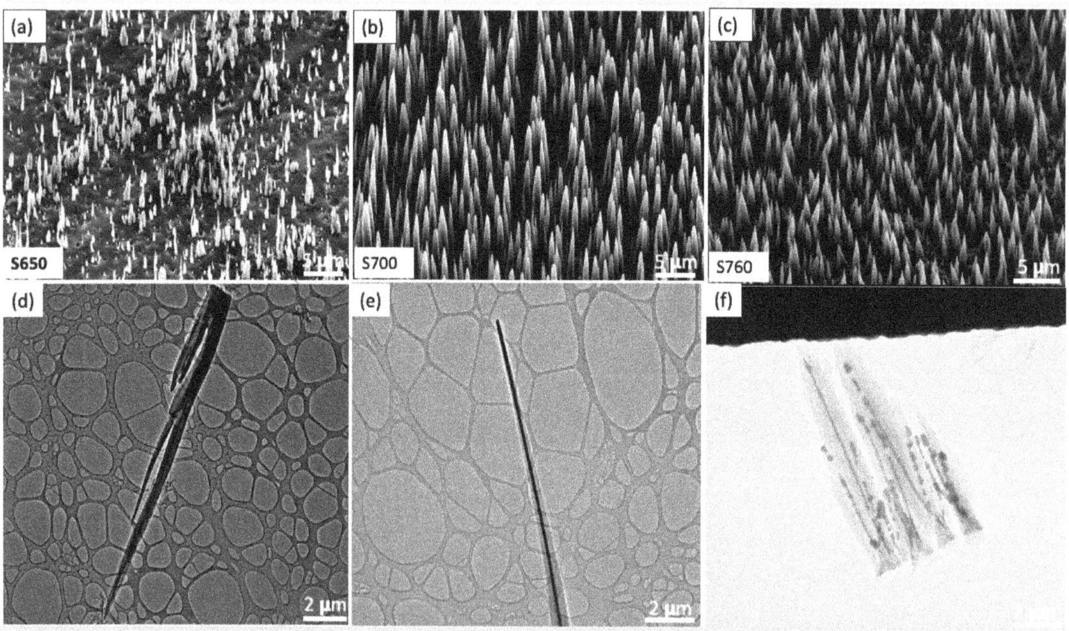


**Figure 1.** (**a**–**c**) SEM images of VACNTs synthesized at 650 °C, 700 °C, and 760 °C at a constant pressure of 7 Torr, plasma power of 70 W, and $NH_3$ and $C_2H_2$ flow rates of 110 sccm and 30 sccm, respectively, for 30 min. (**d**–**f**) Corresponding TEM images of (**a**–**c**).

Figure 1d–f show the corresponding TEM images of the VACNTs filled with foreign materials. Herein, it is evident that all the samples were filled with the guest material from their roots to the tips, creating core–shell structures. It can be seen that all the Cu@VACNT samples possess tapered structures with the number of graphitic layers decreasing from the roots to the tips. As a result, it can be said that the VACNT walls close to the tip of the Cu core were formed at the later stages of the synthesis process [63]. This implies that the growth procedure follows the tip growth mechanism described in our previous report [25], where the Cu tips serve as the catalytically active sites for the decomposition and precipitation of carbon atoms to form VACNT shells. Figure 1f shows that, although the individual VACNTs in the bundle are filled, the Cu nanowires are discontinuous and characterized by segments and dots. This can be attributed to the high internal energy from the very high growth temperature (760 °C). Consequently, the atomic activity of the Cu atoms in the nanowires increases. As a result, the Cu nanowires stretch due to pre-melting [64], decreasing the bonding strength of the Cu atoms and causing atomic gliding dislocations and grain boundary movement to occur [65].

The EDS mappings in Figure 2(a1–c2) reveal that the interior regions of darker contrast identified in the TEM images in Figure 1d–f are Cu nanowires. Figure 2(a1,a2) are the EDS mappings of two carbon nanotubes synthesized at a temperature of 650 °C; the two CNTs are in contact at their roots. Figure 2(a2) shows that the two carbon nanotubes are filled with continuous Cu nanowires. Figure 2(b1,b2) are the EDS mappings of a nanotube synthesized at a temperature of 700 °C; the nanotube is also filled with a long continuous Cu nanowire. From the mapping, we can also deduce that each of the individual VACNTs has Cu nanowires completely encapsulated in them. Figure 2(c1,c2) show the EDS mappings of a bundle of nanotubes synthesized at a temperature of 760 °C. Figure 2(c2) shows Cu nanowires' discontinuity and Cu dots inside the carbon nanotubes.

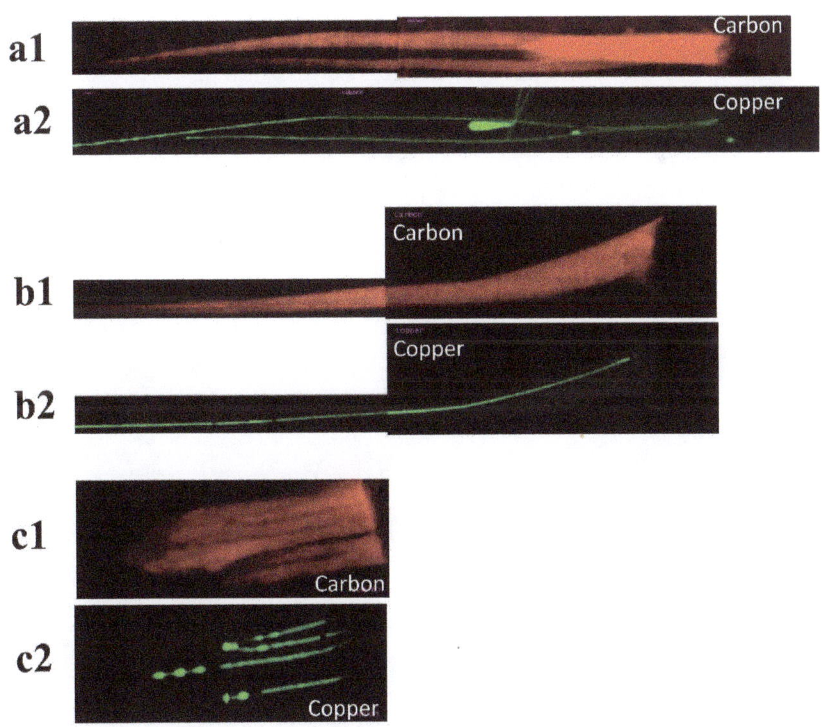


**Figure 2.** TEM-EDS mapping of **(a1,a2)** S650, **(b1,b2)** S700, and **(c1,c2)** S760 Cu@VACNTs.

Figure 3a–c are the HRTEM images of the various Cu@VACNT samples (the insets provide zoomed-out views of the corresponding Cu@VACNTs). In Figure 3a, two sets of regular crystal lattice spacings, 0.21 and 0.34 nm [66,67], related to the separation between two (111) planes of face-centered cubic (fcc) Cu crystal and (002) graphitic carbon planes are observed. Figure 3b,c indicate the presence of the (110) and (111) planes of the fcc Cu crystal with lattice spacings of 0.25 and 0.21 nm, respectively [66]. These observations demonstrate that the Cu nanowires encapsulated inside the S650, S700, and S760 Cu@VACNTs are single crystals with good crystallization.

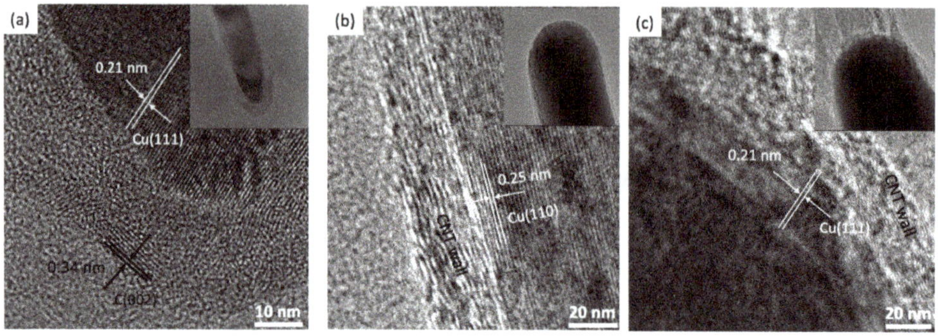


**Figure 3.** HRTEM images of (**a**) S650, (**b**) S700, and (**c**) S760 Cu@VACNTs. The insets provide zoomed-out views of the corresponding Cu@VACNTs.

Figure 4 shows the XRD pattern of the as-synthesized Cu@VACNTs at 600 °C, 700 °C, and 760 °C. Figure 4a shows diffraction peaks (2θ) for all samples at 26° related to the (002) graphitic planes. Figure 4b is the close-up view of Figure 4a in the 2θ range of 40–80°, it shows the diffraction peaks of pure Cu crystals at 43.18°, 50.34°, and 74.06°, which can be indexed as (111), (200), and (220) planes of the fcc Cu phase. However, we observed a broad diffraction peak at 52.30 related to the (020) crystalline phase of copper oxide (CuO). The observed CuO impurity could be from residual Cu nanoparticles attached to the surface of the VACNTs which are susceptible to oxidation upon exposure to atmospheric conditions. During the VACNT growth and Cu filling processes, not all Cu particles are fully encapsulated within the VACNTs. Some residual nanoparticles adhere to the external surfaces of the VACNTs due to surface energy interactions [68]. These surface-bound Cu nanoparticles, when exposed to air, readily oxidize to form CuO [69].

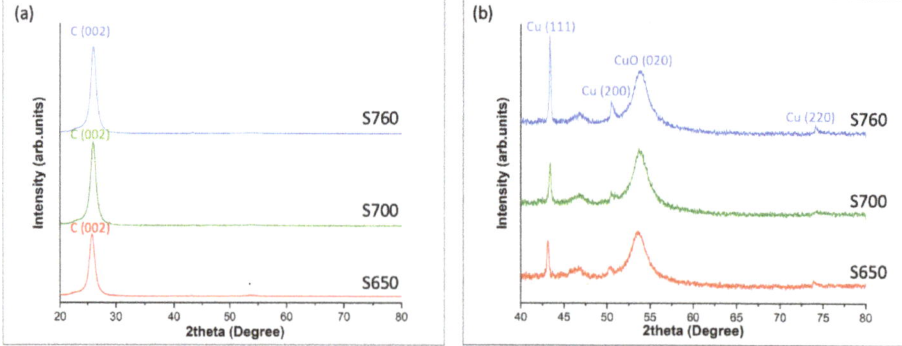


**Figure 4.** XRD patterns of the S650, S700, and S760 samples. (**a**) shows the peak of the graphitic carbon layers, and (**b**), zoom-in of (**a**), shows the peaks of different Cu crystals encapsulated inside the VACNTs.

From these findings, it can be clearly seen that all the growth temperatures resulted in Cu-filled VACNTs, although their densities and structures are different. These results are slightly different from previous findings where we reported discrete Cu nanorod-filled CNTs at 650 °C [25]. Here, due to the thinness of the Cu foil, heating under the reduction environment was able to induce surface breakup of the Cu substrates even at a lower temperature of 650 °C. As a result, the disintegrated solid Cu nanoparticles were transformed into their quasi-liquid state, facilitating the filling of the VACNTs via quasi-liquid capillary adsorption [70,71].

### 3.2. Field Emission (FE) Measurements

#### 3.2.1. Field Emission Theory

Electrons from the surfaces of metals or semiconductors can only be knocked off when an external force supplies additional energy [72]. This extra energy can be produced using a variety of methods, including thermal processes, energy storage in an electric field, using the kinetic energy of charges, or light energy. Depending on the type of source, there are four primary ways to eject electrons from the solid surfaces: (1) cold emission, (2) heated emission, (3) field emission (FE), and (4) secondary emission. The FE method, which involves electron emission from a conductive metal surface by applying a strong electric field, takes advantage of sharp electrodes to enhance the local electric field [73]. In contrast to the other three techniques, the FE process transfers energy to trapped electrons in the material by deforming the potential barrier on its surface.

When an external electric field is applied, the potential barrier changes depending on the strength of the field. Figure 5 illustrates the potential as a simple plane with the electron needing enough energy to exit the material. This energy is known as the work function of the metal and is typically expressed in electron volts (eV). The red dashed line depicts the potential barrier before applying an external electric field. When an electric field is applied (blue dashed line), this potential barrier is distorted, which is shown as a curved dashed line (image potential) in the image. The stronger the applied electric field, the more the barrier is lowered and narrowed (effective barrier) and as a result, cold-trapped electrons near the Fermi energy ($E_F$) level can escape into the vacuum by the 'quantum tunneling' effect, leading to field emission [74].

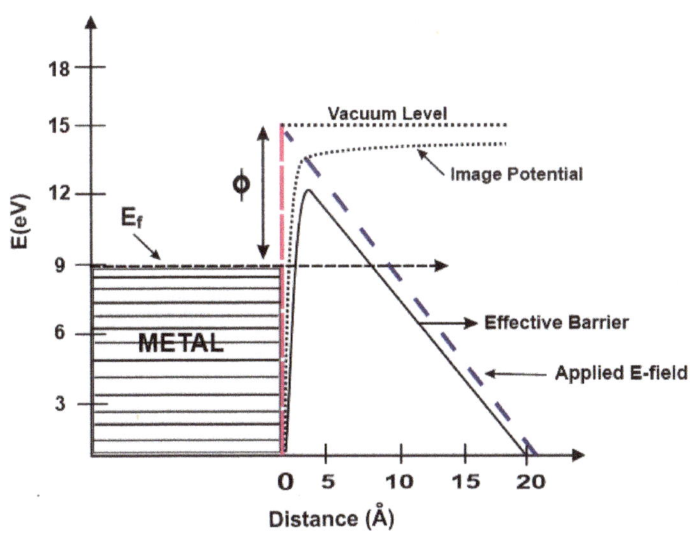


**Figure 5.** Potential energy diagram showing how an exterior electric field affects the energy barrier for electrons at the surface of a metal.

With the advancement of micro and nanofabrication technology in recent decades, generating high electric fields has been made possible by using sharp cathodes with few-hundred-nanometer tip diameters. However, compared to conventional metal emitters, CNTs possess a smaller tip curvature from which electrons are extracted. The reduced tip diameters in CNTs translate to a larger field enhancement factor and a greatly reduced turn-on field. In addition to the nanosized tips of the CNTs, their other properties, such as high mechanical strength, high electronic and heat conductivities, chemical inertness, and high aspect ratio, contribute to the excellent FE properties exhibited by CNTs. The FE behavior strongly depends on the CNTs' morphology, spatial distribution, diameter, degree of alignment, the contact resistance between the CNTs and the substrate, and the nature of the CNTs' tips [75].

3.2.2. Field Emission Results

The emission currents from the Cu@VACNTs were measured at different electric fields to understand the performance of the Cu@VACNT emitters. Figure 6 presents the FE results from this study. To ensure reproducible emission characteristics, a 'cleaning conditioning' procedure was conducted during the first five current density vs. applied electric field (J vs. E) cycles [76]. The conditioning process evaporates surface absorbates from the VACNTs, which could otherwise trigger the FE quicker than their absorbate-free counterparts, resulting in instabilities and significant fluctuations in the measured emission current densities. Thus, the FE results presented in Figure 6a,b were from the subsequent five cycles after the initial cleaning process. Figure 6a,b are the semi-log and linear scale representations of the J vs. E properties of the Cu@VACNTs, respectively. From the curves of the J vs. E plots in Figure 6a, we estimated the turn-on ($E_{turn\text{-}on}$) and threshold ($E_{th}$) electric fields. $E_{turn\text{-}on}$ was defined as the applied electric field required to obtain an emission current density of 10 $\mu A/cm^2$ and $E_{th}$ as the applied electric field required to produce an emission current of 1 $mA/cm^2$. The Cu@VACNTs showed low $E_{turn\text{-}on}$ and $E_{th}$ values of 2.33 V/μm and 3.29 V/μm, respectively. Compared to the $E_{turn\text{-}on}$ and $E_{th}$ (1.77 V/μm and 2.43 V/μm) values previously reported by our group, the slightly larger $E_{turn\text{-}on}$ and $E_{th}$ reported in this work could be a result of increased screening effect, stemming from reduced inter-tube spacing between individual VACNT emitters and lower aspect ratio caused by the larger diameter of the VACNTs. From Figure 6b, the maximum emission current density was found to be in the range of 20.5 to 21.0 $mA/cm^2$ at an applied electric field of 4.5 V/μm.

The emission current density of CNT emitters can be expressed by the following Fowler–Nordheim (F-N) equation [77]:

$$J = \left(\frac{A\beta^2E^2}{\Phi}\right)\exp\left(\frac{-B\Phi^{\frac{3}{2}}}{\beta E}\right) \quad (1)$$

where J is the emission current density, A ($1.56 \times 10^{-6}$ A $V^{-2}$ eV) and B ($6.83 \times 10^{9}$ $eV^{-\frac{3}{2}}$ $Vm^{-1}$) are constants dependent on the work function of the material and the local electric field at the emission tip, Φ (5 eV) is the work function of the VACNT field emitters, E is the applied macroscopic electric field between the VACNT emitters and the anode, and β is the field enhancement factor. Equation (1) can further be transformed as follows:

$$\text{In}\left(\frac{J}{E^2}\right) = \text{In}\left(\frac{A\beta^2}{\Phi}\right) - \frac{B\Phi^{\frac{3}{2}}}{\beta E} \quad (2)$$

From Equation (2), it can be seen that $\ln\left(\frac{J}{E^2}\right)$ is linearly proportional to $\frac{1}{E}$ with a slope of $-\frac{B\Phi^{\frac{3}{2}}}{\beta}$ Hence, the field enhancement factor, β, can be calculated from the slope of the F-N plot shown in Figure 6c. The straight line in the F-N plot indicates the quantum mechanical tunneling process [78]. From the calculation, we obtained a high field enhancement factor of $\beta = 2037$.

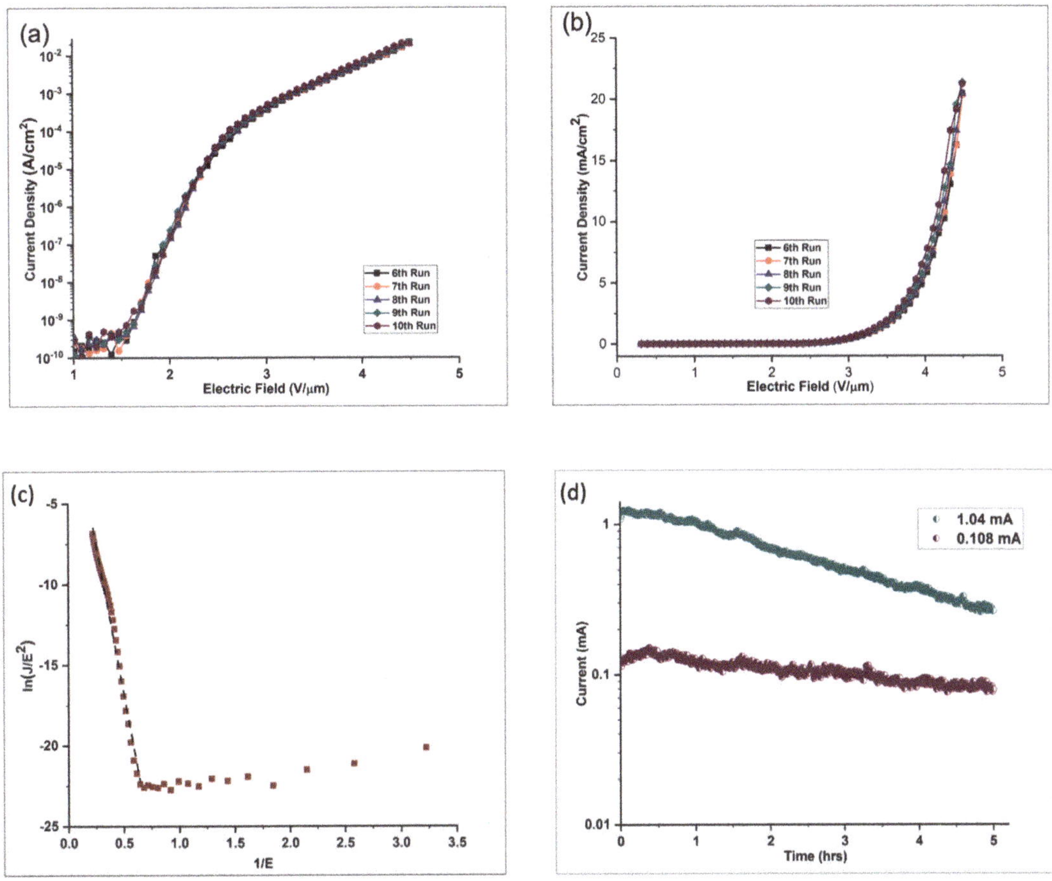


**Figure 6.** Graph of emission current density vs. applied electric field (J vs. E plot) of the S700 sample, (**a**) semi-log and (**b**) linear scale. (**c**) Fowler–Nordheim (F-N) plot of the S700 sample. (**d**) Stability test of the S700 sample at initial currents of 0.108 mA and 1.04 mA.

It is important to note that Equation (1) used in our analysis is a simple approximation of the F-N equation employed by Edgcombe et al. [79] in their study. While the F-N equation is theoretically derived for electron emission from flat surfaces at 0 K [74], it has been found to be experimentally valid at much higher temperatures. This is because field emission is primarily driven by the electric field, and the tunneling mechanism described by the F-N equation continues to be valid and dominant even when thermal effects are present. The simplified version used in our study omits the correction factor v(y), which accounts for exchange and correlation effects for very small tip diameters. Given the relatively large average diameter of our Cu@VACNTs compared to their length, we estimate that the impact of curvature corrections is minimal.

Additionally, the work function of 5 eV used for our analysis is a common value employed in field emission studies involving CNT emitters. This value is well established in the literature for pure CNTs and provides a standardized basis for comparison [80]. Although our CNTs are filled with Cu, which typically has a work function of 4.6 to 4.9 eV [81], the interaction between the Cu and the carbon matrix could alter the Fermi level and affect the overall work function [82]. Thus, choosing 5 eV provides a reasonable approximation, allowing us to maintain consistency with previous studies and ensure ease of comparison.

Stability is another essential factor to be considered in determining the applicability of CNTs as field emission devices. To assess this, we tested the stability of two Cu@VACNTs as shown in Figure 6d. At a low initial test current, the Cu@VACNTs showed good stability, decreasing from 0.108 mA to only about 0.079 mA, which represents about 26% current degradation from the initial test current after the 5 h test. At a higher initial test current of 1.04 mA, the Cu@VACNTs showed good stability in the first 1.5 h with only 19% degradation from the initial test current. Thereafter, the current degraded rapidly from 0.89 mA to a low of 0.264 mA, representing a total degradation of 74% from the initial test current after the 5 h test. The significant current degradation at elevated test current is likely due to joule heating effects exacerbated by the low inter-tube spacing within the arrays. In a closely packed VACNT array, individual VACNTs shield each other from the applied electric field due to electrostatic screening [83]. This leads to a reduction in the effective emission area and the field enhancement factor ($\beta$), as only a fraction of the VACNTs actively participate in the emission process [84]. When this happens, the active VACNT emitters experience increased localized heating (joule heating) impeding effective thermal dissipation. This effect becomes more pronounced at high emission currents as the resulting overheating weakens the bond between the VACNTs and the substrate, leading to the peeling of the VACNTs from the substrate surface [85].

Figure 7a,b show the SEM images of the VACNTs after the stability tests at initial test currents of 0.108 mA and 1.04 mA, respectively. Figure 7b shows that more VACNTs were peeled off from the Cu substrate during the test at 1.04 mA compared to the VACNTs tested at the low current of 0.108 mA (Figure 7a). As discussed earlier, the observed degradation at the higher test current is most likely a result of the increased joule heating at 1.04 mA, exacerbating thermal stress within the VACNT array, and weakening the adhesion between the VACNTs and the substrate, hence the peeling. Figure 7c,d are the TEM images of samples presented in Figure 7a,b, respectively, showing the tip geometry change after the stability test. At the tip of the VACNTs, where the electric field is concentrated, the field emission can lead to the evaporation of carbon atoms [86]. This field-induced evaporation occurs because the strong local electric fields at the tips can overcome the cohesive energy of carbon atoms, causing them to be ejected from the tips. This process gradually reshapes the tip, leading to blunting [87]. Figure 7d reveals that the tip of the VACNT sample, subjected to a high test current of 1.04 mA, not only exhibited blunting but also suffered significant damage to its structural integrity. This deterioration is attributed to the stronger local electric fields and increased joule heating associated with higher test currents, accelerating the degradation of the tips, which serve as the primary emission sites.

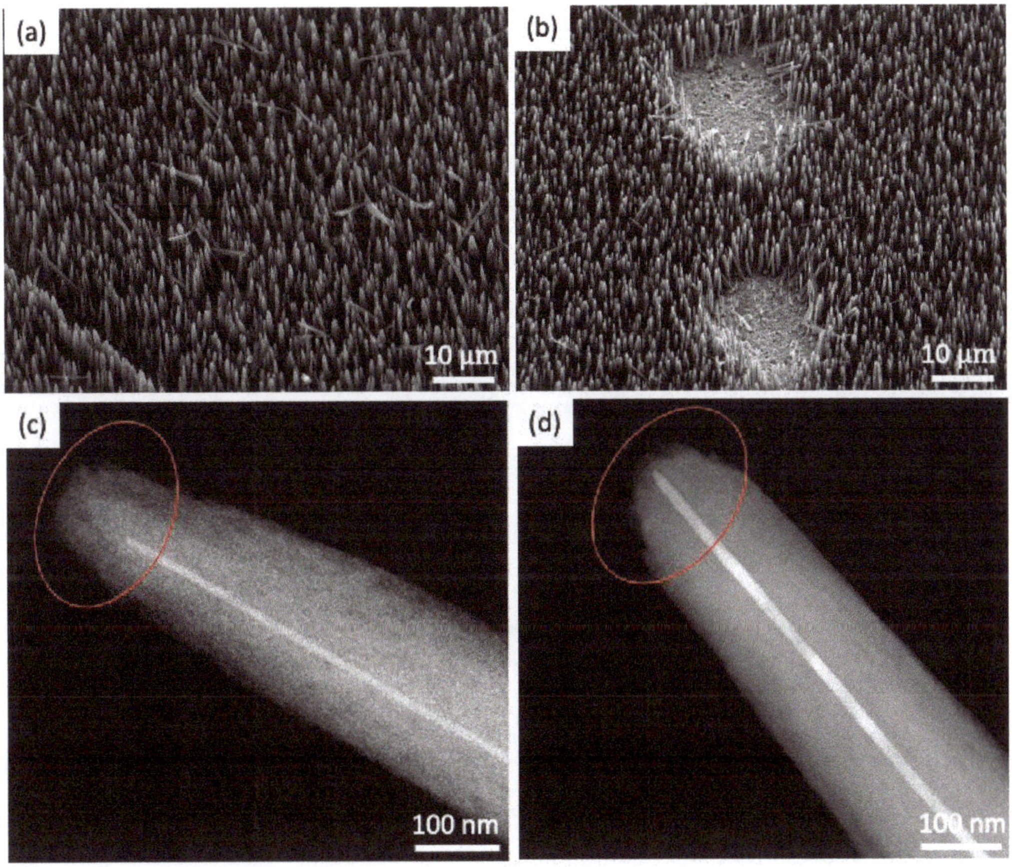


**Figure 7.** (**a**,**b**) SEM images of the VACNTs after emission stability tests at 0.108 mA and 1.04 mA, respectively. (**c**,**d**) TEM images showing the tips of VACNTs in Figures (**a**,**b**).

## 4. Conclusions

In this study, we have successfully synthesized Cu-filled VACNT (Cu@VACNT) arrays on thin Cu foils of 0.1 mm thickness at different temperatures, via the PECVD technique. Our findings highlight the significant role of substrate thickness in promoting the Cu-filling process at lower temperatures. This was evident from the complete filling of the as-synthesized VACNTs with Cu at a temperature as low as 650 °C, which contrasts with our previous studies on thicker Cu substrates where filling occurred only at temperatures ranging from 700 to 750 °C. The detailed SEM and TEM analysis revealed a consistent core structure with the encapsulated Cu nanowires showing good crystallinity, despite the occurrence of some discontinuities and dislocations at a higher growth temperature of 760 °C. The field emission measurements of the S700 Cu@VACNTs indicated favorable field emission properties, with low turn-on and threshold fields. This underscores the huge potential of Cu@VACNT emitters in field emission applications due to the synergetic effect of the sharp tips of the VACNTs and the highly conductive Cu nanowire fillers. The stability test revealed satisfactory performance at a lower emission current, though challenges remain at a much higher emission current likely due to the increased joule heating effect caused by the low inter-tube spacing within the VACNT arrays. To fully utilize the potential of these Cu@VACNTs, further work is required to control the density of the VACNTs and optimize the inter-tube spacing for reduced screening effects.

**Supplementary Materials:** The following supporting information can be downloaded at https://www.mdpi.com/article/10.3390/nano14110988/s1, Figure S1: Schematic illustration of the setup for field emission measurements.

**Author Contributions:** C.E.N.: conceptualization, methodology, investigation, formal analysis, writing—original draft. A.T.: investigation, discussion, review, and editing. J.W.: investigation, discussion, review, and editing. D.S.B.: investigation, discussion, review, and editing. W.L.: conceptualization, methodology, discussion, validation, funding acquisition, supervision, review, and editing. All authors have read and agreed to the published version of the manuscript.

**Funding:** This work was supported by the National Science Foundation under grant 2213923.

**Data Availability Statement:** The data presented in this study are not available due to privacy.

**Acknowledgments:** The authors would also like to acknowledge and appreciate the support from the Advanced Materials Engineering Research Institute (AMERI) at Florida International University. This work was performed, in part, at the Center for Integrated Nanotechnologies, an Office of Science User Facility operated for the U.S. Department of Energy (DOE) Office of Science. Los Alamos National Laboratory, an affirmative action equal opportunity employer, is managed by Triad National Security, LLC for the U.S. Department of Energy's NNSA, under contract 89233218CNA000001. This paper describes objective technical results and analysis. Any subjective views or opinions that might be expressed in the paper do not necessarily represent the views of the U.S. Department of Energy or the United States Government.

**Conflicts of Interest:** The authors do not have any conflicts of interest to declare.

## References

1. Berry, M.J. Helical microtubules of graphitic carbon. *Nature* **1991**, *354*, 737–740. [CrossRef]
2. Yakobson, B.I.; Brabec, C.J.; Bernholc, J. Nanomechanics of carbon tubes: Instabilities beyond linear response. *Phys. Rev. Lett.* **1996**, *76*, 2511–2514. [CrossRef]
3. Abdulhameed, A.; Wahab, N.Z.A.; Mohtar, M.N.; Hamidon, M.N.; Shafie, S.; Halin, I.A. Methods and Applications of Electrical Conductivity Enhancement of Materials Using Carbon Nanotubes. *J. Electron. Mater.* **2021**, *50*, 3207–3221. [CrossRef]
4. Han, Z.; Fina, A. Thermal conductivity of carbon nanotubes and their polymer nanocomposites: A review. *Prog. Polym. Sci.* **2011**, *36*, 914–944. [CrossRef]
5. Inam, F.; Yan, H.; Reece, M.J.; Peijs, T. Structural and chemical stability of multiwall carbon nanotubes in sintered ceramic nanocomposite. *Adv. Appl. Ceram.* **2010**, *109*, 240–245. [CrossRef]
6. Lim, S.C.; Jeong, H.J.; Kim, K.S.; Lee, I.B.; Bae, D.J.; Lee, Y.H. Extracting independently the work function and field enhancement factor from thermal-field emission of multi-walled carbon nanotube tips. *Carbon* **2005**, *43*, 2801–2807. [CrossRef]
7. Okwundu, O.S.; Aniekwe, E.U.; Nwanno, C.E. Unlimited potentials of carbon: Different structures and uses (a Review). *Metall. Mater. Eng.* **2018**, *24*, 145–171. [CrossRef]
8. Kohls, A.; Maurer Ditty, M.; Dehghandehnavi, F.; Zheng, S.Y. Vertically Aligned Carbon Nanotubes as a Unique Material for Biomedical Applications. *ACS Appl. Mater. Interfaces* **2022**, *14*, 6287–6306. [CrossRef]
9. Liu, Q.; Shi, X.; Jiang, Q.; Li, R.; Zhong, S.; Zhang, R. Growth mechanism and kinetics of vertically aligned carbon nanotube arrays. *EcoMat* **2021**, *3*, e12118. [CrossRef]
10. Shi, W.; Plata, D.L. Vertically aligned carbon nanotubes: Production and applications for environmental sustainability. *Green Chem.* **2018**, *20*, 5245–5260. [CrossRef]
11. He, Z.; Zhang, S.; Zheng, L.; Liu, Z.; Zhang, G.; Wu, H.; Wang, B.; Liu, Z.; Jin, Z.; Wang, G. Si-Based NIR Tunneling Heterojunction Photodetector with Interfacial Engineering and 3D-Graphene Integration. *IEEE Electron. Device Lett.* **2022**, *43*, 1818–1821. [CrossRef]
12. Zhang, Z.Y.; Lee, C.S.; Zhang, W.J. Vertically aligned graphene nanosheet arrays: Synthesis, properties and applications in electrochemical energy conversion and storage. *Adv. Energy Mater.* **2017**, *7*, 1700678. [CrossRef]
13. Liang, Q.; Yao, X.; Wang, W.; Liu, Y.; Wong, C.P. A three-dimensional vertically aligned functionalized multilayer graphene architecture: An approach for graphene-based thermal interfacial materials. *ACS Nano* **2011**, *5*, 2392–2401. [CrossRef]
14. Chimowa, G.; Tshabalala, Z.P.; Akande, A.A.; Bepete, G.; Mwakikunga, B.; Ray, S.S.; Benecha, E.M. Improving methane gas sensing properties of multi-walled carbon nanotubes by vanadium oxide filling. *Sens. Actuators B Chem.* **2017**, *247*, 11–18. [CrossRef]
15. Tyagi, P.K.; Singh, M.K.; Misra, A.; Palnitkar, U.; Misra, D.S.; Titus, E.; Ali, N.; Cabral, G.; Gracio, J.; Roy, M.; et al. Preparation of Ni-filled carbon nanotubes for key potential applications in nanotechnology. *Thin Solid Films* **2004**, *469–470*, 127–130. [CrossRef]
16. Dorozhkin, P.S.; Tovstonog, S.V.; Golberg, D.; Zhan, J.; Ishikawa, Y.; Shiozawa, M.; Nakanishi, H.; Nakata, K.; Bando, Y. A liquid-Ga-filled carbon nanotube: A miniaturized temperature sensor and electrical switch. *Small* **2005**, *1*, 1088–1093. [CrossRef] [PubMed]

17. Gao, Y.; Bando, Y.; Liu, Z.; Golberg, D.; Nakanishi, H. Temperature measurement using a gallium-filled carbon nanotube nanothermometer. *Appl. Phys. Lett.* **2003**, *83*, 2913–2915. [CrossRef]
18. Nwanno, C.E.; Li, W. Aligned carbon nanotubes for lithium-ion batteries: A review. *Nano Res.* **2023**, *16*, 12384–12410. [CrossRef]
19. Kiamahalleh, M.; Najafpour, G.; Sata, S.; Buniran, S.; Zein, S. Multiwalled Carbon Nanotubes Filled with Transition Metal Oxides as Supercapacitor Materials. *Carbon* **2010**, *2*, 3.
20. Ugarte, D.; Stöckli, T.; Bonard, J.M.; Châtelain, A.; De Heer, W.A. Filling carbon nanotubes. *Appl. Phys. A Mater. Sci. Process.* **1998**, *67*, 101–105. [CrossRef]
21. Zhang, X.X.; Wen, G.H.; Huang, S.; Dai, L.; Gao, R.; Wang, Z.L. Magnetic properties of Fe nanoparticles trapped at the tips of the aligned carbon nanotubes. *J. Magn. Magn. Mater.* **2001**, *231*, 9–12. [CrossRef]
22. Gautam, U.K.; Costa, P.M.F.J.; Bando, Y.; Fang, X.; Li, L.; Imura, M.; Golberg, D. Recent developments in inorganically filled carbon nanotubes: Successes and challenges. *Sci. Technol. Adv. Mater.* **2010**, *11*, 054501. [CrossRef]
23. Leonhardt, A.; Ritschel, M.; Kozhuharova, R.; Graff, A.; Mühl, T.; Huhle, R.; Mönch, I.; Elefant, D.; Schneider, C.M. Synthesis and properties of filled carbon nanotubes. *Diam. Relat. Mater.* **2003**, *12*, 790–793. [CrossRef]
24. Baro, M.; Pal, A.R. One-step grown multi-walled carbon nanotubes with Ni filling and decoration. *J. Phys. D Appl. Phys.* **2015**, *48*, 225303. [CrossRef]
25. Thapa, A.; Wang, X.; Li, W. Synthesis and field emission properties of Cu-filled vertically aligned carbon nanotubes. *Appl. Surf. Sci.* **2021**, *537*, 148086. [CrossRef]
26. Jin, C.; Wu, Q.; Yang, G.; Zhang, H.; Zhong, Y. Investigation on hybrid nanofluids based on carbon nanotubes filled with metal nanoparticles: Stability, thermal conductivity, and viscosity. *Powder Technol.* **2021**, *389*, 1–10. [CrossRef]
27. Liu, Z.; Han, B. Synthesis of carbon-nanotube composites using supercritical fluids and their potential applications. *Adv. Mater.* **2009**, *21*, 825–829. [CrossRef]
28. Wang, P.; Huang, B.; Wei, J.; Qin, X.; Yao, S.; Zhang, Q. Preparation of Cu nanoparticles on carbon nanotubes by solution infusion method and calcining in ambient atmosphere. *Mater. Lett.* **2007**, *61*, 5255–5257. [CrossRef]
29. Ye, X.R.; Lin, Y.; Wang, C.; Wai, C.M. Supercritical fluid fabrication of metal nanowires and nanorods templated by multiwalled carbon nanotubes. *Adv. Mater.* **2003**, *15*, 316–319. [CrossRef]
30. Arai, S. Fabrication of Metal/Carbon Nanotube Composites by Electrochemical Deposition. *Electrochem* **2021**, *2*, 563–589. [CrossRef]
31. Setlur, A.A.; Lauerhaas, J.M.; Dai, J.Y.; Chang, R.P.H. A method for synthesizing large quantities of carbon nanotubes and encapsulated copper nanowires. *Appl. Phys. Lett.* **1996**, *69*, 345–347. [CrossRef]
32. Zhang, G.Y.; Wang, E.G. Cu-filled carbon nanotubes by simultaneous plasma-assisted copper incorporation. *Appl. Phys. Lett.* **2003**, *82*, 1926–1928. [CrossRef]
33. Poudel, Y.R.; Li, W. Synthesis, properties, and applications of carbon nanotubes filled with foreign materials: A review. *Mater. Today Phys.* **2018**, *7*, 7–34. [CrossRef]
34. Fu, Q.; Weinberg, G.; Su, D.S. Selective filling of carbon nanotubes with metals by selective washing. *New Carbon Mater.* **2008**, *23*, 17–20. [CrossRef]
35. Demoncy, N.; Stéphan, O.; Brun, N.; Colliex, C.; Loiseau, A.; Pascard, H. Filling carbon nanotubes with metals by the arc-discharge method: The key role of sulfur. *Eur. Phys. J. B-Condens. Matter Complex Syst.* **1998**, *4*, 147–157. [CrossRef]
36. Hsin, Y.L.; Hwang, K.C.; Chen, F.R.; Kai, J.J. Production and in-situ metal filling of carbon nanotubes in water. *Adv. Mater.* **2001**, *13*, 830–833. [CrossRef]
37. Ajayan, P.M.; Stephan, O.; Redlich, P.; Colliex, C. Carbon nanotubes as removable templates for metal oxide nanocomposites and nanostructures. *Nature* **1995**, *375*, 564–567. [CrossRef]
38. Liu, X.; Marangon, I.; Melinte, G.; Wilhelm, C.; Ménard-Moyon, C.; Pichon, B.P.; Ersen, O.; Aubertin, K.; Baaziz, W.; Pham-Huu, C.; et al. Design of covalently functionalized carbon nanotubes filled with metal oxide nanoparticles for imaging, therapy, and magnetic manipulation. *ACS Nano* **2014**, *8*, 11290–11304. [CrossRef] [PubMed]
39. Tsang, S.C.; Chen, Y.K.; Harris, P.J.F.; Green, M.L.H. A simple chemical method of opening and filling carbon nanotubes. *Nature* **1994**, *372*, 159–162. [CrossRef]
40. Takenobu, T.; Takano, T.; Shiraishi, M.; Murakami, Y.; Ata, M.; Kataura, H.; Achiba, Y.; Iwasa, Y. Stable and controlled amphoteric doping by encapsulation of organic molecules inside carbon nanotubes. *Nat. Mater.* **2003**, *2*, 683–688. [CrossRef]
41. Lu, J.; Nagase, S.; Yu, D.; Ye, H.; Han, R.; Gao, Z.; Zhang, S.; Peng, L. Amphoteric and controllable doping of carbon nanotubes by encapsulation of organic and organometallic molecules. *Phys. Rev. Lett.* **2004**, *93*, 9–12. [CrossRef]
42. Li, L.J.; Khlobystov, A.N.; Wiltshire, J.G.; Briggs, G.A.D.; Nicholas, R.J. Diameter-selective encapsulation of metallocenes in single-walled carbon nanotubes. *Nat. Mater.* **2005**, *4*, 481–485. [CrossRef] [PubMed]
43. Yanagi, K.; Miyata, Y.; Kataura, H. Highly stabilized β-carotene in carbon nanotubes. *Adv. Mater.* **2006**, *18*, 437–441. [CrossRef]
44. Tschirner, N.; Brose, K.; Maultzsch, J.; Yanagi, K.; Kataura, H.; Thomsen, C. The influence of incorporated β-carotene on the vibrational properties of single wall carbon nanotubes. *Phys. Status Solidi Basic Res.* **2010**, *247*, 2734–2737. [CrossRef]
45. Yanagi, K.; Iakoubovskii, K.; Kazaoui, S.; Minami, N.; Maniwa, Y.; Miyata, Y.; Kataura, H. Light-harvesting function of β-carotene inside carbon nanotubes. *Phys. Rev. B* **2006**, *74*, 155420. [CrossRef]

46. Caillier, C.; Machon, D.; San-Miguel, A.; Arenal, R.; Montagnac, G.; Cardon, H.; Kalbac, M.; Zukalova, M.; Kavan, L. Probing high-pressure properties of single-wall carbon nanotubes through fullerene encapsulation. *Phys. Rev. B* **2008**, *77*, 125418. [CrossRef]
47. Simon, F.; Peterlik, H.; Pfeiffer, R.; Bernardi, J.; Kuzmany, H. Fullerene release from the inside of carbon nanotubes: A possible route toward drug delivery. *Chem. Phys. Lett.* **2007**, *445*, 288–292. [CrossRef]
48. Smith, B.W.; Monthioux, M.; Luzzi, D.E. Carbon nanotube encapsulated fullerenes: A unique class of hybrid materials. *Chem. Phys. Lett.* **1999**, *315*, 31–36. [CrossRef]
49. Kim, B.M.; Qian, S.; Bau, H.H. Filling Carbon Nanotubes with Particles. *Nano Lett.* **2005**, *5*, 873–878. [CrossRef]
50. Wei, X.; Tao, J.; Hu, Y.; Liu, Y.; Bao, R.; Li, F.; Fang, D.; Li, C.; Yi, J. Enhancement of mechanical properties and conductivity in carbon nanotubes (CNTs)/Cu matrix composite by surface and intratube decoration of CNTs. *Mater. Sci. Eng. A* **2021**, *816*, 141248. [CrossRef]
51. Golberg, D.; Costa, P.M.F.J.; Mitome, M.; Hampel, S.; Haase, D.; Mueller, C.; Leonhardt, A.; Bando, Y. Copper-filled carbon nanotubes: Rheostatlike behavior and Femtogram copper mass transport. *Adv. Mater.* **2007**, *19*, 1937–1942. [CrossRef]
52. Sun, Y.; Onwona-Agyeman, B.; Miyasato, T. Controlling the resistivity of multi-walled carbon nanotube networks by copper encapsulation. *Mater. Lett.* **2011**, *65*, 3187–3190. [CrossRef]
53. Kruszka, B.; Terzyk, A.P.; Wiśniewski, M.; Gauden, P.A.; Szybowicz, M. Synthesis of carbon nanotubes and nanotube forests on copper catalyst. *Mater. Res. Express* **2014**, *1*, 035040. [CrossRef]
54. Chen, G.; Neupane, S.; Li, W.Z. Electron field emission properties of vertically aligned carbon nanotube point emitters. *Diam. Relat. Mater.* **2012**, *25*, 134–138. [CrossRef]
55. Atthipalli, G.; Epur, R.; Kumta, P.N.; Yang, M.; Lee, J.K.; Gray, J.L. Nickel catalyst-assisted vertical growth of dense carbon nanotube forests on bulk copper. *J. Phys. Chem. C* **2011**, *115*, 3534–3538. [CrossRef]
56. Yin, X.; Wang, Q.; Lou, C.; Zhang, X.; Lei, W. Growth of multi-walled CNTs emitters on an oxygen-free copper substrate by chemical-vapor deposition. *Appl. Surf. Sci.* **2008**, *254*, 6633–6636. [CrossRef]
57. Rao, L.; Reddy, N.K.; Coulombe, S.; Meunier, J.L.; Munz, R.J. Carbon nanotubes as nanoparticles collector. *J. Nanoparticle Res.* **2007**, *9*, 689–695. [CrossRef]
58. Sepahvand, S.; Safaei, P.; Sanaee, Z. Growth of Carbon Nano Tubes on Copper Substrate Suitable for Lithium Ion Battery Anode. *Procedia Mater. Sci.* **2015**, *11*, 634–638. [CrossRef]
59. Katar, S.L.; González-Berríos, A.; De Jesus, J.; Weiner, B.; Morell, G. Direct deposition of bamboo-like carbon nanotubes on copper substrates by sulfur-assisted HFCVD. *J. Nanomater.* **2008**, *2008*, 515890. [CrossRef]
60. Zhu, J.; Jia, J.; Kwong, F.L.; Ng, D.H.L. Synthesis of bamboo-like carbon nanotubes on a copper foil by catalytic chemical vapor deposition from ethanol. *Carbon* **2012**, *50*, 2504–2512. [CrossRef]
61. Thapa, A.; Neupane, S.; Guo, R.; Jungjohann, K.L.; Pete, D.; Li, W. Direct growth of vertically aligned carbon nanotubes on stainless steel by plasma enhanced chemical vapor deposition. *Diam. Relat. Mater.* **2018**, *90*, 144–153. [CrossRef]
62. Neupane, S.; Lastres, M.; Chiarella, M.; Li, W.; Su, Q.; Du, G. Synthesis and field emission properties of vertically aligned carbon nanotube arrays on copper. *Carbon* **2012**, *50*, 2641–2650. [CrossRef]
63. Tao, X.Y.; Zhang, X.B.; Cheng, J.P.; Liu, F. Synthesis and characterization of Cu filled carbon nanohorns. *Mater. Chem. Phys.* **2007**, *104*, 210–214. [CrossRef]
64. Sun, C.Q. Dominance of broken bonds and nonbonding electrons at the nanoscale? *Nanoscale* **2010**, *2*, 1930–1961. [CrossRef] [PubMed]
65. Huang, P.H.; Kuo, J.K. Thermal stretching of defective nanowires: The coupled effects of vacancy cluster defects, operating temperature, and wire cross-sectional area. *Appl. Phys. A* **2011**, *103*, 1083–1092. [CrossRef]
66. Lin, C.; Lee, C.; Chin, T.; Xiang, R.; Ishikawa, K.; Shiomi, J.; Maruyama, S. Anisotropic electrical conduction of vertically-aligned single-walled carbon nanotube films. *Carbon* **2010**, *49*, 1446–1452. [CrossRef]
67. Liu, J.; Zhang, Y.; Zhang, L.; Xie, F.; Vasileff, A.; Qiao, S.Z. Graphitic Carbon Nitride (g-$C_3N_4$)-Derived N-Rich Graphene with Tuneable Interlayer Distance as a High-Rate Anode for Sodium-Ion Batteries. *Adv. Mater.* **2019**, *31*, 1901261. [CrossRef]
68. Wang, G.; Zhang, Y.; Zhu, L.; Ma, Y.; Zhang, S.; Huang, Z.; Cao, Z.; Meng, X. Theoretical and experimental studies of nucleation and interface structure between carbon nanotubes and metals. *J. Mater. Sci.* **2023**, *58*, 1086–1098. [CrossRef]
69. Jardón-Maximino, N.; Pérez-Alvarez, M.; Sierra-Ávila, R.; Ávila-Orta, C.A.; Jiménez-Regalado, E.; Bello, A.M.; González-Morones, P.; Cadenas-Pliego, G. Oxidation of copper nanoparticles protected with different coatings and stored under ambient conditions. *J. Nanomater.* **2018**, *2018*, 9512768. [CrossRef]
70. Zhao, Y.; Zhang, Y.; Li, Y.; Yan, Z. A flexible chemical vapor deposition method to synthesize copper@carbon core-shell structured nanowires and the study of their structural electrical properties. *New J. Chem.* **2012**, *36*, 1161–1169. [CrossRef]
71. Zhang, Q.; Qian, W.Z.; Yu, H.; Wei, F.; Wen, Q. Synthesis of carbon nanotubes with totally hollow channels and/or with totally copper filled nanowires. *Appl. Phys. A* **2007**, *86*, 265–269. [CrossRef]
72. Awan, T.I.; Bashir, A.; Tehseen, A.; Bibi, S. Electrons in nanostructures. In *Chemistry of Nanomaterials*; Elsevier Inc.: Amsterdam, The Netherlands, 2020. [CrossRef]
73. Huo, C.; Liang, F.; Sun, A.B. Review on development of carbon nanotube field emission cathode for space propulsion systems. *High Volt.* **2020**, *5*, 409–415. [CrossRef]
74. Gadzuk, J.W.; Plummer, E.W. Field emission energy distribution (FEED). *Rev. Mod. Phys.* **1973**, *45*, 487–548. [CrossRef]

75. Zhong, D.Y.; Zhang, G.Y.; Liu, S.; Sakurai, T.; Wang, E.G. Universal field-emission model for carbon nanotubes on a metal tip. *Appl. Phys. Lett.* **2002**, *80*, 506–508. [CrossRef]
76. Semet, V.; Binh, V.T.; Vincent, P.; Guillot, D.; Teo, K.B.K.; Chhowalla, M.; Amaratunga, G.A.J.; Milne, W.I.; Legagneux, P.; Pribat, D. Field electron emission from individual carbon nanotubes of a vertically aligned array. *Appl. Phys. Lett.* **2002**, *81*, 343–345. [CrossRef]
77. Parveen, S.; Kumar, A.; Husain, S.; Husain, M. Fowler Nordheim theory of carbon nanotube based field emitters. *Phys. B Condens. Matter.* **2017**, *505*, 1–8. [CrossRef]
78. Pandey, A.; Prasad, A.; Moscatello, J.P.; Yap, Y.K. Stable electron field emission from PMMA-CNT matrices. *ACS Nano* **2010**, *4*, 6760–6766. [CrossRef] [PubMed]
79. Edgcombe, C.J.; Valdrè, U. Field Emission and Electron Microscopy. *Microsc. Microanal.* **2000**, *6*, 380–387. [CrossRef] [PubMed]
80. Gröning, O.; Küttel, O.M.; Emmenegger, C.; Gröning, P.; Schlapbach, L. Field emission properties of carbon nanotubes. *J. Vac. Sci. Technol. B Microelectron. Nanometer Struct. Process. Meas. Phenom.* **2000**, *18*, 665–678. [CrossRef]
81. Gartland, P.O.; Berge, S.; Slagsvold, B.J. Photoelectric work function of a copper single crystal for the (100), (110), (111), and (112) faces. *Phys. Rev. Lett.* **1972**, *28*, 738–739. [CrossRef]
82. Bonard, J.M.; Croci, M.; Arfaoui, I.; Noury, O.; Sarangi, D.; Châtelain, A. Can we reliably estimate the emission field and field enhancement factor of carbon nanotube film field emitters? *Diam. Relat. Mater.* **2002**, *11*, 763–768. [CrossRef]
83. Nilsson, L.; Groening, O.; Emmenegger, C.; Kuettel, O.; Schaller, E.; Schlapbach, L.; Kind, H.; Bonard, J.M.; Kern, K. Scanning field emission from patterned carbon nanotube films. *Appl. Phys. Lett.* **2000**, *76*, 2071–2073. [CrossRef]
84. Teo, K.B.K.; Chhowalla, M.; Amaratunga, G.A.J.; Milne, W.I.; Pirio, G.; Legagneux, P.; Wyczisk, F.; Pribat, D.; Hasko, D.G. Field emission from dense, sparse, and patterned arrays of carbon nanofibers. *Appl. Phys. Lett.* **2002**, *80*, 2011–2013. [CrossRef]
85. Huang, B.R.; Lin, T.C.; Yang, Y.K.; Tzeng, S. Der: The stability of the CNT/Ni field emission cathode fabricated by the composite plating method. *Diam. Relat. Mater.* **2010**, *19*, 158–161. [CrossRef]
86. Bonard, J.M.; Klinke, C.; Dean, K.A.; Coll, B.F. Degradation and failure of carbon nanotube field emitters. *Phys. Rev. B* **2003**, *67*, 10. [CrossRef]
87. Dean, K.A.; Burgin, T.P.; Chalamala, B.R. Evaporation of carbon nanotubes during electron field emission. *Appl. Phys. Lett.* **2001**, *79*, 1873–1875. [CrossRef]

nanomaterials

MDPI

*Review*

# Electrical Properties of Carbon Nanotubes: From Individual to Assemblies

**Yuxin Xiang, Lili Zhang * and Chang Liu ***

Shenyang National Laboratory for Materials Science, Institute of Metal Research (IMR), Chinese Academy of Sciences, 72 Wenhua Road, Shenyang 110016, China; yxxiang22s@imr.ac.cn
* Correspondence: zhangll@imr.ac.cn (L.Z.); cliu@imr.ac.cn (C.L.)

**Abstract**

Carbon nanotubes (CNTs) have attracted intense research interest owing to their unique one-dimensional structure and exceptional properties. However, when individual CNTs are assembled to macrostructures such as films and fibers, their electrical performance often deteriorates significantly. This review offers a comprehensive look at the recent progress in the electrical properties and measurement techniques of CNTs, ranging from individual nanotubes to their assemblies. Firstly, we explore the methods for measuring the electrical properties of individual CNTs, including scanning tunnelling microscopy, electron microscope-based nanoprobes, and measurements using nanodevices. Secondly, we examine how structural characteristics of CNTs (e.g., chirality, diameter, and defects) influence their electrical behaviors. A critical comparison between individual CNTs and their assemblies reveals the difficulties in transferring the electrical properties from nanoscale to bulk materials. Finally, we put forward strategies to boost the electrical conductivity of CNT assemblies and also sketch out future research and development directions.

**Keywords:** carbon nanotubes; assembly; electrical property; measurements

## 1. Introduction

With the progress of science and technology, there is ever-increasing demand for various electronic devices in the fields of information, display, and sensing. These demands include high-performance, miniaturization, and flexible wearable devices, among others. They present substantial challenges to materials and device design. Since the 1980s, scientists have successively discovered a series of low-dimensional materials such as quantum dots, nanotubes, nanowires, graphene, and graphdiyne. These materials have unique structures and extraordinary properties. It is expected that they will meet the requirements of the post-Moore era in terms of multi-functionality and miniaturization. Among them, carbon nanotubes (CNTs), reported by Sumio Iijima in 1991 [1], are a promising material for achieving large-scale practical applications and taking the lead in future competition within the high-tech industry. CNTs are one-dimensional hollow tubular structures made of $sp^2$-hybridized carbon atoms. More specifically, single-walled CNTs (SWCNTs) are uniquely defined by chiral index (n, m), since they can be taken as rolled graphene sheets at different chiral angles θ, corresponding to specific chiral vectors $C = n\cdot\mathbf{a_1} + m\cdot\mathbf{a_2}$, where $\mathbf{a_1} = a(\sqrt{3},0)$ and $a_2 = a\left(\frac{\sqrt{3}}{2},\frac{3}{2}\right)$ are the basis vectors in graphene lattice (Figure 1a) [2]. As is shown in Figure 1b, SWCNTs will be referred to as zigzag SWCNTs if m = 0, while armchair type if m = n and chiral SWCNTs if n ≠ m. SWCNTs are then categorized electrically into metallic SWCNTs (m-SWCNTs) when $(n - m)/3 = q$ (an integer); otherwise, they


https://doi.org/10.3390/nano15151165

are referred to semiconducting SWCNTs (s-SWCNTs) [3]. Depending on their wall number, CNTs are classified as SWCNTs, double-walled (DWCNTs), or multi-walled (MWCNTs) (Figure 1c).

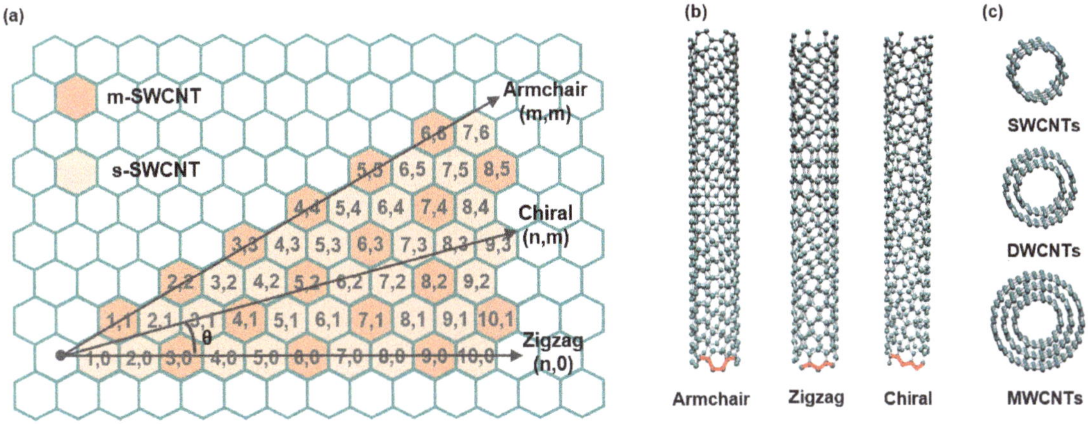


**Figure 1.** (**a**) Schematic diagram of an unrolled SWCNT with chiral vectors and two basis vectors. (**b**) "Ball and stick" models of (5,5) armchair SWCNT, (8,0) zigzag SWCNT, and (8,5) chiral SWCNT (from left to right). (**c**) "Ball and stick" models of SWCNT, DWCNT, and MWCNT.

In general, CNTs possess extraordinary and tunable electrical properties, making them promising candidates for use in transistors, energy storage, transparent conductive films, and other electrical applications [2,4]. A 10 nm long CNT could carry current up to 70 μA [5]. An SWCNT has been demonstrated to withstand a current density of $10^9$ A/cm$^2$ [6], 2–3 orders higher than copper wires. The intrinsic carrier mobility of an intact s-SWCNT can reach $10^5$ cm$^2$ V$^{-1}$ s$^{-1}$ [7,8] at room temperature, which is 100 times higher than that of bulk phase silicon with the dopant concentration of $10^{17}$ cm$^{-3}$ [9]. Furthermore, adjusting the gate voltage of a field-effect transistor (FET) enables the concentration and type of the carriers to be regulated, thereby producing either p-type or n-type semiconductors.

CNTs possess extraordinary ballistic transport characteristics. When the length of a CNT is shorter than the mean free path of electrons, which can extend to approximately a micron [9,10], electron scattering becomes negligible. In this case, the CNTs exhibit high current-carrying capability (~25 μA per tube) and the length of the CNT is independent of its resistance, in line with the law of ballistic transport [6,11]. Nevertheless, ballistic transport can only be maintained when a low bias voltage ($0.1\ \mathrm{V} \leq |V_{ap}| \leq 6\ \mathrm{V}$) is applied to the CNT device; otherwise, scattering between electrons and phonons is significant. When a low bias voltage is applied to a CNT device with a length of several hundred nanometers, the carrier can exhibit ballistic transport characteristics [12].

Under the ideal condition of ballistic transport, the conductance of an SWCNT with two conduction channels is equal to $\frac{4e^2}{h} \approx 77.5$ μS, i.e., a resistance of 6.45 kΩ [13,14], which is a temperature-independent minimum resistance (this value being about $10^7$ higher that clean Cu/Cu contacts). If an electrode with appropriate work function is selected, the on-state conductance of the s-SWCNT device can also be close to $\frac{4e^2}{h}$ [1–3]. When the length of SWCNT continues to increase to several microns, reaching the regime of scattering-dominated electrical conduction, the resistance undergoes a persistent decline.

However, the electrical properties of individual CNTs, as measured experimentally, are much lower than predicted due to the existence of defects and impurities. These properties continually diminish significantly in assemblies due to the introduction of more interfaces and the diverse structures of the comprising CNTs. In other words, the electrical properties of CNTs depend on various structural features, such as chirality, length, defects, and the configuration of CNT assemblies. For example, the highest measured carrier mobility of SWCNT film is about $10^3$ $cm^2/(V{\cdot}s)$ [15], and the highest measured conductivity of SWCNT fibers is about $10^7$ S/m [16,17], which are much lower than those of individual SWCNTs. Recent advancements in flexible electronics and wearable sensors have spurred increased efforts to optimize CNT assemblies. However, it remains a big challenge to achieve the excellent performance of CNTs at macroscopic scale.

In this review, we present an overview of the electrical properties of CNTs from the individual level to macroscopic assemblies, and how these properties are measured. We begin by examining the methods used to measure the electrical properties of CNTs, including scanning tunnelling microscopy (STM), electron microscope-based nanoprobes, and nanodevice-based measurements. The origins of measurement error will be discussed. Next, we summarize the effects of structural features affecting the electrical properties of individual CNTs, CNT bundles, and macroscopic assemblies (Figure 2). It should be noted that the bundles in this review are those comprising countable CNTs with a diameter of 50 nm or less. The macroscopic assemblies of CNTs refer to vertical arrays, aerogel, soot, films, and fibers. The scope of this review is limited to the most widely used assemblies, namely, films and fibers. Furthermore, attention will be paid to the mechanisms underlying the difference in the electrical properties between individual and assembled SWCNTs. Finally, we discuss possible solutions such as doping to improving the electrical properties of individual CNTs and CNT assemblies. Future trends toward bridging the gap between the electrical properties of individual and assembled CNTs are also outlined.

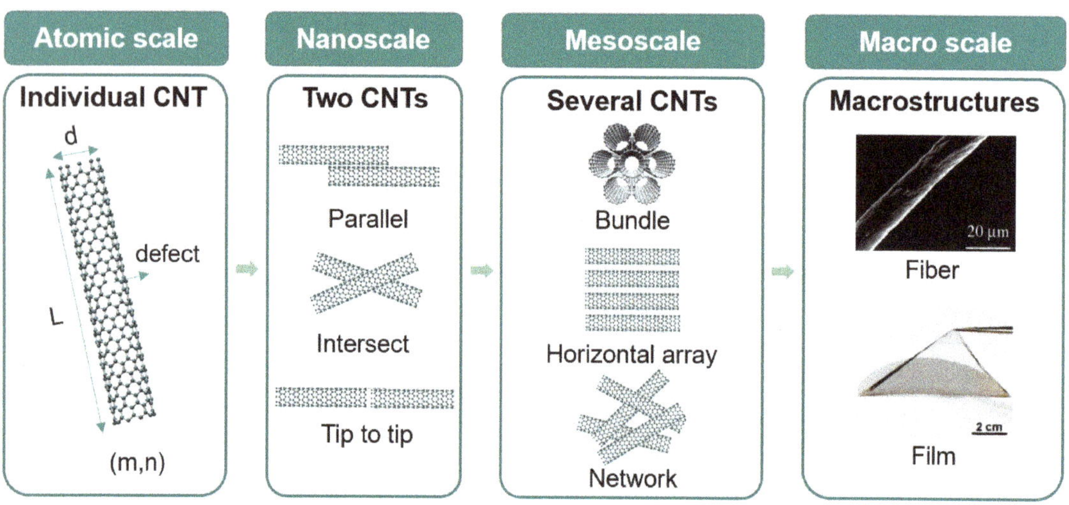


**Figure 2.** Schematic showing the structures of individual CNTs and their assemblies at different scales.

## 2. Measurement of the Electrical Properties of CNTs

### *2.1. Methods for Measuring the Electrical Properties of CNTs*

This section summarizes three typical methodologies for measuring the electrical properties of CNTs: (1) STM, (2) electron microscope-based nanoprobes, and (3) nanodevice-based measurements. These methods can provide critical information on various electrical properties, including the density of states and resistance. Additionally, experimental errors and their potential sources are discussed.

2.1.1. STM

STM is commonly used to characterize the surface structure of CNTs with atomic resolution and to obtain the surface electron state density through scanning tunneling spectroscopy (STS). Figure 3a schematically shows the working mechanism of an STM. A direct current with an additional small alternating voltage is applied between a CNT and a probe. The differential conductance (dI/dV) can then be calculated from tunnelling current versus voltage (I-V) data, which is considered to be proportional to the density of states (DOS) of the examined CNT (Figure 3b) [18]. When the gap value in the differential conductance diagram is 0.5–0.6 V or 1.7–2.0 V, the SWCNT is semiconducting or metallic, respectively [18]. Some differential conductance diagrams of SWCNTs are displayed in Figure 3c.

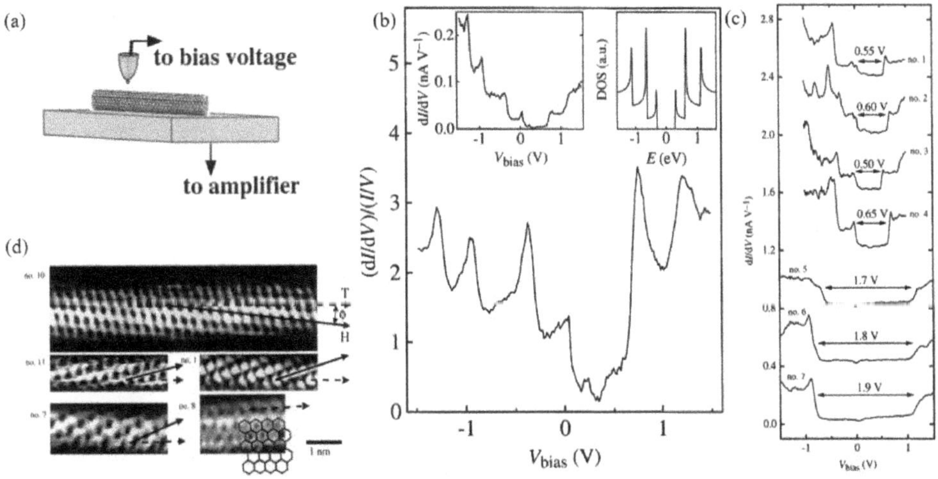


**Figure 3.** (**a**) Schematic diagram of the electrical measurement of individual SWCNTs using STM [19]; (**b**) STS diagram (local DOS at surface via tunneling) and DOS diagram [18]; (**c**) differential conductance (dI/dV) diagram [18]; (**d**) STM image of a SWCNT with atomic resolution [18].

Atomically resolved STM images of individual SWCNTs can also be obtained using the STM technique (Figure 3d). Only a few carbon atoms at the top of the CNT are discernible. It is sufficient to enable the chirality to be identified unambiguously from the chiral angle θ.

2.1.2. Electron Microscope-Based Nanoprobe Measurement

Using an electron microscope-based nanoprobe, it is possible to manipulate, process, and measure samples in situ in a scanning electron microscope (SEM) or a transmission electron microscope (TEM). Data relating to current voltage, current time, temperature profile, and power time can be obtained in conjunction with a series of SEM/TEM images. This makes it possible to establish a correlation between the microstructures and electrical properties of CNTs.

A nanoprobe can manipulate a sample or a nanoprobe electrode, triggering movement in the X, Y, and Z directions at the nanometer scale. In order to manipulate nanomaterials, the probe that grabs the sample must be sharp enough, with the tip size within 100 nm. Compared with mechanical methods of probe fabrication, such as polishing and shearing, electrochemical etching is the most common method due to its good reproducibility and standardization. By finely tuning the applied electrochemical potential, cutoff voltage, etching mode, electrolyte type and concentration, immersion depth, counter electrode type, and probe material, uniform and sharp probes with fewer impurities can be produced [20,21].

A variety of materials have been used for the nanoprobes. The most commonly used materials include tungsten [22–31] with high hardness and gold [32–41] with high chemical stability and good wettability.

Contact resistance between the nanoprobe and CNTs is a key concern in this type of nanoprobe measurement. To minimize its impact on the measured resistance of CNTs, the contact points between the tip and the CNT were irradiated with a focused ion beam (FIB) [42] or heated with a pulsed current [43]. Since CNTs can be welded to the wire to form good physical and electrical contacts, electrical contacts can also be formed by filling the tube with a material that has low vapor pressure, such as a chalcogenide [27], or by depositing nanoparticles that have a low melting point, such as tin [44,45].

Since SWCNTs have much lower structure stability and are more difficult to manipulate under an electron microscope, the nanoprobe technique has mainly been used to investigate the electrical properties of individual MWCNTs. For instance, Huang et al. [39] reported unexpectedly three distinct breakdown sequences in individual MWCNTs, proving that each wall of an MWCNT conducts at high voltage, since a current drop can be observed when the innermost wall was broken. Moreover, the breakdown of each wall initiated in the middle of the tube, proving the transport property is not ballistic (Figure 4a). Also, the thermoelectric power increased as the tube diameter decreased (Figure 4b). Another crucial phenomenon that can be observed is the structure engineering by Joule heating. On exposure to a high current density, resistive hotspots near the contact points later migrated and expanded along the CNT over time, as indicated by the localized sublimation of the encapsulated material (Figure 4c) [27]. The electrical properties of doping systems can also be investigated by this method. It has been observed by Aslam et al. [28] that a large current could induce structural transformation of N-doped MWCNTs, resulting in the removal of the dopant, and a significant change in electrical behavior.

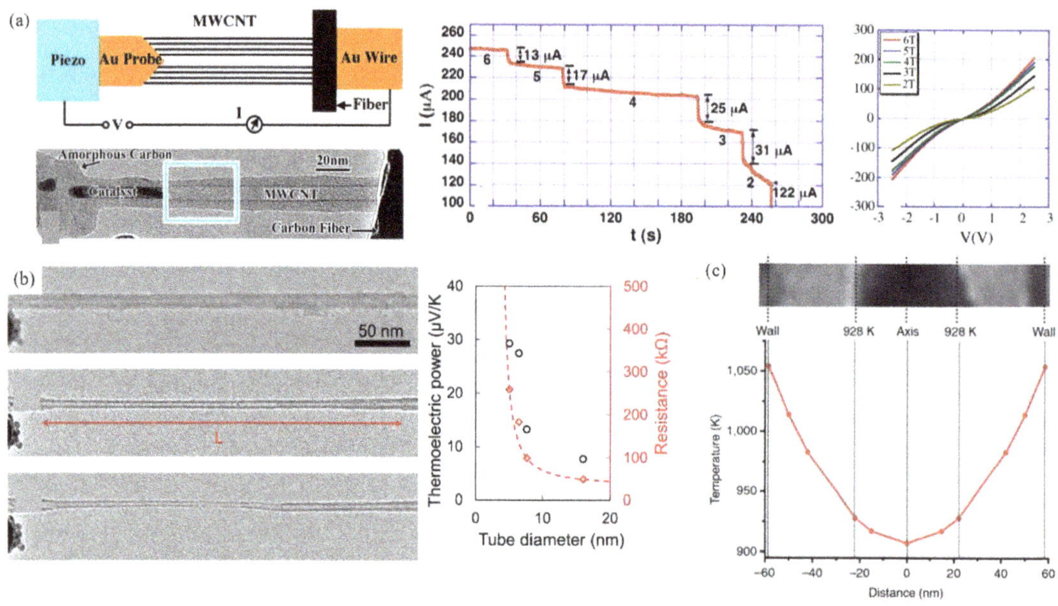


**Figure 4.** (**a**) Schematic of the nanoprobe measurement platform, TEM image of the connection, I-t curve of the breakdown of an MWCNT, and I-V plots with a different number of walls [39]. (**b**) Thermoelectric power and resistance measured during the step-by-step breakdown of an MWCNT wall [44]. (**c**) Estimation of the radial temperature profile in an electrically heated CNT channel [27].

As a conclusion, using this method with SEM/TEM enables more phenomena relating to the Joule heating behavior of CNTs and the doping system to be observed, particularly with regard to the CNT filling system and electromigration processes.

2.1.3. CNT Electronic Devices

A great deal of work has been carried out on designing electronic devices using individual CNTs, CNT arrays, and CNT films, as they are ideal for use in integrated circuits (ICs). One common device is the four-point probe measurement [24,46–49]. This device-based method is often used to measure the electrical resistance of individual CNTs [24,46–48], CNT fibers [15,16,49,50], and CNT films. This can be achieved by passing a current through the two outer terminals ($I_{outer}$) and measuring the voltage using the two inner terminals ($\Delta V_{inner}$). All electrodes must be in contact with the material being measured. This method can mitigate contact resistance errors, and reveal the material's intrinsic resistance, which can be determined by $R = (\Delta V_{inner})/I_{outer}$ [51]. Furthermore, numerous methodologies can be employed to reduce contact resistance between the electrodes and CNTs. Ebbesen et al. [46] reported on four-probe measurements of individual SWCNTs produced by the ion-induced deposition of four 80 nm wide tungsten leads. Wang et al. [52] measured the electrical resistance of 5 cm long single CNT fiber by dipping silver adhesives to four contact regions.

Another widespread electronic device is the CNT-based FET. Due to their ultrahigh carrier mobility and satisfactory carrier concentration, CNTs are a promising material for the fabrication of future ICs. A CNT-based FET can inhibit the short-channel effect in a conventional metal–oxide–semiconductor FET (MOSFET) and can be scaled to a contacted gate pitch of 55 nm, corresponding to a sub-10 nm node. The carrier mobility (1500 $cm^2\ V^{-1}\ s^{-1}$) and Fermi velocity ($12 \times 10^6$ $cm\ s^{-1}$) of this node are higher than those of a 10 nm MOSFET [53]. Also, a CNT-based FET could achieve a remarkable reduction in subthreshold swing (Ss) to ~35 mV/dec, which would significantly reduce power consumption [53,54]. Ultrahigh carrier mobility directly enhances the transconductance ($g_m$) and cutoff frequency ($f_T$), exhibiting favorable high-frequency performance [55]. These parameters are of primary concern in the future development of ICs, and these technological advancements are expected to promote the practical application of carbon-based circuits [56].

The CNT-FET structure comprises five parts. The channels are made of CNTs, either in the form of individual s-SWCNTs (Figure 5a) [11,57], an aligned s-SWCNT array (Figure 5b) [58,59], or a thin film (Figure 5c) [54,60–62]. The source and drain electrodes are made of metals with a high work function, such as palladium [53,63] and platinum [64]. The gate electrode can be either top-gate [65] or back-gate [65,66]. The high-k gate dielectric can be $HfO_2$ [56,65,67,68], $Al_2O_3$ [64,69], or ultrathin $SiO_2$ [70]. The insulating substrate can be $SiO_2$/Si, sapphire [71], or hBN [72], or a flexible organic substrate.

Some special electronic design schemes are also presented to facilitate research into the structure–performance relationship. For example, Oshima et al. [73] proposed an in-situ measurement of the electrical impedance during CNT synthesis by floating catalyst chemical vapor deposition (FCCVD) (see Figure 5d) to monitor CNT film thickness and reveal its correlation with the percolation threshold.

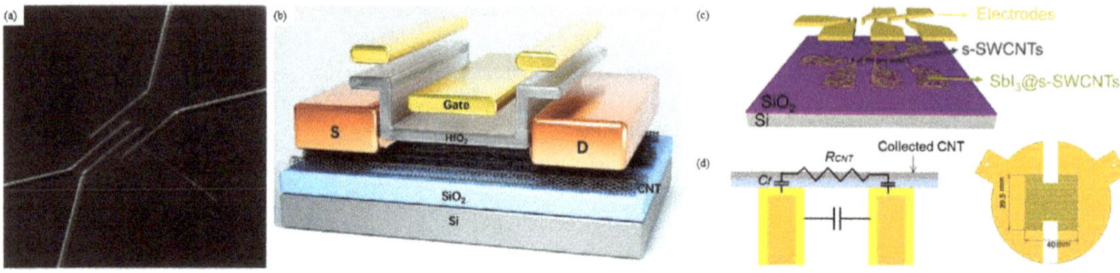


**Figure 5.** (**a**) FIB image of the four-probe measurement on an individual CNT [46]; (**b**) schematic diagram of a top-gate CNT-based FET [65]; (c) schematic diagrams of a thin film transistor device [54]; (**d**) cross-sectional diagrams and partial view of the in situ measurement device placed under a collection filter during CNT collection [73].

## 2.2. Measurement Errors and Their Sources

The previously mentioned methodologies can provide a variety of electrical information, such as band gap, resistance, and thermoelectric power. These approaches have one thing in common: the necessity of establishing contact between CNT and electrodes. In addition to contact resistance, other factors have also been identified as contributors to the diversity of electrical measurement results in different experimental groups. These include variations in measurement environment (such as gas composition, temperature, and humidity) and substrate. Concurrently, a variety of macrostructures with distinct intrinsic and configuration-specific CNT arrangements manifest diverse array electrical properties. These issues pose a significant challenge when it comes to measuring electrical properties.

The primary challenge lies in minimizing the contact resistance. We then model the resistance of an individual CNT contacted at each end of the metal electrodes as the sum of the three parts (Formula (1)).

$$R_{CNT} = R_Q + R_L + R_{CONTACT} \tag{1}$$

where $R_Q$ is called the quantum resistance. For an m-SWCNT with ideal ballistic transport and perfect contacts, $R_Q \approx 6.5\ k\Omega$, which is the minimum theoretical resistance per conduction channel in a 1D system and fundamental limit from 1D quantum transport. This is owing to the inevitable mismatch in the number of conduction channels between the nanotube and the macroscopic metal electrode [74]. $R_L$ is the scattering resistance due to electron–phonon scattering, impurities, and defects. If the nanotube length is shorter than the mean free path, which is about 300–1000 nm, we can usually neglect $R_L$. $R_{CONTACT}$ is the most variable and often dominant component in experiments, which depends on work function difference between the metal and the CNT (the work function of SWCNT is about 4.8 eV [75]), metal wettability, and the bonding configuration [76] (e.g., Pd makes better contacts than Au) and contact geometry (end-bonded vs. side contact) [77]. $R_{CONTACT}$ is about 5–50 kΩ with Pd [78]. As a result, the choice of metal electrode is crucial for reducing contact resistance. For metal electrodes with high work functions, their Fermi levels are close to the valence band of s-SWCNTs. This means holes are more easily injected into the channel than electrons, behaving as p-type semiconductors (Figure 6). Pt, Au, Pd, and Rh [79] are the electrodes of wide use in this category, among which Pd is the most suitable precious metal with a good wettability [11].

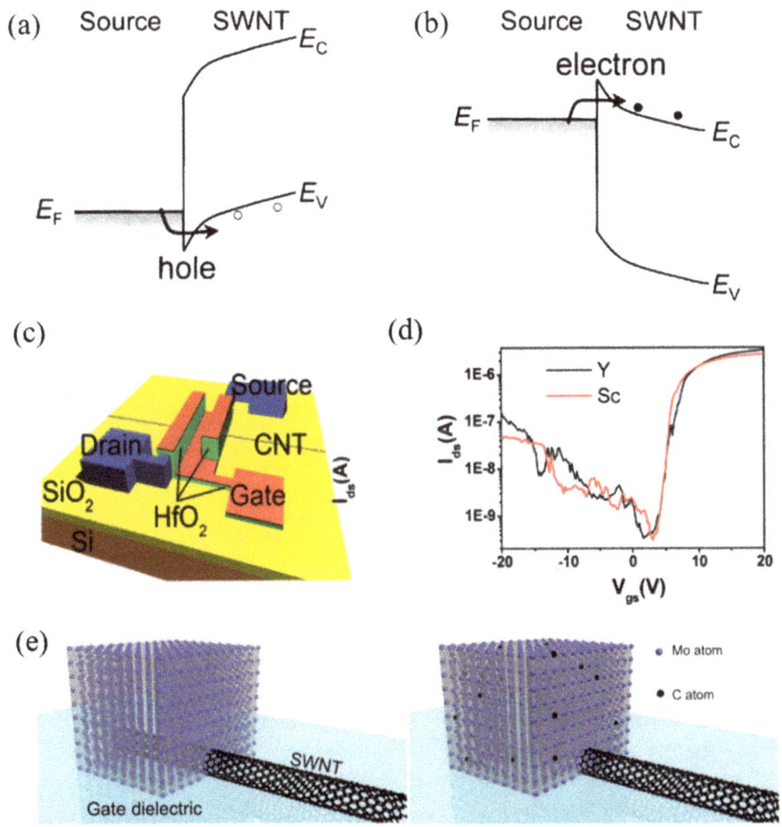


**Figure 6.** Band structure of the contact interface between s-SWCNT and metal electrodes with a (**a**) high work function and (**b**) low work function [80]; (**c**) Y-contacted top-gate CNT-FET device and (**d**) its performance compared to Sc-contacted device; (**e**) schematics showing the conversion from side-bonded contact to end-bonded contact with Mo in SWCNT transistors, and the transfer characteristics of typical Mo end-contacted SWCNT transistors [77].

By contrast, achieving n-type ohmic contact encounters more difficulties, as metals with relatively low work functions tend to oxidize first rather than react with carbon. However, it is still possible to realize end-bonded n-type contacts even with high work function metals through electrostatic doping near the source electrode [81,82]. Moreover, the Fermi level of Al, Sc, and Y is close to the conduction band of s-SWCNTs, also behaving as n-type [78,83,84]. Unlike forming p-type ohmic contact between Pd and CNT, Al can only form Schottky contact with a high barrier [78]. Regarding the economic benefit, Y is the most potential electrode in large-scale CNT ICs [84].

Except for the metal electrodes, graphene [85], hybrid graphene–metal [86], and CNT film [15,87] can also be considered as the electrode material. They can achieve significantly lower resistance than conventional metal electrodes, making them highly beneficial for advanced electrode applications. This stems from graphene's exceptional conductivity, synergistic effects in hybrids or CNTs acting as conductive bridges that enhance charge transfer. These materials also offer superior flexibility, corrosion resistance, thermal stability, and higher surface area—enabling more efficient, durable, and miniaturized devices. Consequently, they outperform metals where low contact resistance, weight savings, or mechanical resilience are critical, though cost and fabrication complexity remain challenges for mass adoption [88].

Reducing contact resistance is essential to achieving desirable performance of FETs, such as high field-effect mobility and on/off ratio. A comparison on the methods for reducing the contact resistance at the individual CNT–electrode interface is shown in Table 1.

**Table 1.** A comparison of the methods for reducing the contact resistance at individual CNT–electrode interface.

| Method | Advantage | Disadvantage |
|---|---|---|
| High-temperature annealing [89] | Batch processing<br>Forming stable low resistance carbides | Large high-temperature zone |
| Local Joule heating [90] | Easy to operate<br>Highly targeted treatment | Poor repeatability |
| Electron beam-induced deposition [91] | Reliable and good electrical contact | Harsh condition<br>Low efficiency |
| Electron beam irradiation [42] | Effective improvement in electrical contact | Small processing area<br>Damage to CNTs |
| Ultrasonic nanowelding [92] | Fast and reliable, normal temperature operation, wide range of adaptation | Difficulty in precise control |

In addition to the selection of electrode materials, the strategies like chemisorption involving carbide formation through high-temperature annealing and local Joule heating with end-bonded type have also been demonstrated to reduce contact resistance. This could form contact with size-independent low resistance and high physical stability [28,29,77]. Such an all-shell connection is demonstrated to prevent the field-emission failure at the CNT–metal interface, promising for real-world applications in CNT-based electronic devices [30]. This chemisorption process has been shown to facilitate substantial interfacial charge transfer via the formation of chemical bonds. This, in turn, results in alterations to the electronic structure, enhancement of durability, and the facilitation of charge injection. The contrary process is known as physical adsorption, which is defined as the process of minimal charge redistribution through weak electrostatic forces, thereby preserving bulk properties and being reversible [93].

Although novel approaches such as ultrasonic nanowelding and electron beam-induced deposition are reliable and repeatable to form good electrical contact, they are hindered by scalability concerns. More efficient methods must therefore be developed.

One impact on the precision of electrical measurement is regulating the measurement environment, including the gas composition, temperature, and humidity. The electrical properties of CNTs have been shown to be highly sensitive to the measurement environment, making this a key consideration in the design of any CNT-based measurement systems. Exposure to air reversibly influences the electrical resistance of the individual SWCNTs, like the conversion from s-SWCNT to m-SWCNT and CNT networks by negligible adsorption of $O_2$ gases [94]. This can be attributed to the hole-doped effect in the presence of adsorbed oxygen or work function change of the metal contact [95]. These results, although demonstrating the potentially viable application on sensitive chemical oxygen detectors [96–98], likewise indicate that many intrinsic properties measured on as-prepared tubes might be substantially compromised by exposure to the external environment. Temperature is also a significant factor. It has been demonstrated that m-SWCNTs exhibit a near-linear increase in the calculated resistivity with temperature over a wide range due to twiston scattering [99]. For s-SWCNT, the on current decreases slightly with rising temperature due to enhanced phonon scattering, while the off current increases significantly due to thermal emission over Schottky barriers. Meanwhile, threshold voltage shifts negatively with temperature, attributed to changes in carrier injection and trap dynamics [65]. The humidity effect is also an essential factor in electrical measurement, especially for flexible electronics applications. Mostafa et al. [100] demonstrated that an increase in relative humidity from 10% to 80% can result in a reduction of approximately 4% in the electrical conductivity of vertically aligned CNTs.

Another impact on the precision of electrical measurement in CNT devices is the substrate. In general, CNTs in FETs are in direct contact with the substrate, which gives rise to several non-ideal behaviors. The additional carriers will be trapped from the substrate, inducing the electrical noise by changing the surface potential [101]. A rough substrate would decrease the mobility and conductivity of the FET devices. Moreover, due to the attachment of CNT and the substrate, a part of the sensing area is lost [102]. As a result, it seems that FETs with suspended CNTs would exhibit higher electrical and sensitive performance. Some efforts were also made to change the type of substrate. Zhang et al. [72] reported on the direct growth of 2D close-packed SWCNT van der Waals crystals with uniform chirality and good alignment on a hexagonal boron nitride substrate. FETs constructed from these SWCNT arrays exhibited excellent electrical performance at room temperature. It demonstrated that atomically flat substrate could lead to a self-assembly growth of the low-dimensional material through van der Waals attraction and ultralow sliding friction.

## 3. Electrical Properties of CNTs

In this section, we discuss the electrical properties of individual CNTs, CNT bundles, and CNT assemblies. The structure-dependent electrical properties include quantitative results such as resistance, conductance, mobility, on/off ratio, and the density of state. Furthermore, the reasons why the electrical properties of individual CNTs do not translate into assemblies will be explained. Certain endeavors like doping to enhance the electrical characteristics of assemblies will also be discussed.

### *3.1. Electrical Properties of Intrinsic Structure of Individual CNTs*

Individual CNTs are believed to exhibit superior electrical properties, which are strongly dependent on their intrinsic structures, including chirality, diameter, number of walls, defects, length, and doping levels.

#### 3.1.1. Chirality and Diameter

The diameter and chirality of CNTs have been identified as the most crucial factors affecting its electrical properties. The one-dimensional dispersion relationship of electron bands in a SWCNT is calculated on the basis of the first principles of graphene band structure and local-density-functional approach. The results indicate that SWCNTs can be categorized into two distinct groups. When the rolling index $n - m = 3q$ (q is an integer), the electron state wave vector k intersects at the primary Brillouin zone boundary. As demonstrated in Figure 7a, the conduction and valence bands intersect at the Fermi level, indicating that the SWCNT exhibits metallic characteristics with a zero-band gap. However, when $(n - m)/3 \neq q$ (q is an integer), the SWCNT behaves as a semiconductor (Figure 7b), and the band gap is inversely proportional to the tube diameter, as outlined in Formula (2) [103]. Consequently, the ratio of m-SWCNT to s-SWCNT is approximately 1:2 [104]. The occurrence of metallic behavior has been observed in armchair and select zigzag tubes, while semiconducting tubes have been shown to exhibit diameter-dependent band gaps ($E_g \propto 1/d$).

$$E_{gap} = \frac{2\gamma_0 a_{c-c}}{d} \tag{2}$$

where $\gamma_0$ stands for the overlap energy of C-C bound with the value of about 347 kJ/mol, $a_{c-c}$ denotes the nearest C-C bond length with the value of 0.142 nm, and d is the tube diameter [105].

However, it should be noted that there are exceptions to this rule in chiral and zigzag SWCNTs with very small diameters (d < 0.6 nm) due to curvature effects. It has been established that a portion of SWCNTs, predicted to be metallic, exhibits semiconducting properties [106]. The finite curvature reduces the overlap between adjacent p orbitals and shifts the band intersections away from the K point, resulting in a curvature-induced band gap in the metallic tube. Furthermore, the width of this band gap is inversely proportional to the square of the tube's diameter [107] as shown in Figure 7c, which has been confirmed experimentally by Ouyang et al. (Figure 7d) [108]. Nonetheless, armchair SWCNTs are still metallic because $k_F$ remains in the sub-band, with both the initial band gap and the curvature-induced band gap being zero [109].

Moreover, s-SWCNTs with small diameters can be divided into two types as shown in Figure 7e,f. For Type I (mod (2n + m, 3) = 1), both the on-state current and the carrier mobility exhibit an increase with an increasing chiral angle within the same family (2n + m = the same integer), whereas the results are the opposite for Type II (mod (2n + m, 3) = 2) [60].

Increases in diameter invariably result in greater numbers of walls being incorporated into current synthesis techniques. There are also a lot of electrical experiment results in MWCNTs [110]. The electrical properties of MWCNTs are inferior to those of SWCNTs, as theoretical explanations posit the occurrence of electronic distortions in MWCNTs, depending on how the lattices are interrelated [111]. In addition, experimental evidence was presented indicating that only the outermost shells contribute to the overall conduction at low bias. However, at high bias voltage, all of the shells contribute to carrier transport [104,105]. This leakage was frozen at a low temperature and low bias limit, indicating that the intershell coupling was thermally activated and tunnel-type [112]. As Liu et al. [113] also concluded in the context of DWCNTs, the only exception is the s/m-DWCNT (where the outer and inner layers are s-SWCNT and m-SWCNT, respectively), which only exhibits semiconducting-like properties and a much lower on/off ratio.

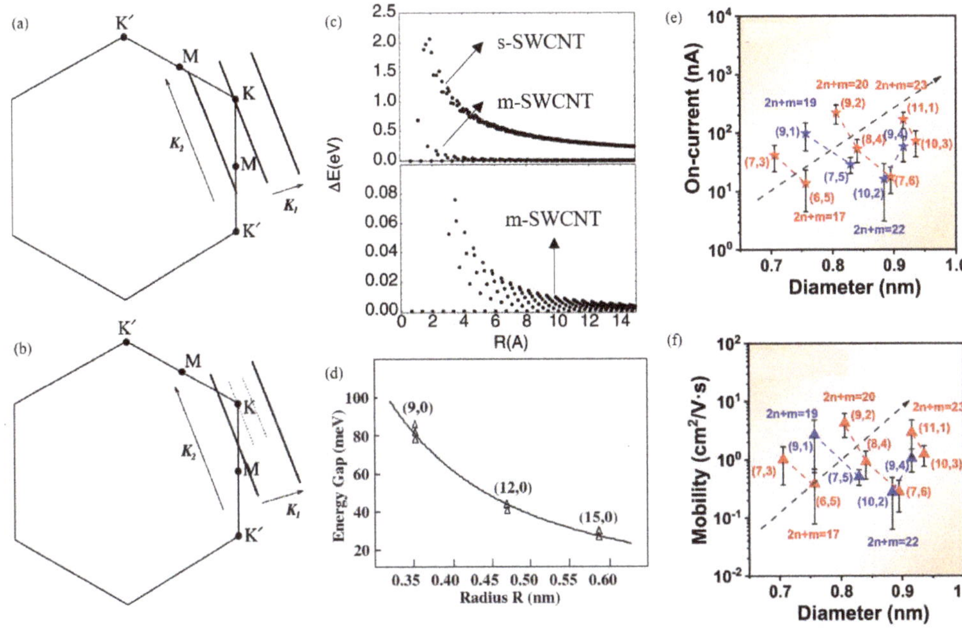


**Figure 7.** The wave vector k is shown as bold lines for (**a**) m-SWCNTs and (**b**) s-SWCNTs in the Brillouin zone of graphite (hexagon) [103]. (**c**) Gaps calculated for SWCNTs with a radius of less than 15 Å. Those with zero primary gap but nonzero curvature induced gaps which scale as $1/R^2$ are shown in the lower curve on top panel, whose expanded scale is shown in the lower panel [107]. (**d**) Curvature-induced gaps in "metallic" zigzag SWCNTs [108]. (**e**,**f**) Statistical distribution of the on-state current and the mobility of the two types. Blue and red symbols represent Type I and Type II, respectively.

### 3.1.2. Defect

Defects in CNTs, such as vacancies and non-hexagonal rings, which may be induced by irradiation or strain, can drastically change their transport properties. This section will discuss how and why the defects influence the electrical properties.

The application of density functional theory in conjunction with Green's function scattering approach facilitates the elucidation of defects' impact on electronic transport properties. The conductance of defective carbon nanotubes (CNTs) has been shown to typically exhibit non-trivial behavior, which has been observed to decrease and affect the band gap. In more precise terms, elevated levels of disorder have been shown to induce a greater degree of localization of DOS (Figure 8a) [114].

Furthermore, experimental results can be obtained. Navarro et al. [115] applied consecutive $Ar^+$ irradiation doses to SWCNTs in order to produce a uniform density of defects. The data demonstrate an exponential dependence of resistance on SWCNT length, thus indicating that the system is within the strong Anderson localization regime (Figure 8b).

A number of strategies have been developed for the purpose of defect healing to restore the conductivity, such as thermal annealing. Khanbolouki et al. [116] applied current-induced annealing (5–60 $MA/m^2$ in high vacuum) at 685–1445 °C to remove impurities, which results in the reduction of sheet resistance by 35–40%. Altuntas et al. [117] found that annealing at 120 °C for 1 h in air led to a reduction in interface traps and contact resistance, from 1.9 MΩ to 12.3 kΩ.

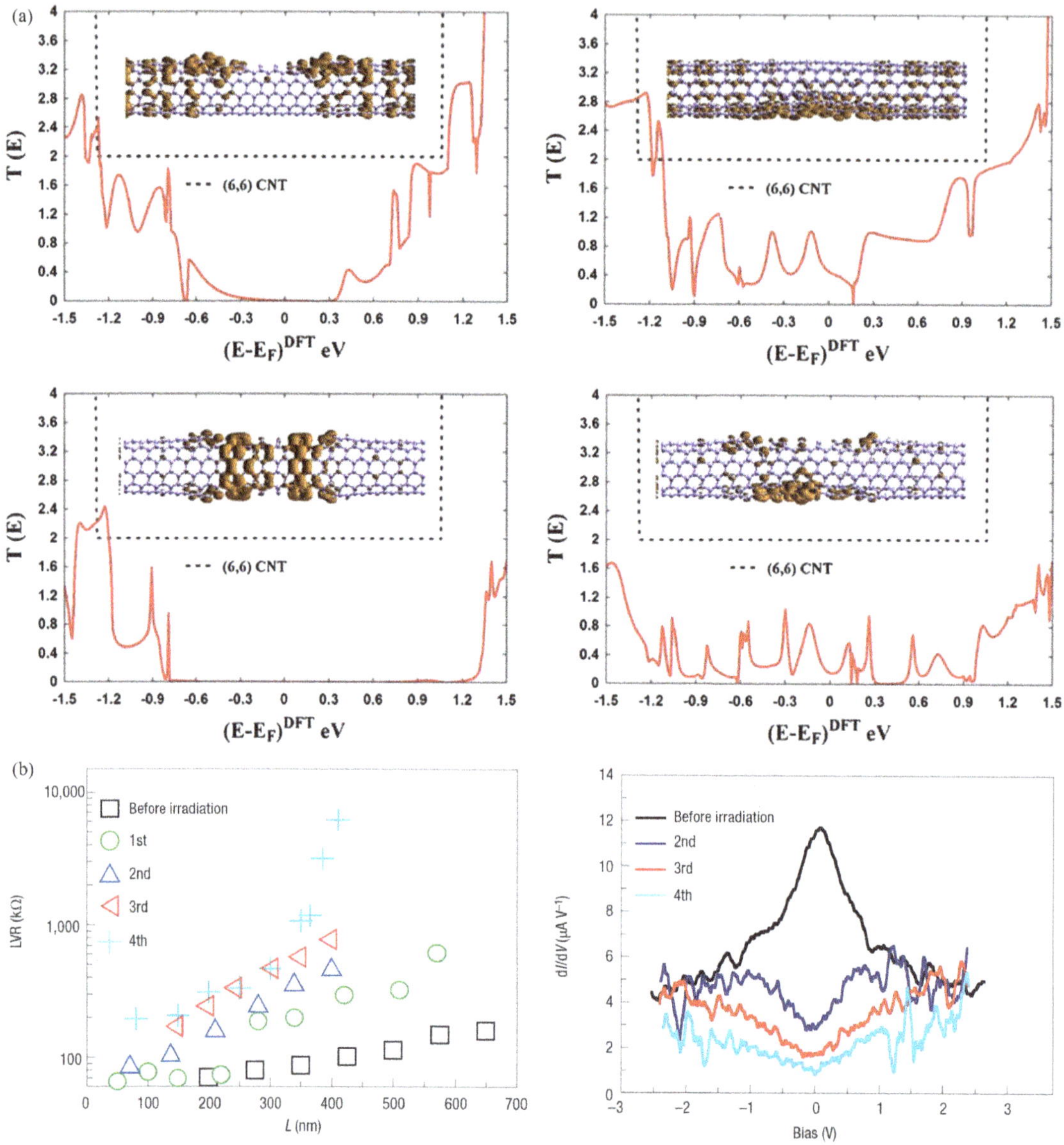


**Figure 8.** (**a**) The zero bias transmission probabilities with different defect types [114]; (**b**) effect of consecutive irradiations on the electrical resistance [115].

## 3.2. Electrical Properties of Several CNTs

It has been demonstrated that due to van der Waals interactions, individual SWCNTs, which are nearly uniform in diameter, readily self-organize into "ropes" [118]. A more common statement is CNT bundles, which also represent the intermediate transitional form from individual tubes to macroscopic CNT assemblies. Consequently, considerable research has been devoted to understanding how the interaction in bundles affects the electrical properties of the CNT assemblies. In this section, an exploration of the primary limitations on electrical properties in multi-tube systems, comprising two CNTs and CNT bundles, will be conducted.

### 3.2.1. Electrical Properties of Two CNTs

The density functional tight (DFT) binding theory and the nonequilibrium Green's functions are frequently utilized to calculate the electrical properties of pairs of CNTs. The template structures utilized in the calculation for two CNTs can be categorized into three distinct types: tip-to-tip, parallel, and perpendicular. Among these studies, two parallel armchair SWCNTs with equal chirality indices are always considered first, as they are the simplest model. However, experimental measurements across pairs of CNTs are much more difficult due to the nanomanipulation involved in SEM or TEM. To date, experimental measurement of junction resistance has been possible only for two MWCNTs.

The following discussion will commence with a consideration of two CNTs arranged in parallel. It has been demonstrated that there is a positive correlation between the overlap length and the electrical current (Figure 9). As demonstrated in Figure 9a, the scatter plot of the calculation results indicates that an overlap length of two armchair SWCNTs in parallel greater than 25 Å can induce a larger current [119]. A more detailed analysis of the results is presented in Figure 9d, where nanotube junctions are formed by two (10,10) SWNCTs and (6,6)/(9,0) SWCNTs. A noteworthy observation is the occurrence of a nonlinear and quasi-periodic relationship between conductance and contact length. These quasi-periods were found to be heavily dependent on the chirality of SWCNTs. For instance, armchair SWCNT and zigzag SWCNT are about $3a_0$ and $a_0$, respectively. This is due to the fact that the Fermi wavelength for armchair tubes is $a_0$, where $a_0$ is the unit cell parameter [111,120]. When contact lengths are compared, the conductance of the armchair/zigzag contact is found to be significantly lower than that of the armchair/armchair contact, by approximately one order of magnitude. The reason for this phenomenon is that two SWCNTs with different chirality are incommensurate and possess mismatched Fermi momenta [121]. It can be concluded that the optimal conductance at a parallel contact can only be achieved when two metallic tubes possess the same chirality. Examples of such chirality include metallic zigzag/zigzag and armchair/armchair types. It is evident that the metallic zigzag tubes possess a gap around the Fermi energy due to the curvature effect. This observation indicates that armchair/armchair tubes in parallel are the most promising geometry for connecting multiple tubes in device applications [111,120].

This similar periodical trend was experimentally observed by Barnett et al., in which two halves of the same MWCNTs were manipulated in a SEM (Figure 9e) [122]. There is an apparent periodicity of 2–3 Å, consistent with the unit cell of armchair SWCNT (2.45 Å). In addition, Hamasaki et al. also reported the in situ measurement of the parallel-contacted resistance between two halves of the same MWCNT in TEM [45]. The hypothesis was that the contact resistance would decrease with increasing contact length within the measurement range of a few hundred nanometers. It is theorized that, under certain conditions, the effect of the interface may become negligible, provided that the contact length is sufficiently long.

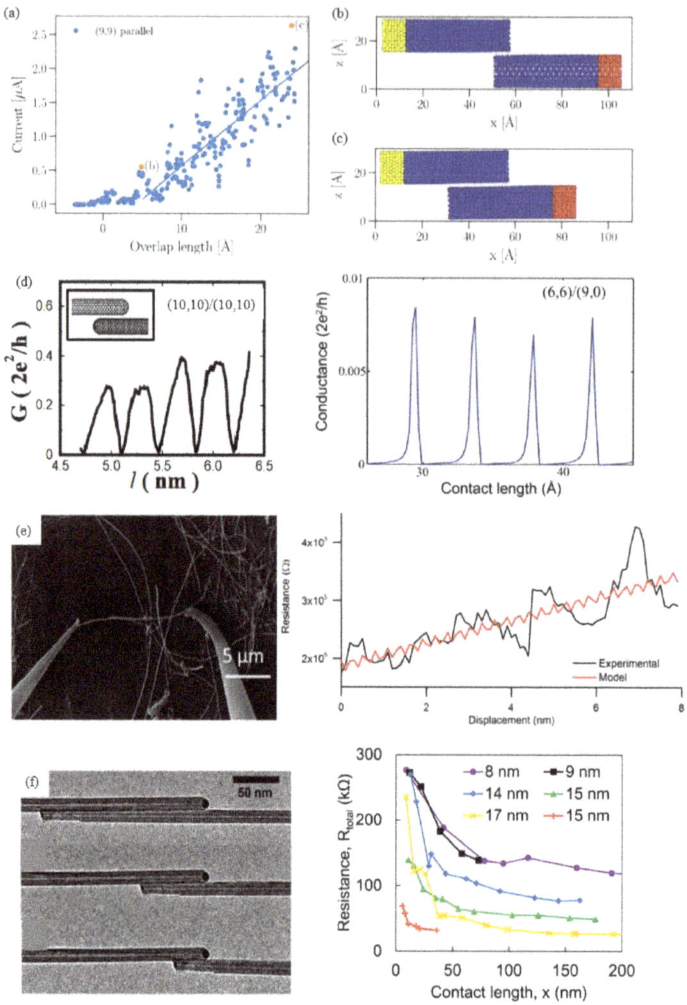

**Figure 9.** (**a**) Relationship between overlap length and current for pairs of parallel (9,9) SWCNTs [119]. (**b**,**c**) Geometries of two highlighted points in orange [119]. (**d**) The calculation results for the variation in conductance at the Fermi energy as a function of overlap length for two (10,10) SWCNTs [123] and a (6,6)/(9,0) contact [124]. (**e**) The SEM image and the schematic of manipulating two halves of a same MWCNT in a parallel manner. The bottom plot is the measured resistance at 1V as a function of the overlap length, where the red line is a simple model result [122]. (**f**) In situ TEM image of two parallel MWCNTs with various contact lengths. The right plot is the electrical resistance as a function of contact length for different MWCNTs. Legend indicates the outer diameter of each MWCNT [45].

Due to the occurrence of angular overlap conduction, the conductive rule is analogous. When one SWCNT is placed on top of another, the conductance between the in-registry tubes is increased. In this instance, the contact region is commensurate, with the carbon atoms arranged in an A-B stacking configuration of graphite [123]. The plot in Figure 10a predicts the resistance variation concerning the rotation angle along the tube axes by calculation. The tubes in the first (18,0)–(10,10) junction are found to be in registry at θ = 30, 90, 150°. The tubes in the second (10,10)–(10,10) junction are arranged in registry at θ = 0, 60, 120, 180°. Lower resistances have been observed in these in-registry orientations. However, even when the tubes are in registry, the low resistance can vary at different rotation angles. The lowest possible resistance is achieved when the carbon atoms in contact area are stacked like A-A stacking of graphite. This variation in electrical properties bears a resemblance to the dependence of mechanical or frictional properties on atomic registry [125,126]. The variation in resistance as a function of the crossing angle between two unparallel tubes is also investigated. As the crossing angle is reduced, the contact resistance decreases because the contact area is reduced; however, for MWCNTs with large diameter and low weight, the change in contact area will be negligible. Barnett et al. conducted a study on the angular dependence of contact resistance of two halves of a single MWCNT in in situ SEM (Figure 10c). The findings demonstrate that the two tubes manifest as "in-registry" at 22° and 44°, exhibiting reduced resistance. This observation signifies that the concept of "in-registry" for MWCNTs is less stringent than the ideal SWCNT. It is conceivable that the minima occur when the overlap of the $\pi$-orbital or electrostatic forces is minimal, leading to reduced repulsion [122].

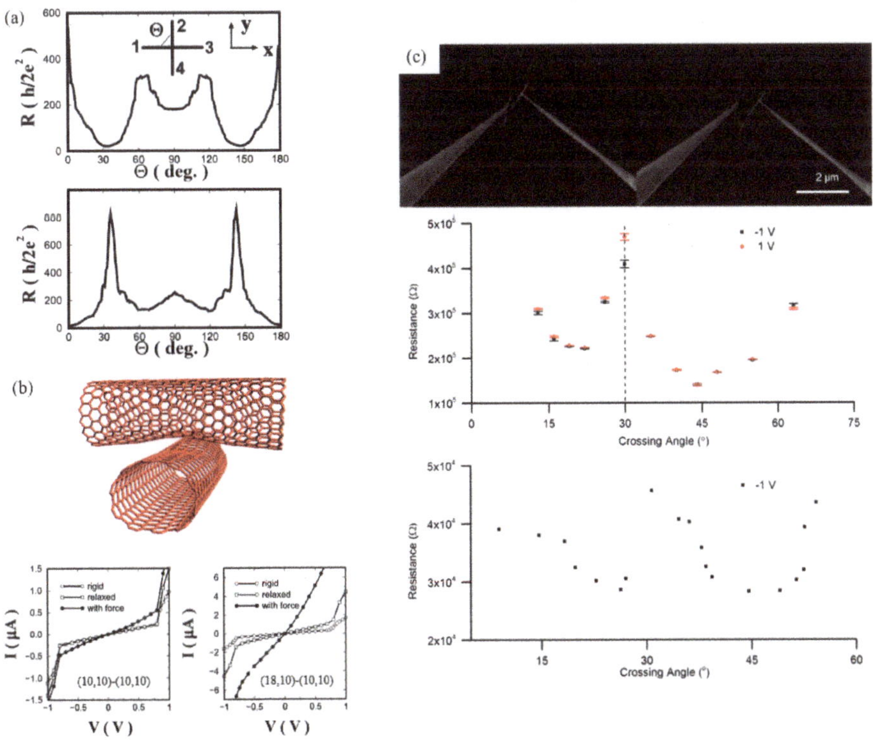


**Figure 10.** (**a**) The variation in contact resistance as a function of the rotation angle [123]. (**b**) The schematic of the cross-junction where the two SWCNTs are perpendicular to each other. I-V characteristics of a (10,10)–(10,10) (out-of-registry) and (18,10)–(10,10) (in-registry) cross-junctions [123]. (**c**) SEM images of manipulating the two halves of an individual MWCNT with different crossing angles. Plots of resistance at ±1 V as a function of junction angle with two different MWCNTs [122].

Furthermore, irrespective of the commensurate status of the contact area, the electronic contact can undergo significant enhancement through the application of forces and relaxation (Figure 10b). However, the mechanism of the transport properties of two CNTs remains unclear because of the lower precision for atomic manipulation and characterization. Further study is required to gain a more profound comprehension of the mechanisms.

#### 3.2.2. Electrical Properties of CNT Bundles

As Thess et al. [118] initially demonstrated, SWCNTs have the capacity to self-assemble with remarkable uniformity into crystalline ropes, which are more commonly referred to as bundles. The electrical properties of CNT bundles are governed by their mixed fine structures, weak inter-tube coupling, and low-dimensional quantum effects.

Bockrath et al. [127] revealed the profound impact of this heterogeneity at low temperatures. Measurements on individual bundles demonstrated a suppression of conductance at low bias (<10 K) and sharp peaks in conductance as a function of gate voltage. These features were interpreted as single-electron charging and resonant tunneling through quantized energy levels within the bundle. Collins et al. [120] addressed the challenge of exploiting s-SWNTs within mixed bundles by introducing "electrical breakdown". The exploitation of high current stress in air, which selectively destroys only the m-SWCNTs within the bundle, results in the conversion of the entire bundle into an effective FET based solely on the remaining s-SWCNTs.

The core limitation for CNT bundle conductivity is inefficient electron transfer at inter-tube junctions. This phenomenon was clearly illustrated through the use of quadrupole STM on herringbone-type CNTs [128]. The junctions were found to adopt a V-, T- and Y-shaped configuration, thereby functioning as ohmic scattering centers. However, they exhibited no gating effect, a phenomenon that can be attributed to the strong inter-layer coupling effect, which effectively suppresses electrostatic modulation.

A number of strategies have been proposed to improve inter-tube coupling. Ha et al. [129] employed laser-induced shockwaves (~3.2 GPa) to densify CNT networks. This process physically transformed CNT bundles into flattened, multi-layered graphene nanoribbons. The combination of pre-stretching for alignment and the other factors under investigation resulted in an approximate fivefold increase in electrical conductivity. This is due to the elimination of voids and the creation of seamless pathways for electron transport via graphitized interfaces. Gong et al. [130] utilized in-situ TEM to form covalent graphitic bonds through Joule-heating-induced welding, thereby augmenting bundle conductivity from approximately $10^2$ S/m to $10^5$ S/m. Qiu et al. [131] employed a two-step strategy, namely interfacial functionalization and densification, to increase the inter-bundle electrical properties. Infiltration with high polarity solvents such as ethylene glycol induced local electrostatic cohesion between bundles via surface dipoles, increasing interfacial electrical conductance by approximately 2.8 times. Acid treatment ($HNO_3$) introduced oxygen-containing functional groups acting as electron relays. This resulted in an additional 95% increase in inter-bundle electrical conductance.

### 3.3. *Electrical Properties of Macroscopic Assemblies of CNTs*

The assembly of CNTs into various macroscopic structures, such as fibers, arrays, and cottons, is contingent upon the specific synthesis process employed. Under these conditions, individual CNTs and CNT bundles agglomerate due to van der Waals interactions. Consequently, introducing inter-tube interfaces and non-uniformity of CNTs has resulted in a challenging task for preserving the unique electrical properties of individual CNTs in CNT macrostructures. In this section, the issue of dependence on the electrical properties of the assemblies of CNTs, such as fibers and films, will be explored in depth. The existing methods for improving their electrical properties will also be employed.

#### 3.3.1. Electrical Properties of CNT Fibers

The electrical properties of CNT fibers are contingent on the intrinsic structure of CNTs present within the macrostructure, including chirality, diameter, and length. Additionally, the extrinsic configuration, such as porosity and the alignment of CNT fibers, must be also considered (Figure 11).

In order to achieve an optimal electrical conductivity, the CNT fibers must consist exclusively of defect-free armchair SWCNTs, characterized by their small diameter, long length, and single chirality. As Sundaram et al. [132] reported, the electrical conductivity of fibers spun from metallic CNTs is expected to be high. The quantity of shells has been demonstrated to influence the conductivity of individual CNTs, as well as the collective conductivity of the fiber. The conductivity of SWCNTs is greater than that of MWCNTs because the mean free path decreases with an increase in the number of walls [133]. It can be posited that even in the context of purely armchair MWCNT fiber, the formation of physical contact between the internal shells and other nanotubes is improbable. Therefore, perfect electronic coupling between the inner and outer shells may prove challenging to achieve [134]. Tajima et al. [135] demonstrated that the specific conductivities of the fibers are almost proportional to the effective length of CNT. It is imperative to note that the increased length of the carbon nanotubes (CNTs) within the fiber will ensure a sufficient number of inter-tube contacts during the passage of a charge along the fiber [134].

It is noteworthy that extrinsic configuration substantially influences the electrical conductivity of CNT fibers. It is hypothesized that if the CNTs in a fiber are well condensed and perfectly axially aligned, it will ensure a long contact area with good physical stability and the shortest pathway for the flow of charge. This will result in a decrease in the overall resistance of the fibers. Miao et al. [136] systematically studied the relationship between the conductivity and the porosity of the CNT fibers. The study's results demonstrated a decreasing conductivity trend with increasing porosity. Francis et al. [137] theoretically suggested that the peak electrical sensitivity is attained within 40% to 60% porosity range. Furthermore, it has been demonstrated that enhancing the degree of alignment represents an effective strategy for enhancing the electrical conductivity and capacity of CNT fibers. The increase in alignment can be achieved through the application of outfield, stretching [138], twisting [139], or a combination of gravity and drawing [52], amongst other methods.

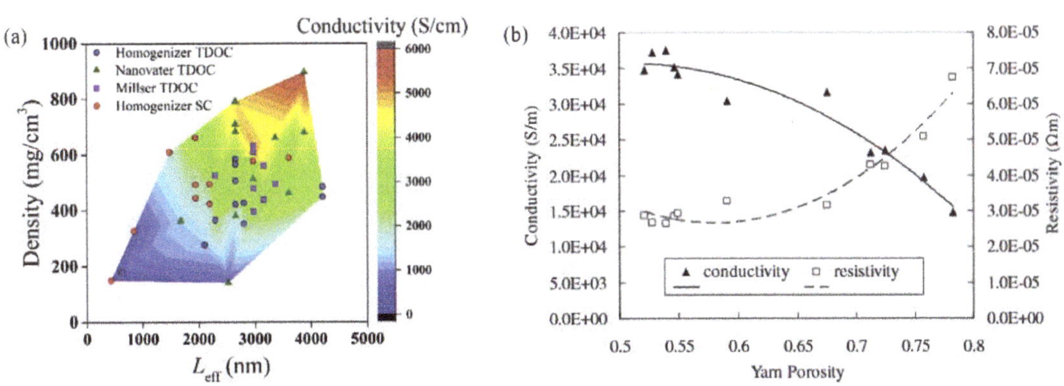


**Figure 11.** (**a**) Contour plot of electrical conductivity as functions of the effective length and fiber density [135]. (**b**) Variation in conductivity and resistivity as a function of the yarn porosity [136].

### 3.3.2. Electrical Properties of CNT Film

The primary factor influencing the electrical performance of CNT transparent conductive film (TCF) is the microstructure of CNTs. It has been demonstrated that long CNTs (about 10–20 μm) [140] with a large diameter (the diameter of the SWCNT and DWCNT is about 1.4 nm and 4nm, respectively) [141,142] and high crystallinity [143] are preferable for the construction of high-performance TCFs. The wall number is also crucial for the performance of TCF. It is evident that SWCNT film is the most desirable material for electronic and optoelectronic devices, given that it contains numerous semiconductive tubes and MWCNT absorbs more photons (Figure 12a) [110,142,144]. Furthermore, the incorporation of DWCNTs with a large diameter (about 4 nm) has been demonstrated to result in a substantial enhancement of the yield threshold of bundling, thereby expanding the contact area at the junctions [142]. The electrical type is also a crucial factor in this regard [145–148]. In a typical SWCNT film, the existence of Schottky barriers between metallic and semiconducting tubes invariably results in the presence of hopping barriers, thereby preventing high conductivity. As revealed by Yanagi et al. [148], quantum transport can be observed and hopping barriers are not present in SWCNT networks formed by high-purity metallic tubes, while a Coulomb-gap-type conduction mode is observed in semiconductive SWCNT networks. Topinka et al. and Yanagi et al. [146,148] provided a summary of the conduction mechanism in SWCNT networks as a function of the metallic/semiconducting ratio (Figure 12c). As the relative content of semiconducting SWCNTs increased, the conduction mechanism underwent a transition from quantum transport to variable range hopping (VRH) and then from VRH to ES-VRH. The VRH model postulated strong localization of electrons, indicating strong disorder in the contact area between bundles. The ES-VRH model indicated the presence of Coulomb interactions between localized electrons, inducing a Coulomb gap in the DOS. It was observed that both m-SWCNT and s-SWCNT networks exhibited the highest conductivity.

It is also imperative to optimize the configuration of CNT networks to enhance the electrical performance of TCFs. Many experiments have revealed that TCF consisting of isolated CNTs exhibits superior conductive properties compared to bundles [149–152]. Moreover, to avoid the introduction of surfactants and defects, it is preferable to synthesize an isolated CNT network by FCCVD directly [132,135] instead of utilizing solution-based techniques [61,153–155]. For instance, Kauppinen et al. [156] obtained SWCNT networks with over 60% isolated tubes by FCCVD method. After a HNO3 doping process, the TCFs demonstrated a comparatively low sheet resistance ($R_s$) of approximately 63 Ω/sq at a transmittance of 90% for 550 nm light. The synthesis of carbon-welded SWCNT film composed of approximately 85% isolated tubes was achieved by Liu et al. using the injection FCCVD method [149]. For pristine and $HNO_3$-doped TCFs, the Rs values were recorded as 41 Ω/sq and 25 Ω/sq, respectively. The high concentration of isolated CNTs in TCF is crucial to its ultrahigh performance. This is due to the provision of sufficient carrier transport paths without the absorption of additional light [145].

The percolation theory is widely used to examine the transport behavior of TCFs [157–159], which is employed to delineate the insulator-to-conductor transition. The percolation threshold $N_C$ is the minimum CNT concentration at which a conductive pathway can be formed, as illustrated in Formula (3).

$$N_C = 4.236^2 \pi L_s^2 \approx \frac{1}{L} \tag{3}$$

$L_s$ and $L$ represent an individual CNT length and aspect ratio, respectively. This indicates that the percolation threshold for SWCNT films with large aspect ratios is low, indicating that a minimal surface coverage is required to establish carrier transport pathways. Concurrently, the film resistance undergoes a precipitous decline above the threshold at elevated CNT densities. In this case, CNT hybrid architectures have been shown to synergize aligned domains (providing directional strength, conductivity, and deformation resistance) with random networks (delivering isotropic resilience, stress absorption, and environmental buffering) to mitigate anisotropy limitations. This balance optimizes mechanical stability, enhances charge transport via directional pathways and percolation networks, and improves environmental adaptability—particularly against humidity or temperature shifts. Hu et al. [160] derived a power exponential relationship between film conductance and thickness by controlling the volume of CNT suspension to form the TCF by vacuum filtration. Furthermore, a reduction in the size of the tube bundle results in a decrease in its square resistance [161]. It has been demonstrated by experimental and theoretical works that the network conductivity of CNTs is approximately linearly related to the mean bundle length and the network thickness, and approximately related to the 0.5 power of the mean bundle diameter (Figure 12b) [162].

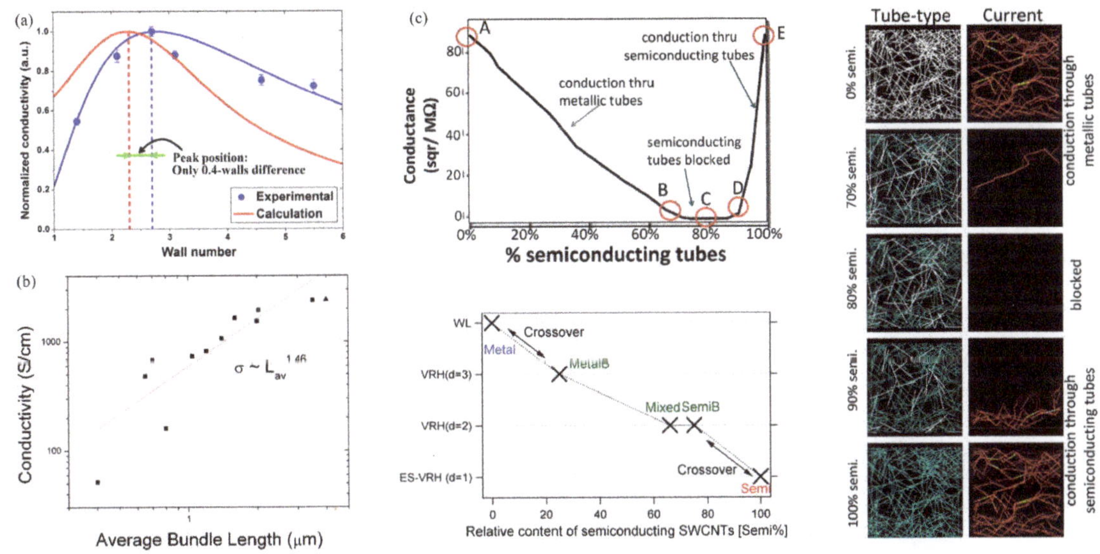


**Figure 12.** (**a**) The variation in calculated electrical conductivity in red line and normalized experimental conductivity in blue line as a function of wall number [110]. (**b**) The relationship between the average bundle length and electrical conductivity of CNT network [162]. (**c**) Electronic phase diagram of conduction mechanisms as a function of semiconducting tube ratio [146,148].

The degree of alignment is another factor that affects the conductivity of TCFs, and the cutoff angle ($\theta_\mu$) is often used to quantify the degree of alignment. If the degree of alignment is excessive or insufficient, the formation of a conductive percolation path is impeded due to anisotropy or inadequate coverage, respectively. In contradistinction, the conductivity of CNT films is elevated by random distribution, a phenomenon attributable to diminished percolation thresholds [163,164]. Furthermore, the presence of amorphous carbon and $sp^3$ bonds within the network leads to carrier scattering, demonstrating that films with a high G/D ratio exhibit enhanced conductivity [165].

It is widely accepted that the junction resistance (Figure 13a) [147] is the primary contributor to the Rs of a TCF, suggesting that enhancing the performance of TCFs is more effectively achieved by focusing on the control of inter-tube connections rather than the structure of composing CNTs. A considerable amount of effort has been invested into the enhancement of junction resistance. The integration of graphitic carbon within the junctions has been demonstrated to facilitate a transformation of the Schottky contacts between s-SWCNT and m-SWCNT, converting these contacts into near-ohmic configurations (Figure 13b) [149]. Also, controlling the unique morphology of the network, characterized by a preponderance of Y-junctions over X-junctions, which possess larger contact areas, has the potential to yield reduced junction resistances [61]. Moreover, densification through the application of acid treatments, such as $HNO_3$ [166] and chlorosulfonic acid (CSA) [167] or employing simple compression [166–168] could also enhance the conductivity of TCFs. It enhances the zipping effect between tubes, leading to highly packed and oriented CNTs in the film, with a concomitant decrease in voids (Figure 13c).

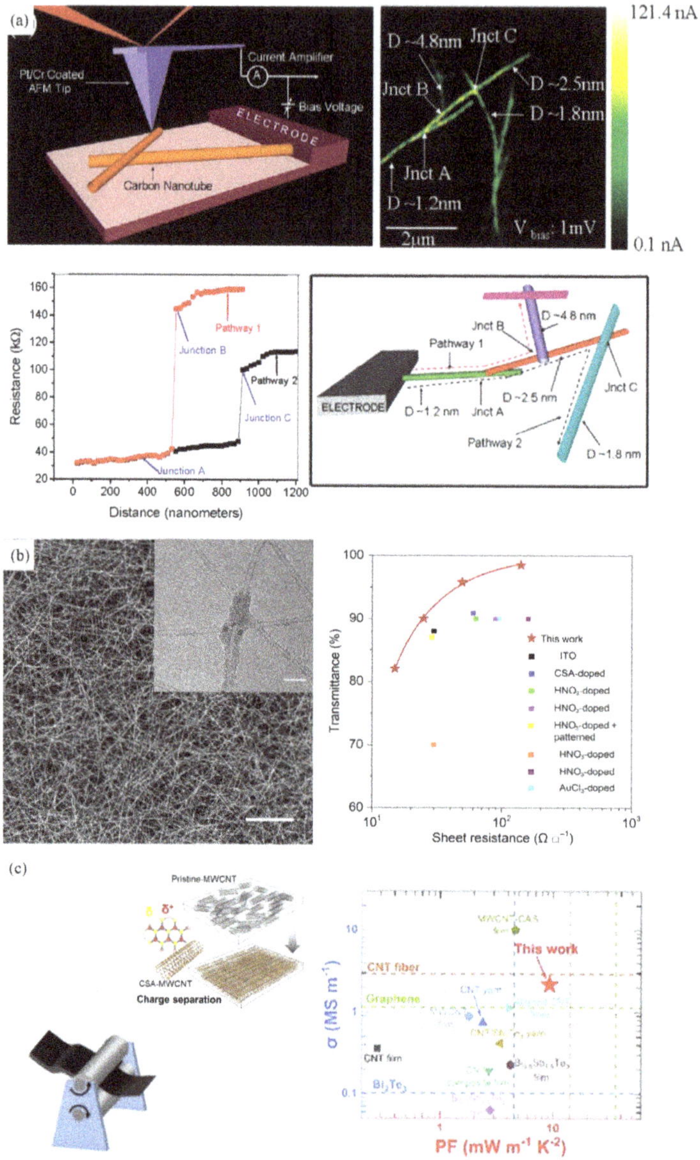


**Figure 13.** (**a**) Schematic of conductive mode in atomic force microscope, and one of the current maps and resistance analysis of interconnected tubes [147]. (**b**) The left are the SEM and TEM images of SWCNTs with carbon-welded joints. The right is the transmittance (550 nm light) versus Rs of doped SWCNT TCFs and other reported doping results [149]. (**c**) Schematic of compression and acid treatment for densification, SEM images of pristine and compressed CNT films, and the thermal and electrical performance compared to various materials [157,158].

### 3.4. Discrepancy in the Electrical Property of Individual CNTs and Their Assemblies

It has been demonstrated that individual CNTs, especially individual SWCNTs, exhibit excellent electrical properties. The conductivities of SWCNTs can be comparable to, or even superior to those of highly conductive metals. In an ideal condition, the carrier mobility and current-carrying capacity of a SWCNT with a defect-free structure can be extremely high. However, when assembled into macrostructures, the distinct transport properties of the individual tubes do not translate into the assemblies. Table 2 summarizes experimental results of the electrical properties of individual CNTs and their assemblies.

**Table 2.** Experimental result of the electrical properties of individual CNTs and their assemblies.

| | Individual CNTs | Assemblies of CNTs |
|---|---|---|
| Conductivity | $10^6$–$10^7$ S/m (m-SWCNT)<br>$10^2$–$10^3$ S/m (s-SWCNT) | $10^3$–$10^5$ S/m |
| Carrier mobility | $10^5$ $cm^2/(V{\cdot}s)$ (m-SWCNT)<br>$10^4$ $cm^2/(V{\cdot}s)$ (s-SWCNT) | ~$10^3$ $cm^2/(V{\cdot}s)$ |
| Current carrying capacity | ~$10^9$ $A/cm^2$ | ~$10^6$–$10^7$ $A/cm^2$ |
| Anisotropy | One-dimensional(1D) conductor<br>Extremely anisotropic | Alignment CNT: anisotropic<br>Disordered CNT: ~isotropy |
| Stability and environmental sensitivity | Susceptible to surface adsorbents (such as oxygen, water), resulting in electrical performance fluctuations | Higher structural stability, but the interface oxidation or mechanical deformation may occur for a long time. |
| Application | Nanoelectronic devices (transistors, sensors), quantum wires, scanning probes, etc. | Flexible conductors, electromagnetic shielding materials, battery electrodes, composite reinforcement phase |

The significant difference in the electrical properties of individual CNTs and their assemblies can be attributed to the following factors: First and foremost, is the cross junctions exhibited in assemblies. Since the inter-tube contact at cross junctions is achieved by van der Waals interaction, it may be difficult to achieve an ohmic contact. The second factor is a mixture of nanotubes that vary in chirality, diameter, number of walls, and length in the assemblies. It is still challenging to prepare pure SWCNTs with identical structures. Thirdly, some overall structural factors like tube alignment, densification, and porosity also significantly influence the properties. If the arrangement is highly disordered, the electron transport paths become zigzag, which leads to a decrease in the conductivity.

As a result, a large number of strategies for optimizing the electrical performance of assemblies have been proposed. On orientation control, CNTs are arranged by drawing [52,169], applying electric field [170,171], or using template method [172] to reduce the interface resistance. Substitutional or surface charge transfer doping was also performed to improve the electrical conductivity (Section 3.5). Additionally, in order to obtain reliable conductivity measurements, it is necessary to undertake pre-treatments like acid treatment to remove residual impurities and post-treatments like high-temperature annealing to reduce defects. It should also be noted that trade-offs between mechanical and electrical properties invariably occur in CNT assemblies. Increasing the density and alignment of CNTs improves the electrical properties and also promotes the mechanical properties by improving the interfacial resistance. However, certain methods, such as high-temperature treatment, have been observed to enhance electrical properties at the expense of toughness due to reduced inter-tube overlap [173].

### 3.5. Electrical Property of Doped CNTs

As mentioned above, the electrical properties of individual CNTs are strongly dependent on their atomic structures, such as chirality and defects. In contrast, the electrical properties of assemblies are also influenced by inter-tube junctions, extrinsic configurations such as porosity, and alignment. A number of strategies are proposed to enhance the conductivity in both Sections 3.2 and 3.3. However, it is worth noting that the atomically controlled growth of CNTs and charge transfers in CNT assemblies have yet to be fully achieved. Doping is a critical strategy for tailoring the electrical properties of CNTs and has broad application prospects in electronic devices. There are mainly two primary doping methods: substitutional doping and surface charge transfer doping.

#### 3.5.1. Substitutional Doping

Substitutional doping in CNTs incorporates foreign atoms into the carbon lattice. This is similar to the bulk doping process used in conventional semiconductors. Owing to their similar atomic radii to carbon (C) atoms, nitrogen (N) [174–178] and boron (B) [179–181] are the most commonly used doping elements, which release excess electrons to CNTs and create holes in the valence band in CNTs, respectively.

There are basically two approaches to prepare the substitutionally doped CNT: in situ doping during synthesis and post-synthesis doping. For the former, precursors containing the dopant atoms are used during CNT growth, by arc discharge [182,183], laser ablation [184–186], chemical vapor deposition (CVD) [187,188], spray-pyrolysis [177,189], or plasma-assisted CVD [190]. For the later, post-synthesis chemical reaction was performed to replace carbon atoms in CNTs with doping atoms, which can be realized by heat treatment [191,192], plasma treatment [193,194], or ball-milling [195,196].

B-doped and N-doped CNTs exhibit unique electronic and transport properties. At low concentrations of B or N, the dopants are uniformly distributed within the CNT and the impurity states appear in the band gap, which can be detected by STM. The donor state induced by N doping appears near the conduction band minimum, while the acceptor state in the case of B is close to the valence band maximum. Moreover, some calculation results suggested that the localization of the extra charge associated with the impurity differs considerably between s-CNTs and m-CNTs. Compared to an s-CNT, the defect state is spatially more localized for a metallic one, resulting in a significant difference in electrical and chemical properties [197].

With a high concentration of B or N, several studies have revealed that N-doped and B-doped CNTs exhibit metal-like behavior [198] because of the presence of dopant-rich islands like $BC_3$ nanodomains [199] and $CN_x$ tubes [200].

Owing to their excellent conductivity and chemical stability along with unique electronic structure and high electron density, N-doped and B-doped CNTs have been widely used in flexible micro-supercapacitors, batteries, and catalysis.

#### 3.5.2. Surface Charge Transfer Doping

Surface charge transfer doping is an unconventional way to achieve doping through surface engineering, which is commonly used in the doping of CNT assemblies. Take p-type doping as an example, acceptors are known to possess unoccupied molecular orbitals for electrons (UMOs). In the event of the energetically lowest of these orbitals (LUMO) being in proximity to the valence band maximum of the CNT, an electron will be transferred from the CNT. Consequently, holes will form in the CNTs, and negative charge will be localized on the surface acceptors. The process of charge separation is accompanied by the establishment of an electrostatic potential that confines the holes in a perpendicular direction. However, the holes are still able to move in a parallel direction to the surface [201].

Ionic dopants form a heterogeneous interface between CNT and dopants after CNT synthesis. N-type dopants like alkali metals (Li [202], K [95,203–208]) and p-type dopants like halogen elements ($Br_2$ [208], $I_2$ [209–213]), strong acid (CSA [167,209]), and transition metal halides (CuI [214], Ibr [209,215], Icl [215], $AuCl_3$ [216,217], $FeCl_3$ [203,218], $PbI_2$ [219]) are the most commonly used dopants. Regarding the location of the dopants, the doping type can be classified as exohedral (intercalation and deposition) and endohedral (filling) [220].

A substantial body of research remains on exohedral doping (Figure 14). Madrona et al. [221] demonstrated that the intercalation of long-range ordered Br into DWCNT fibers reduces the inter-tube resistance, and increases the conductance of individual DWCNTs. Madrona et al. [222] introduced $FeCl_3$ intercalation into fibers consisted of collapsed CNTs, which leads to a stable increase in conductivity measured at a factor of six. Qiu et al. [223] deposited Au nanoparticles in CNT fibers to enhance the interfacial electrical transport, inducing an equivalent p-type doping.

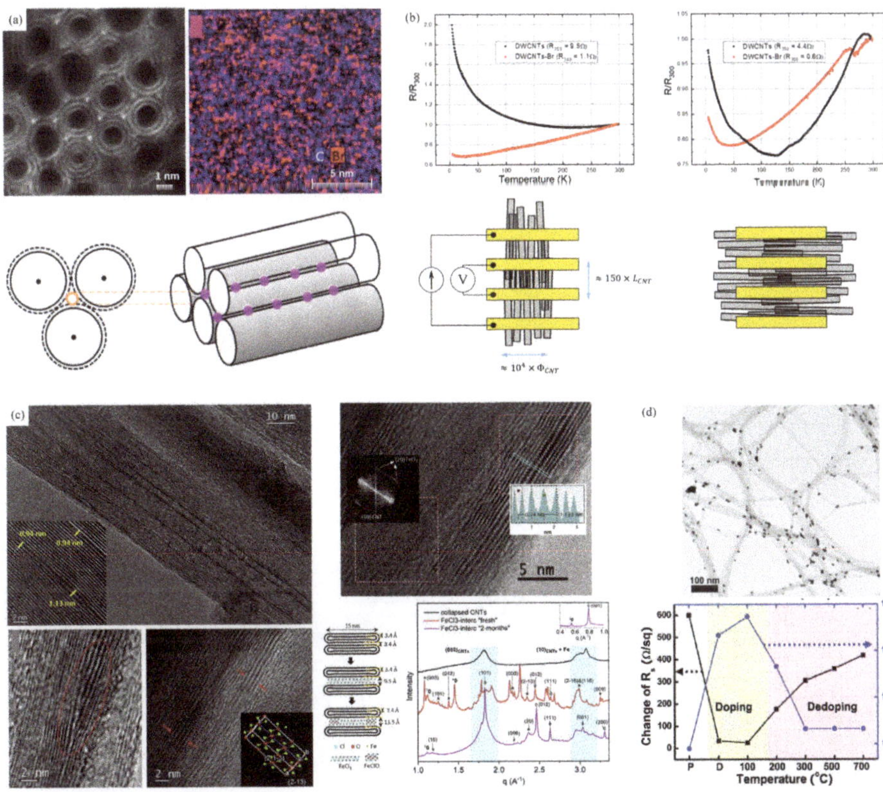


**Figure 14.** (**a**) High-angle annular dark-field (HAADF)–scanning transmission electron microscopy (STEM) image, energy dispersive X-ray chemical map, and illustrations of the interstitial doping of the cross section of a Br-intercalated DWCNT fiber [221]. (**b**) Low-temperature transport properties before and after Br-intercalation with different transport directions [221]. (**c**) High-resolution TEM image of $FeCl_3$-intercalated collapsed CNT fiber and evidence of the existence of FeClO and $FeCl_3$ [222]. (**d**) TEM image of $AuCl_3$-doping SWCNT film and the variation of Rs and G peak position as a function of temperature [217].

Recently, considerable attention has been paid to the construction of one-dimensional van der Waals heterostructures by endohedral doping for the purpose of enhancing the electrical properties. Due to the confinement protection of CNTs, they exhibit good stability. Zhang et al. [224] prepared CuI@SWCNT networks with an electrical conductivity of 32 kS $m^{-1}$ and a current carrying capacity of $2 \times 10^7$ A $cm^{-2}$ (Figure 15a). Teng et al. [219] prepared a $PbI_2$@SWCNTs hybrid and fabricated self-powered photodetectors that exhibit exceptional photocurrent and a switching ratio of three orders of magnitude (Figure 15b). Du et al. [225] developed a synthesizing method named comelting–filling–freezing–modification for simultaneously encapsulating various high-entropy metal phosphide (HEP) into SWCNTs. The SWCNT could protect the 1D HEP and donate $\pi$ electrons to the HEP for enhanced electron delocalization, thus promoting high electrocatalytic activity and stability (Figure 15c).

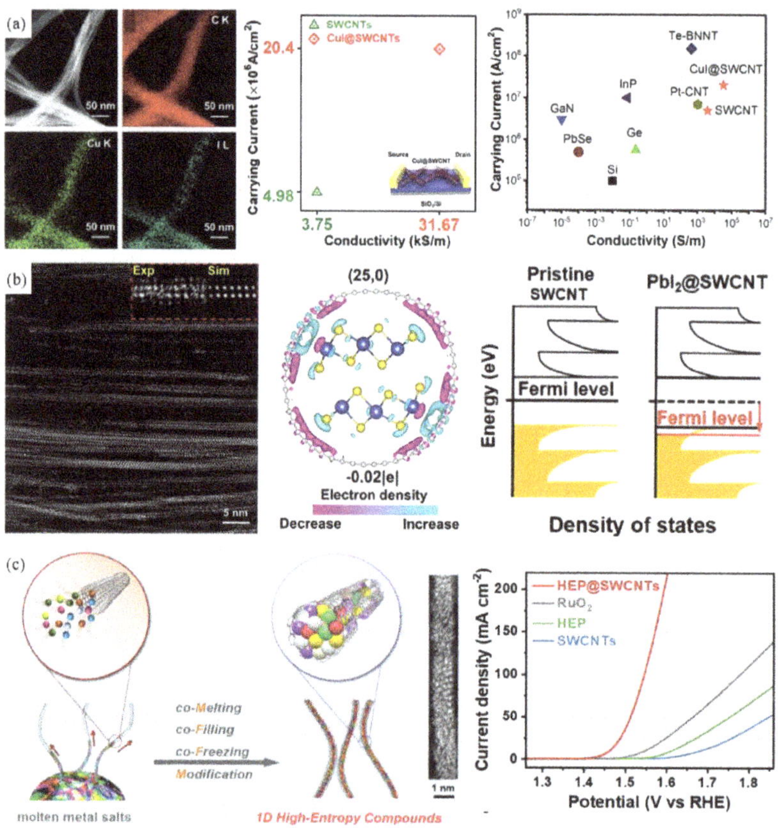


**Figure 15.** (**a**) Elemental mapping images and electrical performance of CuI@SWCNT networks [224]. (**b**) HAADF STEM image, structural models, DFT calculations showing the electron density difference and Scheme showing the downshift of the Fermi level of $PbI_2$@SWCNT [219]. (**c**) Schematic illustration for the synthesis and the electrochemical characterization of HEP@SWCNTs [225].

It is important to note that the substitutional doping always demonstrates superior thermal and chemical stability compared to the surface charge transfer doping because the substitutional dopants exhibit superior thermal resilience via strong chemical bonds. The dopants employed in surface charge transfer doping, instead, are susceptible to de-doping in high temperature or humid/ionic environments. This limitation restricts the applicability of this method to high-temperature applications like cables.

## 4. Conclusions and Outlook

In summary, CNTs are a promising candidate for use in electronics, energy storage, transparent conductive films, and other electrical applications due to their exceptional electrical properties. This review briefly introduces the methods explored for measuring the electrical properties of CNTs, such as STM, electron microscope-based nanoprobes, CNT electronic devices. These methodologies can be used to obtain a variety of electrical information, such as electronic band structure, resistance, conductance, and thermoelectric power. Most approaches require contact between CNTs and electrodes, inevitably leading to measurement errors. In addition to contact resistance, other factors such as measurement environment and substrate have also been identified as contributors to the diversity of electrical measurement results in different experimental groups. Moreover, the effects of intrinsic structures of CNTs such as chirality, diameter, defects, and length on the electrical properties of individual CNTs at the atomic scale are discussed. More importantly, the primary limitations on electrical properties in multi-tube systems have been explored by considering two CNTs, CNT bundles, fibers, and films. The core limitations for charge transfer are clarified as inter-tube junctions and extrinsic configurations of CNT assemblies, such as porosity, alignment, etc. Several strategies such as acid treatment, application of outfields and doping have been developed to modify the alignment, van der Waals interactions, and contact resistance of CNTs in macroscopic assemblies.

Despite the development made in improving the electrical properties of individual CNTs and CNT assemblies, many challenges still remain. Firstly, the electrical transfer mechanism from individuals to assemblies remains unclear. Although a general understanding is that the inter-tube interfaces is the main origin of the degraded electrical properties, there has been no general theory, principle, and method to solve this problem. Secondly, bridging the gap between theory and experiment is still challenging. The structure of CNTs cannot be precisely controlled and measurement errors do exist.

Based on the current research status and the aforementioned challenges, we propose the following strategies to address these issues: First of all, defects are inevitably introduced during CNT synthesis and subsequent processing, and these defects act as scattering centers of electron transport. It is important to develop defect engineering techniques, such as laser annealing and chemical functionalization, which enables vacancy healing in a gentler and more controllable way. Secondly, the transport properties of CNTs are strongly dependent on their structures, especially their chirality. Consequently, realizing chirality-specific synthesis is an important way to boost the electrical properties of individual and assembled SWCNTs. Thirdly, advanced doping methods need developed to achieve more stable n/p-type CNT assemblies. Fourthly, modifying the interface structure between CNTs in their macroscopic assemblies is highly important. It is expected that efficient electrical conduction will be realized through the optimization of the assembly configuration and the interaction between the comprising CNTs. The application of machine learning to guide assembly design [226] and conduct high-throughput synthesis screening for CNT assembly optimization [227,228] is also a viable option.

**Author Contributions:** Writing—original draft preparation, Y.X.; writing—review and editing, L.Z. and C.L. All authors have read and agreed to the published version of the manuscript.

**Funding:** This project was supported by National Natural Science Foundation of China (Nos. 52130209, 51927803, 52188101, 52372054, and 22003074), National Key R&D Program of China (No. 2022YFA1203302), Guangdong Provincial Key Laboratory Program of the Guangdong Science and Technology Department (No. 2021B1212040001), Youth Innovation Promotion Association CAS (No. 2022366), and the Shenzhen Science and Technology Program (No. JCYJ20240813154813018).

**Conflicts of Interest:** The authors declare no conflict of interest.

## Abbreviations

The following abbreviations are used in this manuscript:

| | |
|---|---|
| CNTs | Carbon nanotubes |
| SWCNTs | Single-walled carbon nanotubes |
| m-SWCNTs | Metallic single-walled carbon nanotubes |
| s-SWCNTs | Semiconducting single-walled carbon nanotubes |
| DWCNTs | Double-walled carbon nanotubes |
| MWCNTs | Multi-walled carbon nanotubes |
| FET | Field-effect transistor |
| STM | Scanning tunnelling microscopy |
| STS | Scanning tunneling spectroscopy |
| DOS | Density of states |
| SEM | Scanning electron microscope |
| TEM | Transmission electron microscope |
| FIB | Focused ion beam |
| ICs | Integrated circuits |
| MOSFET | Metal–oxide–semiconductor field-effect transistor |
| FCCVD | Floating catalyst chemical vapor deposition |
| TCF | Transparent conductive film |
| VRH | Variable range hopping |
| CSA | Chlorosulfonic acid |
| 1D | One-dimensional |
| HAADF | High-angle annular dark-field |
| STEM | Scanning transmission electron microscopy |
| HEP | High-entropy metal phosphide |
| DFT | Density functional tight |

## References

1. Iijima, S. Helical Microtubules of Graphitic Carbon. *Nature* **1991**, *354*, 56–58. [CrossRef]
2. Venkataraman, A.; Amadi, E.V.; Chen, Y.; Papadopoulos, C. Carbon Nanotube Assembly and Integration for Applications. *Nanoscale Res. Lett.* **2019**, *14*, 220. [CrossRef]
3. Rathinavel, S.; Priyadharshini, K.; Panda, D. A Review on Carbon Nanotube: An Overview of Synthesis, Properties, Functionalization, Characterization, and the Application. *Mater. Sci. Eng. B* **2021**, *268*, 115095. [CrossRef]
4. Shooshtari, M. Ammonia Gas Sensors Based on Multi-Wall Carbon Nanofiber Field Effect Transistors by Using Gate Modulation. *Colloids Surf. Physicochem. Eng. Asp.* **2025**, *704*, 135563. [CrossRef]
5. Javey, A.; Guo, J.; Paulsson, M.; Wang, Q.; Mann, D.; Lundstrom, M.; Dai, H. High-Field Quasiballistic Transport in Short Carbon Nanotubes. *Phys. Rev. Lett.* **2004**, *92*, 106804. [CrossRef] [PubMed]
6. Liang, W.J.; Bockrath, M.; Bozovic, D.; Hafner, J.H.; Tinkham, M.; Park, H. Fabry-Perot Interference in a Nanotube Electron Waveguide. *Nature* **2001**, *411*, 665–669. [CrossRef] [PubMed]
7. Perebeinos, V.; Tersoff, J.; Avouris, P. Mobility in Semiconducting Carbon Nanotubes at Finite Carrier Density. *Nano Lett.* **2006**, *6*, 205–208. [CrossRef] [PubMed]
8. Durkop, T.; Getty, S.A.; Cobas, E.; Fuhrer, M.S. Extraordinary Mobility in Semiconducting Carbon Nanotubes. *Nano Lett.* **2004**, *4*, 35–39. [CrossRef]
9. Arora, N.D.; Hauser, J.R.; Roulston, D.J. Electron and Hole Mobilities in Silicon as a Function of Concentration and Temperature. *IEEE Trans. Electron Devices* **1982**, *29*, 292–295. [CrossRef]
10. McEuen, P.L.; Bockrath, M.; Cobden, D.H.; Yoon, Y.-G.; Louie, S.G. Disorder, Pseudospins, and Backscattering in Carbon Nanotubes. *Phys. Rev. Lett.* **1999**, *83*, 5098–5101. [CrossRef]

11. Javey, A.; Guo, J.; Wang, Q.; Lundstrom, M.; Dai, H. Ballistic Carbon Nanotube Field-Effect Transistors. *Nature* **2003**, *424*, 654–657. [CrossRef]
12. Frank, S.; Poncharal, P.; Wang, Z.L.; de Heer, W.A. Carbon Nanotube Quantum Resistors. *Science* **1998**, *280*, 1744–1746. [CrossRef]
13. Kong, J.; Yenilmez, E.; Tombler, T.W.; Kim, W.; Dai, H.; Laughlin, R.B.; Liu, L.; Jayanthi, C.S.; Wu, S.Y. Quantum Interference and Ballistic Transmission in Nanotube Electron Waveguides. *Phys. Rev. Lett.* **2001**, *87*, 106801. [CrossRef]
14. White, C.T.; Todorov, T.N. Carbon Nanotubes as Long Ballistic Conductors. *Nature* **1998**, *393*, 240–242. [CrossRef]
15. Sun, D.-M.; Timmermans, M.Y.; Kaskela, A.; Nasibulin, A.G.; Kishimoto, S.; Mizutani, T.; Kauppinen, E.I.; Ohno, Y. Mouldable All-Carbon Integrated Circuits. *Nat. Commun.* **2013**, *4*, 2302. [CrossRef] [PubMed]
16. Jiao, X.; Shi, C.; Zhao, Y.; Xu, L.; Liu, S.; Hou, P.-X.; Liu, C.; Cheng, H.-M. Efficient Fabrication of High-Quality Single-Walled Carbon Nanotubes and Their Macroscopic Conductive Fibers. *ACS Nano* **2022**, *16*, 20263–20271. [CrossRef] [PubMed]
17. Xu, L.; Jiao, X.; Shi, C.; Wu, A.-P.; Hou, P.-X.; Liu, C.; Cheng, H.-M. Single-Walled Carbon Nanotube/Copper Core–Shell Fibers with a High Specific Electrical Conductivity. *ACS Nano* **2023**, *17*, 9245–9254. [CrossRef]
18. Wilder, J.W.G.; Venema, L.C.; Rinzler, A.G.; Smalley, R.E.; Dekker, C. Electronic Structure of Atomically Resolved Carbon Nanotubes. *Nature* **1998**, *391*, 59–62. [CrossRef]
19. Odom, T.W.; Huang, J.-L.; Lieber, C.M. STM Studies of Single-Walled Carbon Nanotubes. *J. Phys. Condens. Matter* **2002**, *14*, R145. [CrossRef]
20. Jobbins, M.M.; Raigoza, A.F.; Kandel, S.A. Note: Circuit Design for Direct Current and Alternating Current Electrochemical Etching of Scanning Probe Microscopy Tips. *Rev. Sci. Instrum.* **2012**, *83*, 1115. [CrossRef]
21. Zhang, J.; Wang, P.; Zhang, X.; Ji, H.; Luo, J.; Wang, H.; Wang, J. Systematic Electrochemical Etching of Various Metal Tips for Tunneling Spectroscopy and Scanning Probe Microscopy. *Rev. Sci. Instrum.* **2021**, *92*, 015124. [CrossRef] [PubMed]
22. Liu, Z.; Zhang, Y.; Zhao, P.; Ye, Y.; Chen, J.; Xu, N.; Deng, S. In Situ Study of Field Emission Vacuum Breakdown of Individual Multi-wall Carbon Nanotube. *Micro Nano Lett.* **2019**, *14*, 206–210. [CrossRef]
23. Wei, X.; Chen, Q.; Peng, L.; Cui, R.; Li, Y. In Situ Measurements on Individual Thin Carbon Nanotubes Using Nanomanipulators inside a Scanning Electron Microscope. *Ultramicroscopy* **2010**, *110*, 182–189. [CrossRef] [PubMed]
24. Feng, W.; Hayama, K.; Akinaga, H. Investigations of Electrical Transport Properties of Individual Carbon Nanotubes with Nanoprober. *Jpn. J. Appl. Phys.* **2016**, *55*, 06JH01. [CrossRef]
25. Jeong, S.; Higashitani, K.; Kaneko, T.; Yamada, T.; Li, Z.; Fujimori, T.; Tejima, S.; Fujita, J. Elucidating Slipping Behaviors Between Carbon Nanotubes: Using Nitrogen Doping and Electron Irradiation to Suppress Slippage. *Carbon* **2024**, *231*, 119693. [CrossRef]
26. Tang, D.-M.; Kvashnin, D.G.; Najmaei, S.; Bando, Y.; Kimoto, K.; Koskinen, P.; Ajayan, P.M.; Yakobson, B.I.; Sorokin, P.B.; Lou, J.; et al. Nanomechanical Cleavage of Molybdenum Disulphide Atomic Layers. *Nat. Commun.* **2014**, *5*, 3631. [CrossRef]
27. Costa, P.M.F.J.; Gautam, U.K.; Bando, Y.; Golberg, D. Direct Imaging of Joule Heating Dynamics and Temperature Profiling inside a Carbon Nanotube Interconnect. *Nat. Commun.* **2011**, *2*, 421. [CrossRef]
28. Aslam, Z.; Nicholls, R.; Koós, A.; Nicolosi, V.; Grobert, N. Current-Induced Restructuring and Chemical Modification of N-Doped Multi-Walled Carbon Nanotubes. *Adv. Funct. Mater.* **2011**, *21*, 3933–3937. [CrossRef]
29. Wang, M.-S.; Golberg, D.; Bando, Y. Interface Dynamic Behavior Between a Carbon Nanotube and Metal Electrode. *Adv. Mater.* **2010**, *22*, 93–98. [CrossRef]
30. Wang, M.; Golberg, D.; Bando, Y. Superstrong Low-Resistant Carbon Nanotube-Carbide-Metal Nanocontacts. *Adv. Mater.* **2010**, *22*, 5350–5355. [CrossRef]
31. Barnett, C.J.; Gowenlock, C.E.; Welsby, K.; Orbaek White, A.; Barron, A.R. Spatial and Contamination-Dependent Electrical Properties of Carbon Nanotubes. *Nano Lett.* **2018**, *18*, 695–700. [CrossRef]
32. Tuchband, M.; He, J.; Huang, S.; Lindsay, S. Insulated Gold Scanning Tunneling Microscopy Probes for Recognition Tunneling in an Aqueous Environment. *Rev. Sci. Instrum.* **2012**, *83*, 15102. [CrossRef]
33. Gingery, D.; Bühlmann, P. Single-Step Electrochemical Method for Producing Very Sharp Au Scanning Tunneling Microscopy Tips. *Rev. Sci. Instrum.* **2007**, *78*, 113703. [CrossRef]
34. Libioulle, L.; Houbion, Y.; Gilles, J.-M. Very Sharp Gold and Platinum Tips to Modify Gold Surfaces in Scanning Tunneling Microscopy. *J. Vac. Sci. Technol. B Microelectron. Nanometer Struct. Process. Meas. Phenom.* **1995**, *13*, 1325–1331. [CrossRef]
35. Lopes, M.; Toury, T.; De La Chapelle, M.L.; Bonaccorso, F.; Giuseppe Gucciardi, P. Fast and Reliable Fabrication of Gold Tips with Sub-50 Nm Radius of Curvature for Tip-Enhanced Raman Spectroscopy. *Rev. Sci. Instrum.* **2013**, *84*, 73702. [CrossRef]
36. Xu, G.; Liu, Z.; Xu, K.; Zhang, Y.; Zhong, H.; Fan, Y.; Huang, Z. Constant Current Etching of Gold Tips Suitable for Tip-Enhanced Raman Spectroscopy. *Rev. Sci. Instrum.* **2012**, *83*, 103708. [CrossRef] [PubMed]
37. Lee, C.; Kim, S.T.; Jeong, B.G.; Yun, S.J.; Song, Y.J.; Lee, Y.H.; Park, D.J.; Jeong, M.S. Tip-Enhanced Raman Scattering Imaging of Two-Dimensional Tungsten Disulfide with Optimized Tip Fabrication Process. *Sci. Rep.* **2017**, *7*, 40810. [CrossRef] [PubMed]
38. Narasiwodeyar, S.; Dwyer, M.; Liu, M.; Park, W.K.; Greene, L.H. Two-Step Fabrication Technique of Gold Tips for Use in Point-Contact Spectroscopy. *Rev. Sci. Instrum.* **2015**, *86*, 33903. [CrossRef] [PubMed]

39. Huang, J.Y.; Chen, S.; Jo, S.H.; Wang, Z.; Han, D.X.; Chen, G.; Dresselhaus, M.S.; Ren, Z.F. Atomic-Scale Imaging of Wall-by-Wall Breakdown and Concurrent Transport Measurements in Multiwall Carbon Nanotubes. *Phys. Rev. Lett.* **2005**, *94*, 236802. [CrossRef]
40. Bajpai, A.; Aslam, Z.; Hampel, S.; Klingeler, R.; Grobert, N. A Carbon-Nanotube Based Nano-Furnace for in-Situ Restructuring of a Magnetoelectric Oxide. *Carbon* **2017**, *114*, 291–300. [CrossRef]
41. Lord, A.M.; Ramasse, Q.M.; Kepaptsoglou, D.M.; Periwal, P.; Ross, F.M.; Wilks, S.P. Stability of Schottky and Ohmic Au Nanocatalysts to ZnO Nanowires. *Nano Lett.* **2017**, *17*, 6626–6636. [CrossRef]
42. Bachtold, A.; Henny, M.; Terrier, C.; Strunk, C.; Schönenberger, C.; Salvetat, J.-P.; Bonard, J.-M.; Forró, L. Contacting Carbon Nanotubes Selectively with Low-Ohmic Contacts for Four-Probe Electric Measurements. *Appl. Phys. Lett.* **1998**, *73*, 274–276. [CrossRef]
43. Dong, L.; Youkey, S.; Bush, J.; Jiao, J.; Dubin, V.M.; Chebiam, R.V. Effects of Local Joule Heating on the Reduction of Contact Resistance between Carbon Nanotubes and Metal Electrodes. *J. Appl. Phys.* **2007**, *101*, 024320. [CrossRef]
44. Hamasaki, H.; Li, Y.; Ohnishi, M.; Shiomi, J.; Yanagi, K.; Hirahara, K. Thermoelectric Power of a Single van Der Waals Interface between Carbon Nanotubes. *ACS Nano* **2024**, *18*, 612–617. [CrossRef] [PubMed]
45. Hamasaki, H.; Nagahama, S.; Hirahara, K. Electrical Conductivity of a Single Parallel Contact between Carbon Nanotubes. *Nanoscale* **2022**, *14*, 11529–11534. [CrossRef]
46. Ebbesen, T.W.; Lezec, H.J.; Hiura, H.; Bennett, J.W.; Ghaemi, H.F.; Thio, T. Electrical Conductivity of Individual Carbon Nanotubes. *Nature* **1996**, *382*, 54–56. [CrossRef]
47. Collins, P.G.; Hersam, M.; Arnold, M.; Martel, R.; Avouris, P. Current Saturation and Electrical Breakdown in Multiwalled Carbon Nanotubes. *Phys. Rev. Lett.* **2001**, *86*, 3128–3131. [CrossRef]
48. de Picciotto, R.; Stormer, H.L.; Pfeiffer, L.N.; Baldwin, K.W.; West, K.W. Four-Terminal Resistance of a Ballistic Quantum Wire. *Nature* **2001**, *411*, 51–54. [CrossRef]
49. Lüpke, F.; Cuma, D.; Korte, S.; Cherepanov, V.; Voigtländer, B. Four-Point Probe Measurements Using Current Probes with Voltage Feedback to Measure Electric Potentials. *J. Phys. Condens. Matter* **2018**, *30*, 54004. [CrossRef]
50. Zhaoqing, G.; Xu, L.; No, J.; Li, X.; He, C.; Hao-ZiKe, W.; Sun, C.; Hou, P.-X.; Liu, C.; Cheng, H.-M. Strong Connection of Single-Wall Carbon Nanotube Fibers with a Copper Substrate Using an Intermediate Nickel Layer. *ACS Nano* **2023**, *17*, 18290–18298. [CrossRef]
51. Park, J.Y.; Cho, J.; Jun, S.C. Review of Contact-Resistance Analysis in Nano-Material. *J. Mech. Sci. Technol.* **2018**, *32*, 539–547. [CrossRef]
52. Wang, H.-Z.; Jiao, X.-Y.; Gao, Z.-Q.; Hou, P.-X.; Xu, L.-L.; Shi, C.; Liang, Y.; Wang, Y.-P.; Liu, C. Highly Conductive Double-Wall Carbon Nanotube Fibers Produced by Dry-Jet Wet Spinning. *Adv. Funct. Mater.* **2024**, *34*, 2404538. [CrossRef]
53. Lin, Y.; Cao, Y.; Ding, S.; Zhang, P.; Xu, L.; Liu, C.; Hu, Q.; Jin, C.; Peng, L.-M.; Zhang, Z. Scaling Aligned Carbon Nanotube Transistors to a Sub-10 Nm Node. *Nat. Electron.* **2023**, *6*, 506–515. [CrossRef]
54. Guo, Q.; Wang, X.; Zhao, P.; Zhang, Z.; Geng, L.; Liu, Y.; Teng, Y.; Zhong, Y.; Kang, L. Performance Enhancement of Carbon Nanotube Network Transistors via SbI3 Inner-Doping in Selected Regions. *Adv. Mater.* **2025**, *37*, 2415442. [CrossRef]
55. Gwyther, R.E.A.; Côté, S.; Lee, C.-S.; Miao, H.; Ramakrishnan, K.; Palma, M.; Dafydd Jones, D. Optimising CNT-FET Biosensor Design through Modelling of Biomolecular Electrostatic Gating and Its Application to β-Lactamase Detection. *Nat. Commun.* **2024**, *15*, 7482. [CrossRef]
56. Ze, Y.; Liu, Y.; Wang, B.; Yin, H.; Jin, C.; Zhang, Z. Carbon Nanotube Materials for Future Integrated Circuit Applications. *Mater. Today* **2024**, *79*, 97–111. [CrossRef]
57. Shim, M.; Javey, A.; Shi Kam, N.W.; Dai, H. Polymer Functionalization for Air-Stable n-Type Carbon Nanotube Field-Effect Transistors. *J. Am. Chem. Soc.* **2001**, *123*, 11512–11513. [CrossRef]
58. Lin, S.; Ye, T.; Zhang, X.; Zuo, H.; Zhu, L.; Wang, X.; Li, C.; Yang, Z.; Du, R.; Lin, D.; et al. Arrays of Horizontal Semiconducting Carbon Nanotubes Grown from Non-Metal Catalysts Prepared by a "Thermophoresis-Anchoring" Synergistic Strategy. *Nano Today* **2025**, *61*, 102562. [CrossRef]
59. Franklin, A.D. Creating High-Performance Transistors by Coating Carbon Nanotube Arrays. *Nat. Electron.* **2024**, *7*, 1068–1069. [CrossRef]
60. Su, W.; Li, X.; Li, L.; Yang, D.; Wang, F.; Wei, X.; Zhou, W.; Kataura, H.; Xie, S.; Liu, H. Chirality-Dependent Electrical Transport Properties of Carbon Nanotubes Obtained by Experimental Measurement. *Nat. Commun.* **2023**, *14*, 1672. [CrossRef]
61. Sun, D.; Timmermans, M.Y.; Tian, Y.; Nasibulin, A.G.; Kauppinen, E.I.; Kishimoto, S.; Mizutani, T.; Ohno, Y. Flexible High-Performance Carbon Nanotube Integrated Circuits. *Nat. Nanotechnol.* **2011**, *6*, 156–161. [CrossRef]
62. LeMieux, M.C.; Roberts, M.; Barman, S.; Jin, Y.W.; Kim, J.M.; Bao, Z. Self-Sorted, Aligned Nanotube Networks for Thin-Film Transistors. *Science* **2008**, *321*, 101–104. [CrossRef] [PubMed]
63. Foradori, S.M.; Prussack, B.; Berson, A.; Arnold, M.S. Assembly and Alignment of High Packing Density Carbon Nanotube Arrays Using Lithographically Defined Microscopic Water Features. *ACS Nano* **2024**, *18*, 8259–8269. [CrossRef] [PubMed]

64. Hills, G.; Lau, C.; Wright, A.; Fuller, S.; Bishop, M.D.; Srimani, T.; Kanhaiya, P.; Ho, R.; Amer, A.; Stein, Y.; et al. Modern Microprocessor Built from Complementary Carbon Nanotube Transistors. *Nature* **2019**, *572*, 595–602. [CrossRef] [PubMed]
65. Sun, Y.; Lu, P.; Zhang, L.; Cao, Y.; Bai, L.; Ding, L.; Han, J.; Zhang, C.; Zhu, M.; Zhang, Z. Investigation and Improvement of the Bias Temperature Instability in Carbon Nanotube Transistors. *Adv. Electron. Mater.* **2024**, *11*, 2400464. [CrossRef]
66. Zamzami, M.A.; Rabbani, G.; Ahmad, A.; Basalah, A.A.; Al-Sabban, W.H.; Ahn, S.N.; Choudhry, H. Fabrication and Characterization of Field Effect Transistor Based on Single Walled Carbon Nanotubes. *J. King Saud Univ. Sci.* **2022**, *34*, 102137. [CrossRef]
67. Zhu, M.; Yin, H.; Cao, J.; Xu, L.; Lu, P.; Liu, Y.; Ding, L.; Fan, C.; Liu, H.; Zhang, Y.; et al. Inner Doping of Carbon Nanotubes with Perovskites for Ultralow Power Transistors. *Adv. Mater.* **2024**, *36*, 2403743. [CrossRef]
68. Han, J.; Xu, X.; Zhang, Z. Removing Conjugated Polymers from Aligned Carbon Nanotube Arrays. *Small* **2024**, *20*, 2309654. [CrossRef]
69. Cao, Q.; Tersoff, J.; Farmer, D.B.; Zhu, Y.; Han, S.-J. Carbon Nanotube Transistors Scaled to a 40-Nanometer Footprint. *Science* **2017**, *356*, 1369–1372. [CrossRef]
70. Zhu, Z.; Wei, N.; Cheng, W.; Shen, B.; Sun, S.; Gao, J.; Wen, Q.; Zhang, R.; Xu, J.; Wang, Y.; et al. Rate-Selected Growth of Ultrapure Semiconducting Carbon Nanotube Arrays. *Nat. Commun.* **2019**, *10*, 4467. [CrossRef]
71. Hu, Y.; Kang, L.; Zhao, Q.; Zhong, H.; Zhang, S.; Yang, L.; Wang, Z.; Lin, J.; Li, Q.; Zhang, Z.; et al. Growth of High-Density Horizontally Aligned SWNT Arrays Using Trojan Catalysts. *Nat. Commun.* **2015**, *6*, 6099. [CrossRef]
72. Zhang, Z.; Chen, Y.; Shen, P.; Chen, J.; Wang, S.; Wang, B.; Ma, S.; Lyu, B.; Zhou, X.; Lou, S.; et al. Homochiral Carbon Nanotube van Der Waals Crystals. *Science* **2025**, *387*, 1310–1316. [CrossRef] [PubMed]
73. Oshima, H.; Iwase, K.; Ohno, Y. In Situ Monitoring of the Electrical Property of Carbon Nanotube Thin Film in Floating Catalyst Chemical Vapor Deposition. *Jpn. J. Appl. Phys.* **2022**, *61*, 38002. [CrossRef]
74. Nessim, G.D. Properties, Synthesis, and Growth Mechanisms of Carbon Nanotubes with Special Focus on Thermal Chemical Vapor Deposition. *Nanoscale* **2010**, *2*, 1306. [CrossRef] [PubMed]
75. Suzuki, S.; Bower, C.; Watanabe, Y.; Zhou, O. Work Functions and Valence Band States of Pristine and Cs-Intercalated Single-Walled Carbon Nanotube Bundles. *Appl. Phys. Lett.* **2000**, *76*, 4007–4009. [CrossRef]
76. Mei, H.; Cui, J.; Cheng, Y.; He, X.; Mei, X. Heterogeneous Connection of Carbon Nanotubes with Metal Electrodes and Its Electrical Properties. *Chin. J. Lasers* **2021**, *48*, 802023. [CrossRef]
77. Cao, Q.; Han, S.; Tersoff, J.; Franklin, A.; Zhu, Y.; Zhang, Z.; Tulevski, G.; Tang, J.; Haensch, W. End-Bonded Contacts for Carbon Nanotube Transistors with Low, Size-Independent Resistance. *Science* **2015**, *350*, 68–72. [CrossRef]
78. Javey, A.; Wang, Q.; Kim, W.; Dai, H.J. Advancements in Complementary Carbon Nanotube Field-Effect Transistors. In Proceedings of the 2003 IEEE International Electron Devices Meeting, Technical Digest, Washington, DC, USA, 8–10 December 2003; IEEE: New York, NY, USA, 2003; pp. 741–744.
79. Kim, W.; Javey, A.; Tu, R.; Cao, J.; Wang, Q.; Dai, H.J. Electrical Contacts to Carbon Nanotubes down to 1 Nm in Diameter. *Appl. Phys. Lett.* **2005**, *87*, 173101. [CrossRef]
80. Nosho, Y.; Ohno, Y.; Kishimoto, S.; Mizutani, T. *N*-Type Carbon Nanotube Field-Effect Transistors Fabricated by Using Ca Contact Electrodes. *Appl. Phys. Lett.* **2005**, *86*, 073105. [CrossRef]
81. Zhang, J.; Wang, C.; Fu, Y.; Che, Y.; Zhou, C. Air-Stable Conversion of Separated Carbon Nanotube Thin-Film Transistors from p-Type to n-Type Using Atomic Layer Deposition of High-κ Oxide and Its Application in CMOS Logic Circuits. *ACS Nano* **2011**, *5*, 3284–3292. [CrossRef]
82. Franklin, A.D.; Koswatta, S.O.; Farmer, D.B.; Smith, J.T.; Gignac, L.; Breslin, C.M.; Han, S.-J.; Tulevski, G.S.; Miyazoe, H.; Haensch, W.; et al. Carbon Nanotube Complementary Wrap-Gate Transistors. *Nano Lett.* **2013**, *13*, 2490–2495. [CrossRef] [PubMed]
83. Zhang, Z.; Liang, X.; Wang, S.; Yao, K.; Hu, Y.; Zhu, Y.; Chen, Q.; Zhou, W.; Li, Y.; Yao, Y.; et al. Doping-Free Fabrication of Carbon Nanotube Based Ballistic CMOS Devices and Circuits. *Nano Lett.* **2007**, *7*, 3603–3607. [CrossRef]
84. Ding, L.; Wang, S.; Zhang, Z.; Zeng, Q.; Wang, Z.; Pei, T.; Yang, L.; Liang, X.; Shen, J.; Chen, Q.; et al. Y-Contacted High-Performance n-Type Single-Walled Carbon Nanotube Field-Effect Transistors: Scaling and Comparison with Sc-Contacted Devices. *NANO Lett.* **2009**, *9*, 4209–4214. [CrossRef] [PubMed]
85. Liu, N.; Chortos, A.; Lei, T.; Jin, L.; Kim, T.R.; Bae, W.-G.; Zhu, C.; Wang, S.; Pfattner, R.; Chen, X.; et al. Ultratransparent and Stretchable Graphene Electrodes. *Sci. Adv.* **2017**, *3*, e1700159. [CrossRef]
86. Lee, D.; Lee, H.; Ahn, Y.; Jeong, Y.; Lee, D.-Y.; Lee, Y. Highly Stable and Flexible Silver Nanowire–Graphene Hybrid Transparent Conducting Electrodes for Emerging Optoelectronic Devices. *Nanoscale* **2013**, *5*, 7750–7755. [CrossRef]
87. Zhang, L.; Sun, D.-M.; Hou, P.-X.; Liu, C.; Liu, T.; Wen, J.; Tang, N.; Luan, J.; Shi, C.; Li, J.-C.; et al. Selective Growth of Metal-Free Metallic and Semiconducting Single-Wall Carbon Nanotubes. *Adv. Mater.* **2017**, *29*, 1605719. [CrossRef]
88. Chen, Y.; Yue, Y.-Y.; Wang, S.-R.; Zhang, N.; Feng, J.; Sun, H.-B. Graphene as a Transparent and Conductive Electrode for Organic Optoelectronic Devices. *Adv. Electron. Mater.* **2019**, *5*, 1900247. [CrossRef]

89. Zhang, Y.; Ichihashi, T.; Landree, E.; Nihey, F.; Iijima, S. Heterostructures of Single-Walled Carbon Nanotubes and Carbide Nanorods. *Science* **1999**, *285*, 1719–1722. [CrossRef]
90. Asaka, K.; Karita, M.; Saito, Y. Modification of Interface Structure and Contact Resistance between a Carbon Nanotube and a Gold Electrode by Local Melting. *Appl. Surf. Sci.* **2011**, *257*, 2850–2853. [CrossRef]
91. Madsen, D.N.; Mølhave, K.; Mateiu, R.; Rasmussen, A.M.; Brorson, M.; Jacobsen, C.J.H.; Bøggild, P. Soldering of Nanotubes onto Microelectrodes. *Nano Lett.* **2003**, *3*, 47–49. [CrossRef]
92. Chen, C.X.; Yan, L.J.; Kong, E.S.W.; Zhang, Y.F. Ultrasonic Nanowelding of Carbon Nanotubes to Metal Electrodes. *Nanotechnology* **2006**, *17*, 2192–2197. [CrossRef]
93. Long, R. Understanding the Electronic Structures of Graphene Quantum Dot Physisorption and Chemisorption onto the $TiO_2$ (110) Surface: A First-Principles Calculation. *Chemphyschem Eur. J. Chem. Phys. Phys. Chem.* **2013**, *14*, 579–582. [CrossRef]
94. Collins, P.G.; Bradley, K.; Ishigami, M.; Zettl, A. Extreme Oxygen Sensitivity of Electronic Properties of Carbon Nanotubes. *Science* **2000**, *287*, 1801–1804. [CrossRef]
95. Heinze, S.; Tersoff, J.; Martel, R.; Derycke, V.; Appenzeller, J.; Avouris, P. Carbon Nanotubes as Schottky Barrier Transistors. *Phys. Rev. Lett.* **2002**, *89*, 106801. [CrossRef] [PubMed]
96. Wettstein, L.; Specht, J.; Kesselring, V.; Sieben, L.; Pan, Y.; Käch, D.; Baster, D.; Krumeich, F.; El Kazzi, M.; Bezdek, M.J. A Dye-Sensitized Sensor for Oxygen Detection under Visible Light. *Adv. Sci.* **2024**, *11*, 2405694. [CrossRef] [PubMed]
97. Kauffman, D.R.; Shade, C.M.; Uh, H.; Petoud, S.; Star, A. Decorated Carbon Nanotubes with Unique Oxygen Sensitivity. *Nat. Chem.* **2009**, *1*, 500–506. [CrossRef] [PubMed]
98. Rajavel, K.; Lalitha, M.; Radhakrishnan, J.K.; Senthilkumar, L.; Rajendra Kumar, R.T. Multiwalled Carbon Nanotube Oxygen Sensor: Enhanced Oxygen Sensitivity at Room Temperature and Mechanism of Sensing. *ACS Appl. Mater. Interfaces* **2015**, *7*, 23857–23865. [CrossRef]
99. Kane, C.L.; Mele, E.J.; Lee, R.S.; Fischer, J.E.; Petit, P.; Dai, H.; Thess, A.; Smalley, R.E.; Verschueren, A.R.M.; Tans, S.J.; et al. Temperature-Dependent Resistivity of Single-Wall Carbon Nanotubes. *Europhys. Lett.* **1998**, *41*, 683–688. [CrossRef]
100. Shooshtari, M.; Salehi, A.; Vollebregt, S. Effect of Humidity on Gas Sensing Performance of Carbon Nanotube Gas Sensors Operated at Room Temperature. *IEEE Sens. J.* **2021**, *21*, 5763–5770. [CrossRef]
101. Lin, Y.-M.; Tsang, J.C.; Freitag, M.; Avouris, P. Impact of Oxide Substrate on Electrical and Optical Properties of Carbon Nanotube Devices. *Nanotechnology* **2007**, *18*, 295202. [CrossRef]
102. Sun, Y.; Peng, Z.; Li, H.; Wang, Z.; Mu, Y.; Zhang, G.; Chen, S.; Liu, S.; Wang, G.; Liu, C.; et al. Suspended CNT-Based FET Sensor for Ultrasensitive and Label-Free Detection of DNA Hybridization. *Biosens. Bioelectron.* **2019**, *137*, 255–262. [CrossRef] [PubMed]
103. Saito, R.; Dresselhaus, G.; Dresselhaus, M.S. Trigonal Warping Effect of Carbon Nanotubes. *Phys. Rev. B* **2000**, *61*, 2981–2990. [CrossRef]
104. Saito, R.; Fujita, M.; Dresselhaus, G.; Dresselhaus, M.S. Electronic Structure of Chiral Graphene Tubules. *Appl. Phys. Lett.* **1992**, *60*, 2204–2206. [CrossRef]
105. Dresselhaus, M.S.; Dresselhaus, G.; Eklund, P.C. *Science of Fullerenes and Carbon Nanotubes*; Elsevier: Amsterdam, The Netherlands, 1996.
106. Dresselhaus, M.S.; Eklund, P.C. Phonons in Carbon Nanotubes. *Adv. Phys.* **2000**, *49*, 705–814. [CrossRef]
107. Kane, C.L.; Mele, E.J. Size, Shape, and Low Energy Electronic Structure of Carbon Nanotubes. *Phys. Rev. Lett.* **1997**, *78*, 1932–1935. [CrossRef]
108. Ouyang, M.; Huang, J.-L.; Cheung, C.L.; Lieber, C.M. Energy Gaps in "Metallic" Single-Walled Carbon Nanotubes. *Science* **2001**, *292*, 702–705. [CrossRef]
109. Crespi, V.H.; Cohen, M.L.; Rubio, A. In Situ Band Gap Engineering of Carbon Nanotubes. *Phys. Rev. Lett.* **1997**, *79*, 2093–2096. [CrossRef]
110. Chen, G.; Futaba, D.N.; Sakurai, S.; Yumura, M.; Hata, K. Interplay of Wall Number and Diameter on the Electrical Conductivity of Carbon Nanotube Thin Films. *Carbon* **2014**, *67*, 318–325. [CrossRef]
111. Kwon, Y.-K.; Tománek, D. Electronic and Structural Properties of Multiwall Carbon Nanotubes. *Phys. Rev. B* **1998**, *58*, R16001–R16004. [CrossRef]
112. Bourlon, B.; Miko, C.; Forró, L.; Glattli, D.C.; Bachtold, A. Determination of the Intershell Conductance in Multiwalled Carbon Nanotubes. *Phys. Rev. Lett.* **2004**, *93*, 176806. [CrossRef]
113. Liu, K.; Wang, W.; Xu, Z.; Bai, X.; Wang, E.; Yao, Y.; Zhang, J.; Liu, Z. Chirality-Dependent Transport Properties of Double-Walled Nanotubes Measured in Situ on Their Field-Effect Transistors. *J. Am. Chem. Soc.* **2009**, *131*, 62–63. [CrossRef] [PubMed]
114. Algharagholy, L.A. Defects in Carbon Nanotubes and Their Impact on the Electronic Transport Properties. *J. Electron. Mater.* **2019**, *48*, 2301–2306. [CrossRef]
115. Gómez-Navarro, C.; Pablo, P.J.D.; Gómez-Herrero, J.; Biel, B.; Garcia-Vidal, F.J.; Rubio, A.; Flores, F. Tuning the Conductance of Single-Walled Carbon Nanotubes by Ion Irradiation in the Anderson Localization Regime. *Nat. Mater.* **2005**, *4*, 534–539. [CrossRef]

116. Khanbolouki, P.; Tehrani, M. Purification, Structural Evolutions, and Electrical Properties of Carbon Nanotube Yarns Processed via Incandescent Annealing. *Carbon* **2020**, *168*, 710–718. [CrossRef]
117. Altuntas, H.; Oke-Altuntas, F.; Silva, S.R.P. The Effect of Low-Temperature Annealing on the Electrical Characteristics of Carbon Nanotube Network Field-Effect Transistors. *J. Electron. Mater.* **2024**, *53*, 2104–2114. [CrossRef]
118. Thess, A.; Lee, R.; Nikolaev, P.; Dai, H.; Petit, P.; Robert, J.; Xu, C.; Lee, Y.H.; Kim, S.G.; Rinzler, A.G.; et al. Crystalline Ropes of Metallic Carbon Nanotubes. *Science* **1996**, *273*, 483–487. [CrossRef]
119. Durrant, T.R.; El-Sayed, A.-M.; Gao, D.Z.; Rueckes, T.; Bersuker, G.; Shluger, A.L. Atomistic Modeling of the Electrical Conductivity of Single-Walled Carbon Nanotube Junctions. *Phys. Status Solidi RRL Rapid Res. Lett.* **2022**, *16*, 2200118. [CrossRef]
120. Collins, P.G.; Arnold, M.S.; Avouris, P. Engineering Carbon Nanotubes and Nanotube Circuits Using Electrical Breakdown. *Science* **2001**, *292*, 706–709. [CrossRef]
121. Tunney, M.A.; Cooper, N.R. Effects of Disorder and Momentum Relaxation on the Intertube Transport of Incommensurate Carbon Nanotube Ropes and Multiwall Nanotubes. *Phys. Rev. B* **2006**, *74*, 075406. [CrossRef]
122. Barnett, C.J.; Evans, C.; McCormack, J.E.; Gowenlock, C.E.; Dunstan, P.; Adams, W.; Orbaek White, A.; Barron, A.R. Experimental Measurement of Angular and Overlap Dependence of Conduction between Carbon Nanotubes of Identical Chirality and Diameter. *Nano Lett.* **2019**, *19*, 4861–4865. [CrossRef] [PubMed]
123. Buldum, A.; Lu, J.P. Contact Resistance between Carbon Nanotubes. *Phys. Rev. B* **2001**, *63*, 161403. [CrossRef]
124. Xu, F.; Sadrzadeh, A.; Xu, Z.; Yakobson, B.I. Can Carbon Nanotube Fibers Achieve the Ultimate Conductivity?—Coupled-Mode Analysis for Electron Transport through the Carbon Nanotube Contact. *J. Appl. Phys.* **2013**, *114*, 063714. [CrossRef]
125. Imamog¯lu, A.; Awschalom, D.D.; Burkard, G.; DiVincenzo, D.P.; Loss, D.; Sherwin, M.; Small, A. Quantum Information Processing Using Quantum Dot Spins and Cavity QED. *Phys. Rev. Lett.* **1999**, *83*, 4204–4207. [CrossRef]
126. Falvo, M.R.; Taylor, R.M., II; Helser, A.; Chi, V.; Brooks, F.P., Jr.; Washburn, S.; Superfine, R. Nanometre-Scale Rolling and Sliding of Carbon Nanotubes. *Nature* **1999**, *397*, 236–238. [CrossRef]
127. Bockrath, M.; Cobden, D.H.; McEuen, P.L.; Chopra, N.G.; Zettl, A.; Thess, A.; Smalley, R.E. Single-Electron Transport in Ropes of Carbon Nanotubes. *Science* **1997**, *275*, 1922–1925. [CrossRef]
128. Kim, T.-H.; Wendelken, J.F.; Li, A.-P.; Du, G.; Li, W. Probing Electrical Transport in Individual Carbon Nanotubes and Junctions. *Nanotechnology* **2008**, *19*, 485201. [CrossRef]
129. Ha, J.; Noh, J.; Jung, Y.J.; Liang, W.; Choi, T.-Y.; Kim, J.; Kim, D. High Compaction and Physical Graphitization of CNT Bundles and Network via Extreme-Load Compression Using Laser-Induced Shockwave. *Carbon* **2023**, *213*, 118219. [CrossRef]
130. Gong, X.; Zhang, H.; Sun, Z.; Zhang, X.; Xu, J.; Chu, F.; Sun, L.; Ramakrishna, S. A Viable Method to Enhance the Electrical Conductivity of CNT Bundles: Direct in Situ TEM Evaluation. *Nanoscale* **2020**, *12*, 13095–13102. [CrossRef]
131. Qiu, L.; Wang, X.; Tang, D.; Zheng, X.; Norris, P.M.; Wen, D.; Zhao, J.; Zhang, X.; Li, Q. Functionalization and Densification of Inter-Bundle Interfaces for Improvement in Electrical and Thermal Transport of Carbon Nanotube Fibers. *Carbon* **2016**, *105*, 248–259. [CrossRef]
132. Sundaram, R.M.; Koziol, K.K.K.; Windle, A.H. Continuous Direct Spinning of Fibers of Single-walled Carbon Nanotubes with Metallic Chirality. *Adv. Mater.* **2011**, *23*, 5064–5068. [CrossRef]
133. Dariyal, P.; Arya, A.K.; Singh, B.P.; Dhakate, S.R. A Review on Conducting Carbon Nanotube Fibers Spun via Direct Spinning Technique. *J. Mater. Sci.* **2021**, *56*, 1087–1115. [CrossRef]
134. Lekawa-Raus, A.; Patmore, J.; Kurzepa, L.; Bulmer, J.; Koziol, K. Electrical Properties of Carbon Nanotube Based Fibers and Their Future Use in Electrical Wiring. *Adv. Funct. Mater.* **2014**, *24*, 3661–3682. [CrossRef]
135. Tajima, N.; Watanabe, T.; Morimoto, T.; Kobashi, K.; Mukai, K.; Asaka, K.; Okazaki, T. Nanotube Length and Density Dependences of Electrical and Mechanical Properties of Carbon Nanotube Fibres Made by Wet Spinning. *Carbon* **2019**, *152*, 1–6. [CrossRef]
136. Miao, M. Electrical Conductivity of Pure Carbon Nanotube Yarns. *Carbon* **2011**, *49*, 3755–3761. [CrossRef]
137. Avilés, F.; Cabañas-Valdovinos, S.; Oliva-Avilés, A.I. Finite Element Investigation of the Influence of Porosity and Resin Infiltration on the Electrical Conductivity of Carbon Nanotube Yarns. *Adv. Eng. Mater.* **2025**, *27*, 2402714. [CrossRef]
138. Mukai, K.; Asaka, K.; Wu, X.; Morimoto, T.; Okazaki, T.; Saito, T.; Yumura, M. Wet Spinning of Continuous Polymer-Free Carbon-Nanotube Fibers with High Electrical Conductivity and Strength. *Appl. Phys. Express* **2016**, *9*, 55101. [CrossRef]
139. Headrick, R.J.; Tsentalovich, D.E.; Berdegue, J.; Bengio, E.A.; Liberman, L.; Kleinerman, O.; Lucas, M.S.; Talmon, Y.; Pasquali, M. Structure-Property Relations in Carbon Nanotube Fibers by Downscaling Solution Processing. *Adv. Mater.* **2018**, *30*, 1704482. [CrossRef]
140. Wang, Y.; Fugetsu, B. Mono-Dispersed Ultra-Long Single-Walled Carbon Nanotubes as Enabling Components in Transparent and Electrically Conductive Thin Films. *Carbon* **2015**, *82*, 152–160. [CrossRef]
141. Battie, Y.; Broch, L.; En Naciri, A.; Lauret, J.-S.; Guézo, M.; Loiseau, A. Diameter Dependence of the Optoelectronic Properties of Single Walled Carbon Nanotubes Determined by Ellipsometry. *Carbon* **2015**, *83*, 32–39. [CrossRef]

142. Zhang, Q.; Nam, J.-S.; Han, J.; Datta, S.; Wei, N.; Ding, E.-X.; Hussain, A.; Ahmad, S.; Skakalova, V.; Khan, A.T.; et al. Large-Diameter Carbon Nanotube Transparent Conductor Overcoming Performance–Yield Tradeoff. *Adv. Funct. Mater.* **2022**, *32*, 2103397. [CrossRef]
143. Ansón-Casaos, A.; Mis-Fernández, R.; López-Alled, C.M.; Almendro-López, E.; Hernández-Ferrer, J.; González-Domínguez, J.M.; Martínez, M.T. Transparent Conducting Films Made of Different Carbon Nanotubes, Processed Carbon Nanotubes, and Graphene Nanoribbons. *Chem. Eng. Sci.* **2015**, *138*, 566–574. [CrossRef]
144. Li, Z.; Kandel, H.R.; Dervishi, E.; Saini, V.; Biris, A.S.; Biris, A.R.; Lupu, D. Does the Wall Number of Carbon Nanotubes Matter as Conductive Transparent Material? *Appl. Phys. Lett.* **2007**, *91*, 053115. [CrossRef]
145. Jiang, S.; Hou, P.-X.; Liu, C.; Cheng, H.-M. High-Performance Single-Wall Carbon Nanotube Transparent Conductive Films. *J. Mater. Sci. Technol.* **2019**, *35*, 2447–2462. [CrossRef]
146. Topinka, M.A.; Rowell, M.W.; Goldhaber-Gordon, D.; McGehee, M.D.; Hecht, D.S.; Gruner, G. Charge Transport in Interpenetrating Networks of Semiconducting and Metallic Carbon Nanotubes. *Nano Lett.* **2009**, *9*, 1866–1871. [CrossRef] [PubMed]
147. Nirmalraj, P.N.; Lyons, P.E.; De, S.; Coleman, J.N.; Boland, J.J. Electrical Connectivity in Single-Walled Carbon Nanotube Networks. *Nano Lett.* **2009**, *9*, 3890–3895. [CrossRef] [PubMed]
148. Yanagi, K.; Udoguchi, H.; Sagitani, S.; Oshima, Y.; Takenobu, T.; Kataura, H.; Ishida, T.; Matsuda, K.; Maniwa, Y. Transport Mechanisms in Metallic and Semiconducting Single-Wall Carbon Nanotube Networks. *ACS Nano* **2010**, *4*, 4027–4032. [CrossRef] [PubMed]
149. Jiang, S.; Hou, P.-X.; Chen, M.-L.; Wang, B.-W.; Sun, D.-M.; Tang, D.-M.; Jin, Q.; Guo, Q.-X.; Zhang, D.-D.; Du, J.-H.; et al. Ultrahigh-Performance Transparent Conductive Films of Carbon-Welded Isolated Single-Wall Carbon Nanotubes. *Sci. Adv.* **2018**, *4*, eaap9264. [CrossRef]
150. Song, Y.I.; Lee, J.W.; Kim, T.Y.; Jung, H.J.; Jung, Y.C.; Suh, S.J.; Yang, C.-M. Performance-Determining Factors in Flexible Transparent Conducting Single-Wall Carbon Nanotube Film. *Carbon Lett.* **2013**, *14*, 255–258. [CrossRef]
151. Kaskela, A.; Nasibulin, A.G.; Timmermans, M.Y.; Aitchison, B.; Papadimitratos, A.; Tian, Y.; Zhu, Z.; Jiang, H.; Brown, D.P.; Zakhidov, A.; et al. Aerosol-Synthesized SWCNT Networks with Tunable Conductivity and Transparency by a Dry Transfer Technique. *Nano Lett.* **2010**, *10*, 4349–4355. [CrossRef]
152. Chen, Z.-K.; Hu, X.-G.; Guo, S.-Y.; Hou, P.-X.; Liu, C. Air-Stable Room-Temperature Photodetector Based on Large-Diameter Small-Bundle Single-Wall Carbon Nanotube Films. *J. Mater. Sci. Technol.* **2021**, *73*, 205. [CrossRef]
153. Tenent, R.C.; Barnes, T.M.; Bergeson, J.D.; Ferguson, A.J.; To, B.; Gedvilas, L.M.; Heben, M.J.; Blackburn, J.L. Ultrasmooth, Large-Area, High-Uniformity, Conductive Transparent Single-Walled-Carbon-Nanotube Films for Photovoltaics Produced by Ultrasonic Spraying. *Adv. Mater.* **2009**, *21*, 3210–3216. [CrossRef]
154. Du, J.; Pei, S.; Ma, L.; Cheng, H.-M. 25th Anniversary Article: Carbon Nanotube- and Graphene-Based Transparent Conductive Films for Optoelectronic Devices. *Adv. Mater.* **2014**, *26*, 1958–1991. [CrossRef] [PubMed]
155. Tung, V.C.; Chen, L.-M.; Allen, M.J.; Wassei, J.K.; Nelson, K.; Kaner, R.B.; Yang, Y. Low-Temperature Solution Processing of Graphene−carbon Nanotube Hybrid Materials for High-Performance Transparent Conductors. *Nano Lett.* **2009**, *9*, 1949–1955. [CrossRef] [PubMed]
156. Mustonen, K.; Laiho, P.; Kaskela, A.; Zhu, Z.; Reynaud, O.; Houbenov, N.; Tian, Y.; Susi, T.; Jiang, H.; Nasibulin, A.G.; et al. Gas Phase Synthesis of Non-Bundled, Small Diameter Single-Walled Carbon Nanotubes with near-Armchair Chiralities. *Appl. Phys. Lett.* **2015**, *107*, 13106. [CrossRef]
157. Wang, Y.; Weng, G.J.; Meguid, S.A.; Hamouda, A.M. A Continuum Model with a Percolation Threshold and Tunneling-Assisted Interfacial Conductivity for Carbon Nanotube-Based Nanocomposites. *J. Appl. Phys.* **2014**, *115*, 193706. [CrossRef]
158. Huang, J.; Mao, C.; Zhu, Y.; Jiang, W.; Yang, X. Control of Carbon Nanotubes at the Interface of a Co-Continuous Immiscible Polymer Blend to Fabricate Conductive Composites with Ultralow Percolation Thresholds. *Carbon* **2014**, *73*, 267–274. [CrossRef]
159. Tian, X.; Moser, M.L.; Pekker, A.; Sarkar, S.; Ramirez, J.; Bekyarova, E.; Itkis, M.E.; Haddon, R.C. Effect of Atomic Interconnects on Percolation in Single-Walled Carbon Nanotube Thin Film Networks. *Nano Lett.* **2014**, *14*, 3930–3937. [CrossRef]
160. Hu, L.; Hecht, D.S.; Grüner, G. Percolation in Transparent and Conducting Carbon Nanotube Networks. *Nano Lett.* **2004**, *4*, 2513–2517. [CrossRef]
161. Shin, D.H.; Shim, H.C.; Song, J.-W.; Kim, S.; Han, C.-S. Conductivity of Films Made from Single-Walled Carbon Nanotubes in Terms of Bundle Diameter. *Scr. Mater.* **2009**, *60*, 607–610. [CrossRef]
162. Ho, X.; Wei, J. Films of Carbon Nanomaterials for Transparent Conductors. *Materials* **2013**, *6*, 2155–2181. [CrossRef]
163. Du, F.; Fischer, J.E.; Winey, K.I. Effect of Nanotube Alignment on Percolation Conductivity in Carbon Nanotube/Polymer Composites. *Phys. Rev. B* **2005**, *72*, 121404. [CrossRef]
164. Prikhod'ko, A.V.; Kon'kov, O.I. Modification of the Structure and Percolation of Current in an Array of Single-Walled Carbon Nanotubes. *Phys. Solid State* **2014**, *56*, 1463–1466. [CrossRef]

165. Geng, H.-Z.; Kim, K.K.; Lee, K.; Kim, G.Y.; Choi, H.K.; Lee, D.S.; An, K.H.; Lee, Y.H.; Chang, Y.; Lee, Y.S.; et al. Dependence of Material Quality on Performance of Flexible Transparent Conducting Films with Single-Walled Carbon Nanotubes. *Nano* **2007**, *2*, 157–167. [CrossRef]
166. Geng, H.-Z.; Kim, K.K.; So, K.P.; Lee, Y.S.; Chang, Y.; Lee, Y.H. Effect of Acid Treatment on Carbon Nanotube-Based Flexible Transparent Conducting Films. *J. Am. Chem. Soc.* **2007**, *129*, 7758–7759. [CrossRef]
167. Wang, H.; Sun, X.; Wang, Y.; Li, K.; Wang, J.; Dai, X.; Chen, B.; Chong, D.; Zhang, L.; Yan, J. Acid Enhanced Zipping Effect to Densify MWCNT Packing for Multifunctional MWCNT Films with Ultra-High Electrical Conductivity. *Nat. Commun.* **2023**, *14*, 380. [CrossRef]
168. Wang, H.; Li, K.; Hao, X.; Pan, J.; Zhuang, T.; Dai, X.; Wang, J.; Chen, B.; Chong, D. Capillary Compression Induced Outstanding N-Type Thermoelectric Power Factor in CNT Films towards Intelligent Temperature Controller. *Nat. Commun.* **2024**, *15*, 5617. [CrossRef]
169. Guo, Y.; Shi, E.; Zhu, J.; Shen, P.-C.; Wang, J.; Lin, Y.; Mao, Y.; Deng, S.; Li, B.; Park, J.-H.; et al. Soft-Lock Drawing of Super-Aligned Carbon Nanotube Bundles for Nanometre Electrical Contacts. *Nat. Nanotechnol.* **2022**, *17*, 278–284. [CrossRef]
170. Chen, Z.; Yang, Y.; Wu, Z.; Luo, G.; Xie, L.; Liu, Z.; Ma, S.; Guo, W. Electric-Field-Enhanced Assembly of Single-Walled Carbon Nanotubes on a Solid Surface. *J. Phys. Chem. B* **2005**, *109*, 5473–5477. [CrossRef]
171. Karimzad Ghavidel, A.; Zadshakoyan, M.; Arjmand, M. Mechanical Analysis of Aligned Carbon Nanotube Bundles under Electric Field. *Int. J. Mech. Sci.* **2021**, *196*, 106289. [CrossRef]
172. Xu, J.; Zhang, X.; Chen, F.; Li, T.; Li, Y.; Tao, X.; Wang, Y.; Wu, X. Preparation and Modification of Well-Aligned CNTs Grown on AAO Template. *Appl. Surf. Sci.* **2005**, *239*, 320–326. [CrossRef]
173. Zhang, X.; Volder, M.D.; Zhou, W.; Issman, L.; Wei, X.; Kaniyoor, A.; Portas, J.T.; Smail, F.; Wang, Z.; Wang, Y.; et al. Simultaneously Enhanced Tenacity, Rupture Work, and Thermal Conductivity of Carbon Nanotube Fibers by Raising Effective Tube Portion. *Sci. Adv.* **2022**, *8*, eabq3515. [CrossRef]
174. Suslova, E.V.; Arkhipova, E.A.; Kalashnik, A.V.; Ivanov, A.S.; Savilov, S.V.; Xia, H.; Lunin, V.V. Effect of the Functionalization of Nitrogen-Doped Carbon Nanotubes on Electrical Conductivity. *Russ. J. Phys. Chem. A* **2019**, *93*, 1952–1956. [CrossRef]
175. Lee, W.J.; Maiti, U.N.; Lee, J.M.; Lim, J.; Han, T.H.; Kim, S.O. Nitrogen-Doped Carbon Nanotubes and Graphene Composite Structures for Energy and Catalytic Applications. *Chem. Commun.* **2014**, *50*, 6818–6830. [CrossRef]
176. Gong, K.; Du, F.; Xia, Z.; Durstock, M.; Dai, L. Nitrogen-Doped Carbon Nanotube Arrays with High Electrocatalytic Activity for Oxygen Reduction. *Science* **2009**, *323*, 760–764. [CrossRef]
177. Cruz-Silva, E.; Cullen, D.A.; Gu, L.; Romo-Herrera, J.M.; Muñoz-Sandoval, E.; López-Urías, F.; Sumpter, B.G.; Meunier, V.; Charlier, J.-C.; Smith, D.J.; et al. Heterodoped Nanotubes: Theory, Synthesis, and Characterization of Phosphorus−Nitrogen Doped Multiwalled Carbon Nanotubes. *ACS Nano* **2008**, *2*, 441–448. [CrossRef]
178. Lee, J.M.; Park, J.S.; Lee, S.H.; Kim, H.; Yoo, S.; Kim, S.O. Selective Electron- or Hole-Transport Enhancement in Bulk-Heterojunction Organic Solar Cells with N- or B-Doped Carbon Nanotubes. *Adv. Mater.* **2011**, *23*, 629–633. [CrossRef] [PubMed]
179. Sawant, S.V.; Patwardhan, A.W.; Joshi, J.B.; Dasgupta, K. Boron Doped Carbon Nanotubes: Synthesis, Characterization and Emerging Applications—A Review. *Chem. Eng. J.* **2022**, *427*, 131616. [CrossRef]
180. Fujisawa, K.; Hayashi, T.; Endo, M.; Terrones, M.; Kim, J.H.; Kim, Y.A. Effect of Boron Doing on the Electrical Conductivity of Metallicity-Separated Single Walled Carbon Nanotubes. *Nanoscale* **2018**, *10*, 12723–12733. [CrossRef] [PubMed]
181. Wei, B.; Spolenak, R.; Kohler-Redlich, P.; Rühle, M.; Arzt, E. Electrical Transport in Pure and Boron-Doped Carbon Nanotubes. *Appl. Phys. Lett.* **1999**, *74*, 3149–3151. [CrossRef]
182. Glerup, M.; Steinmetz, J.; Samaille, D.; Stéphan, O.; Enouz, S.; Loiseau, A.; Roth, S.; Bernier, P. Synthesis of N-Doped SWNT Using the Arc-Discharge Procedure. *Chem. Phys. Lett.* **2004**, *387*, 193–197. [CrossRef]
183. Stephan, O.; Ajayan, P.M.; Colliex, C.; Redlich, P.; Lambert, J.M.; Bernier, P.; Lefin, P. Doping Graphitic and Carbon Nanotube Structures with Boron and Nitrogen. *Science* **1994**, *266*, 1683–1685. [CrossRef] [PubMed]
184. Gai, P.L.; Stephan, O.; McGuire, K.; Rao, A.M.; Dresselhaus, M.S.; Dresselhaus, G.; Colliex, C. Structural Systematics in Boron-Doped Single Wall Carbon Nanotubes. *J. Mater. Chem.* **2004**, *14*, 669–675. [CrossRef]
185. Zhang, Y.; Gu, H.; Suenaga, K.; Iijima, S. Heterogeneous Growth of B-C-N Nanotubes by Laser Ablation. *Chem. Phys. Lett.* **1997**, *279*, 264–269. [CrossRef]
186. Ayala, P.; Reppert, J.; Grobosch, M.; Knupfer, M.; Pichler, T.; Rao, A.M. Evidence for Substitutional Boron in Doped Single-Walled Carbon Nanotubes. *Appl. Phys. Lett.* **2010**, *96*, 183110. [CrossRef]
187. Terrones, M.; Redlich, P.; Grobert, N.; Trasobares, S.; Hsu, W.K.; Terrones, H.; Zhu, Y.Q.; Hare, J.P.; Reeves, C.L.; Cheetham, A.K.; et al. Carbon Nitride Nanocomposites: Formation of Aligned $C_xN_y$ Nanofibers. *Adv. Mater.* **1999**, *11*, 655–658. [CrossRef]
188. Terrones, M.; Grobert, N.; Olivares, J.; Zhang, J.P.; Terrones, H.; Kordatos, K.; Hsu, W.K.; Hare, J.P.; Townsend, P.D.; Prassides, K.; et al. Controlled Production of Aligned-Nanotube Bundles. *Nature* **1997**, *388*, 52–55. [CrossRef]

189. Pinault, M.; Mayne-L'Hermite, M.; Reynaud, C.; Pichot, V.; Launois, P.; Ballutaud, D. Growth of Multiwalled Carbon Nanotubes during the Initial Stages of Aerosol-Assisted CCVD. *Carbon* **2005**, *43*, 2968–2976. [CrossRef]
190. Teo, K.B.K.; Hash, D.B.; Lacerda, R.G.; Rupesinghe, N.L.; Bell, M.S.; Dalal, S.H.; Bose, D.; Govindan, T.R.; Cruden, B.A.; Chhowalla, M.; et al. The Significance of Plasma Heating in Carbon Nanotube and Nanofiber Growth. *Nano Lett.* **2004**, *4*, 921–926. [CrossRef]
191. Golberg, D.; Bando, Y.; Bourgeois, L.; Kurashima, K.; Sato, T. Large-Scale Synthesis and HRTEM Analysis of Single-Walled B- and N-Doped Carbon Nanotube Bundles. *Carbon* **2000**, *38*, 2017–2027. [CrossRef]
192. Endo, M.; Muramatsu, H.; Hayashi, T.; Kim, Y.-A.; Lier, G.V.; Charlier, J.-C.; Terrones, H.; Terrones, M.; Dresselhaus, M.S. Atomic Nanotube Welders: Boron Interstitials Triggering Connections in Double-Walled Carbon Nanotubes. *Nano Lett.* **2005**, *5*, 1099–1105. [CrossRef]
193. Jeong, H.M.; Lee, J.W.; Shin, W.H.; Choi, Y.J.; Shin, H.J.; Kang, J.K.; Choi, J.W. Nitrogen-Doped Graphene for High-Performance Ultracapacitors and the Importance of Nitrogen-Doped Sites at Basal Planes. *Nano Lett.* **2011**, *11*, 2472–2477. [CrossRef]
194. Evlashin, S.A.; Maksimov, Y.M.; Dyakonov, P.V.; Pilevsky, A.A.; Maslakov, K.I.; Mankelevich, Y.A.; Voronina, E.N.; Vavilov, S.V.; Pavlov, A.A.; Zenova, E.V.; et al. N-Doped Carbon NanoWalls for Power Sources. *Sci. Rep.* **2019**, *9*, 6716. [CrossRef]
195. Soares, O.S.G.P.; Rocha, R.P.; Gonçalves, A.G.; Figueiredo, J.L.; Órfão, J.J.M.; Pereira, M.F.R. Easy Method to Prepare N-Doped Carbon Nanotubes by Ball Milling. *Carbon* **2015**, *91*, 114–121. [CrossRef]
196. Jeon, I.-Y.; Bae, S.-Y.; Seo, J.-M.; Baek, J.-B. Scalable Production of Edge-Functionalized Graphene Nanoplatelets via Mechanochemical Ball-Milling. *Adv. Funct. Mater.* **2015**, *25*, 6961–6975. [CrossRef]
197. Nevidomskyy, A.H.; Csányi, G.; Payne, M.C. Chemically Active Substitutional Nitrogen Impurity in Carbon Nanotubes. *Phys. Rev. Lett.* **2003**, *91*, 105502. [CrossRef] [PubMed]
198. Pan, Z.; Ren, J.; Guan, G.; Fang, X.; Wang, B.; Doo, S.-G.; Son, I.H.; Huang, X.; Peng, H. Synthesizing Nitrogen-Doped Core–Sheath Carbon Nanotube Films for Flexible Lithium Ion Batteries. *Adv. Energy Mater.* **2016**, *6*, 1600271. [CrossRef]
199. Carroll, D.L.; Redlich, P.; Blase, X.; Charlier, J.-C.; Curran, S.; Ajayan, P.M.; Roth, S.; Rühle, M. Effects of Nanodomain Formation on the Electronic Structure of Doped Carbon Nanotubes. *Phys. Rev. Lett.* **1998**, *81*, 2332–2335. [CrossRef]
200. Czerw, R.; Terrones, M.; Charlier, J.-C.; Blase, X.; Foley, B.; Kamalakaran, R.; Grobert, N.; Terrones, H.; Tekleab, D.; Ajayan, P.M.; et al. Identification of Electron Donor States in N-Doped Carbon Nanotubes. *Nano Lett.* **2001**, *1*, 457–460. [CrossRef]
201. Ristein, J. Surface Transfer Doping of Semiconductors. *Science* **2006**, *313*, 1057–1058. [CrossRef]
202. Shimoda, H.; Gao, B.; Tang, X.P.; Kleinhammes, A.; Fleming, L.; Wu, Y.; Zhou, O. Lithium Intercalation into Opened Single-Wall Carbon Nanotubes: Storage Capacity and Electronic Properties. *Phys. Rev. Lett.* **2001**, *88*, 015502. [CrossRef]
203. Mordkovich, V.Z.; Baxendale, M.; Chang, R.P.H.; Yoshimura, S. Intercalation into Carbon Nanotubes without Breaking the Tubular Structure. *Synth. Met.* **1997**, *86*, 2049–2050. [CrossRef]
204. Zhou, C.; Kong, J.; Yenilmez, E.; Dai, H. Modulated Chemical Doping of Individual Carbon Nanotubes. *Science* **2000**, *290*, 1552–1555. [CrossRef] [PubMed]
205. Lee, R.S.; Kim, H.J.; Fischer, J.E.; Lefebvre, J.; Radosavljević, M.; Hone, J.; Johnson, A.T. Transport Properties of a Potassium-Doped Single-Wall Carbon Nanotube Rope. *Phys. Rev. B* **2000**, *61*, 4526–4529. [CrossRef]
206. Kong, J.; Zhou, C.; Yenilmez, E.; Dai, H. Alkaline Metal-Doped *n* -Type Semiconducting Nanotubes as Quantum Dots. *Appl. Phys. Lett.* **2000**, *77*, 3977–3979. [CrossRef]
207. Bockrath, M.; Hone, J.; Zettl, A.; McEuen, P.L.; Rinzler, A.G.; Smalley, R.E. Chemical Doping of Individual Semiconducting Carbon-Nanotube Ropes. *Phys. Rev. B* **2000**, *61*, R10606–R10608. [CrossRef]
208. Lee, R.S.; Kim, H.J.; Fischer, J.E.; Thess, A.; Smalley, R.E. Conductivity Enhancement in Single-Walled Carbon Nanotube Bundles Doped with K and Br. *Nature* **1997**, *388*, 255–257. [CrossRef]
209. Puchades, I.; Lawlor, C.C.; Schauerman, C.M.; Bucossi, A.R.; Rossi, J.E.; Cox, N.D.; Landi, B.J. Mechanism of Chemical Doping in Electronic-Type-Separated Single Wall Carbon Nanotubes towards High Electrical Conductivity. *J. Mater. Chem. C* **2015**, *3*, 10256–10266. [CrossRef]
210. Fan, L.; Xu, X. A Stable Iodine-Doped Multi-Walled Carbon Nanotube-Polypyrrole Composite with Improved Electrical Property. *Compos. Sci. Technol.* **2015**, *118*, 264–268. [CrossRef]
211. Zhao, Y.; Wei, J.; Vajtai, R.; Ajayan, P.M.; Barrera, E.V. Iodine Doped Carbon Nanotube Cables Exceeding Specific Electrical Conductivity of Metals. *Sci. Rep.* **2011**, *1*, 83. [CrossRef]
212. Sankapal, B.R.; Setyowati, K.; Chen, J.; Liu, H. Electrical Properties of Air-Stable, Iodine-Doped Carbon-Nanotube–Polymer Composites. *Appl. Phys. Lett.* **2007**, *91*, 173103. [CrossRef]
213. Grigorian, L.; Williams, K.A.; Fang, S.; Sumanasekera, G.U.; Loper, A.L.; Dickey, E.C.; Pennycook, S.J.; Eklund, P.C. Reversible Intercalation of Charged Iodine Chains into Carbon Nanotube Ropes. *Phys. Rev. Lett.* **1998**, *80*, 5560–5563. [CrossRef]
214. Zhou, Y.; Shimada, S.; Saito, T.; Azumi, R. Understanding the Doping Effects on the Structural and Electrical Properties of Ultrathin Carbon Nanotube Networks. *J. Appl. Phys.* **2015**, *118*, 215305. [CrossRef]
215. Janas, D.; Milowska, K.Z.; Bristowe, P.D.; Koziol, K.K. Improving the Electrical Properties of Carbon Nanotubes with Interhalogen Compounds. *Nanoscale* **2017**, *9*, 3212–3221. [CrossRef]

216. Yoon, S.-M.; Kim, U.J.; Benayad, A.; Lee, I.H.; Son, H.; Shin, H.-J.; Choi, W.M.; Lee, Y.H.; Jin, Y.W.; Lee, E.-H.; et al. Thermal Conversion of Electronic and Electrical Properties of AuCl3-Doped Single-Walled Carbon Nanotubes. *ACS Nano* **2011**, *5*, 1353–1359. [CrossRef]
217. Kim, S.M.; Kim, K.K.; Jo, Y.W.; Park, M.H.; Chae, S.J.; Duong, D.L.; Yang, C.W.; Kong, J.; Lee, Y.H. Role of Anions in the AuCl3-Doping of Carbon Nanotubes. *ACS Nano* **2011**, *5*, 1236–1242. [CrossRef] [PubMed]
218. Naruse, J.; Yokoi, T.; Ishino, K.; Hikita, Y.; Iwase, K. High-Temperature Electrical Conductivity and Thermal Stability of FeCl3 Doping in Carbon Nanotube Tapes. *Appl. Phys. Express* **2021**, *14*, 095002. [CrossRef]
219. Teng, Y.; Zhang, Y.; Xie, X.; Yao, J.; Zhang, Z.; Geng, L.; Zhao, P.; Yang, C.; Gong, W.; Wang, X.; et al. Interfacial Electron Transfer in PbI2@Single-Walled Carbon Nanotube van Der Waals Heterostructures for High-Stability Self-Powered Photodetectors. *J. Am. Chem. Soc.* **2024**, *146*, 6231–6239. [CrossRef] [PubMed]
220. Souza Filho, A.G.; Terrones, M. Properties and Applications of Doped Carbon Nanotubes. In *B-C-N Nanotubes and Related Nanostructures*; Springer: New York, NY, USA, 2009; pp. 223–269. ISBN 978-1-4419-0086-9.
221. Madrona, C. Continuous Intercalation Compound Fibers of Bromine Wires and Aligned CNTs for High-Performance Conductors. *Carbon* **2023**, *204*, 211–218. [CrossRef]
222. Madrona, C.; Vila, M.; Oropeza, F.E.; De La Peña O'Shea, V.A.; Vilatela, J.J. Macroscopic Yarns of FeCl3-Intercalated Collapsed Carbon Nanotubes with High Doping and Stability. *Carbon* **2021**, *173*, 311–321. [CrossRef]
223. Qiu, L.; Zou, H.; Wang, X.; Feng, Y.; Zhang, X.; Zhao, J.; Zhang, X.; Li, Q. Enhancing the Interfacial Interaction of Carbon Nanotubes Fibers by Au Nanoparticles with Improved Performance of the Electrical and Thermal Conductivity. *Carbon* **2019**, *141*, 497–505. [CrossRef]
224. Zhang, R.; Wang, X.; Zhang, Z.; Zhang, W.; Lai, J.; Zhu, S.; Li, Y.; Zhang, Y.; Cao, K.; Qiu, S.; et al. CuI Encapsulated within Single-Walled Carbon Nanotube Networks with High Current Carrying Capacity and Excellent Conductivity. *Adv. Funct. Mater.* **2023**, *33*, 2301864. [CrossRef]
225. Du, J.; Liu, S.; Liu, Y.; Wu, G.; Liu, X.; Zhang, W.; Zhang, Y.; Li, Q.; Kang, L. One-Dimensional High-Entropy Compounds. *J. Am. Chem. Soc.* **2024**, *146*, 8464–8471. [CrossRef]
226. Ji, Z.-H.; Zhang, L.; Tang, D.-M.; Chen, C.-M.; Nordling, T.E.M.; Zhang, Z.-D.; Ren, C.-L.; Da, B.; Li, X.; Guo, S.-Y.; et al. High-Throughput Screening and Machine Learning for the Efficient Growth of High-Quality Single-Wall Carbon Nanotubes. *Nano Res.* **2021**, *14*, 4610–4615. [CrossRef]
227. Gao, Z.-D.; Ji, Z.-H.; Zhang, L.; Tang, D.-M.; Zou, M.-K.; Xie, R.-H.; Liu, S.-K.; Liu, C. Optimizing the Growth of Vertically Aligned Carbon Nanotubes by Literature Mining and High-Throughput Experiments. *New Carbon Mater.* **2023**, *38*, 887–897. [CrossRef]
228. Ji, Z.-H.; Zhang, L.; Tang, D.-M.; Zhao, Y.-M.; Zou, M.-K.; Xie, R.-H.; Liu, C.; Cheng, H.-M. Statistical Patterns in High-Throughput Growth of Single-Wall Carbon Nanotubes from Co/Pt/Mo Ternary Catalysts. *Carbon* **2023**, *210*, 118073. [CrossRef]

*nanomaterials*

MDPI

*Article*

# Structural, Electrical, and Optical Properties of Single-Walled Carbon Nanotubes Synthesized through Floating Catalyst Chemical Vapor Deposition

**Melorina Dolafi Rezaee [1], Biplav Dahal [1], John Watt [2], Mahir Abrar [3], Deidra R. Hodges [3] and Wenzhi Li [1,*]**

1 Department of Physics, Florida International University, Miami, FL 33199, USA; mdola008@fiu.edu (M.D.R.); bdaha002@fiu.edu (B.D.)
2 Center for Integrated Nanotechnologies, Los Alamos National Laboratory, Los Alamos, NM 87545, USA; watt@lanl.gov
3 Department of Electrical & Computer Engineering, Florida International University, Miami, FL 33174, USA; mabrar@fiu.edu (M.A.); dhodges@fiu.edu (D.R.H.)
* Correspondence: wenzhi.li@fiu.edu

**Abstract:** Single-walled carbon nanotube (SWCNT) thin films were synthesized by using a floating catalyst chemical vapor deposition (FCCVD) method with a low flow rate (200 sccm) of mixed gases (Ar and $H_2$). SWCNT thin films with different thicknesses can be prepared by controlling the collection time of the SWCNTs on membrane filters. Transmission electron microscopy (TEM) showed that the SWCNTs formed bundles and that they had an average diameter of 1.46 nm. The Raman spectra of the SWCNT films suggested that the synthesized SWCNTs were very well crystallized. Although the electrical properties of SWCNTs have been widely studied so far, the Hall effect of SWCNTs has not been fully studied to explore the electrical characteristics of SWCNT thin films. In this research, Hall effect measurements have been performed to investigate the important electrical characteristics of SWCNTs, such as their carrier mobility, carrier density, Hall coefficient, conductivity, and sheet resistance. The samples with transmittance between 95 and 43% showed a high carrier density of $10^{21}$–$10^{23}$ cm$^{-3}$. The SWCNTs were also treated using Brønsted acids (HCl, $HNO_3$, $H_2SO_4$) to enhance their electrical properties. After the acid treatments, the samples maintained their p-type nature. The carrier mobility and conductivity increased, and the sheet resistance decreased for all treated samples. The highest mobility of 1.5 cm$^2$/Vs was obtained with the sulfuric acid treatment at 80 °C, while the highest conductivity (30,720 S/m) and lowest sheet resistance (43 ohm/square) were achieved with the nitric acid treatment at room temperature. Different functional groups were identified in our synthesized SWCNTs before and after the acid treatments using Fourier-Transform Infrared Spectroscopy (FTIR).

**Keywords:** SWCNTs; FCCVD method; Raman spectroscopy; Hall effect measurements; acid treatments; FTIR

## 1. Introduction

Carbon nanotubes (CNTs) are cylindrical molecules made of rolled-up sheets of single-layer carbon atoms (graphene) [1]. SWCNTs have diameters in the range of 1–2 nanometers (nm), double-walled carbon nanotubes (DWCNTs) have diameters in the range of 2–4 nm, and multi-walled carbon nanotubes (MWCNTs) have diameters in the range of a few nanometers to one hundred nanometers. MWCNTs consist of many concentrically stacked nanotubes with lengths of several micrometers or millimeters [2]. The credit for discovering MWCNTs and SWCNTs is given to Ijima [3,4] in 1991 and 1993, respectively.

Due to their lightweight and one-dimensional structure, SWCNTs show unique mechanical, electrical, optical, and thermal properties [5–7]. For example, their electrical conductivity is estimated theoretically to be $10^8$ S/m, and they exhibit a high thermal conductivity of 3500 W/k.m. The theoretical tensile strength of individual SWCNTs is around 100 Gpa, with a Young's modulus of 1 TPa [8–10]. Because of these distinctive properties, SWCNTs can be utilized in a wide range of applications, such as flexible and transparent microelectronics, energy storage and conversion devices, multifunctional composites, aerospace devices, electric conductors, etc. [11,12].

Compared to other methods like arc discharge or laser ablation, chemical vapor deposition (CVD) is one of the techniques most used to synthesize SWCNTs due to its high yield, low impurity, and gentle synthesis conditions [13]. So far, different CVD processes have been introduced to fabricate SWCNTs, such as aerosol-assisted CVD [14], microwave-plasma-enhanced CVD [15], hot-filament CVD [16], oxygen-assisted CVD [17], FCCVD, etc. Compared to the other CVD techniques, FCCVD is more appropriate for the mass production of SWCNTs because of its low cost, good flexibility, scalability, and controllability [18,19]. The properties of SWCNT films strongly depend on the morphology of the SWCNTs' bundles, like their length, diameter, density, etc. [20], which can be efficiently controlled in the FCCVD technique.

In the FCCVD method, the appropriate choice of growth parameters plays an important role in controlling the growth of SWCNTs. Thiophene is commonly used to deliver the sulfur components used in SWCNT synthesis [21]. Sulfur plays a vital role in nucleation, renucleation, surface chemistry, and the aerogel formation of the catalyst nanoparticles. On the other hand, the catalyst affects the SWCNTs' morphology and helps the growth of the SWCNTs by reducing the synthesis energy [22]. Because of the low bonding energy between the sulfur and carbon atoms, thiophene decomposes earlier than ferrocene (catalyst) in the initial phase of the FCCVD reaction. The results of this process are hydrocarbon species and liberated sulfur atoms. A thin layer of coating is formed on the surface of the metallic catalyst by these liberated sulfur atoms. This is advantageous to the growth of SWCNTs because the sulfur surfactant can prevent the encapsulation of carbon particles. Moreover, it can stop agglomeration between the sulfur-coated catalyst particles [23].

The type, morphology, and crystallinity of the synthesized SWCNTs can be determined by carbon precursors such as ethanol. The carbon precursor is dissociated after being absorbed on the surface of the catalyst. A SWCNT cap is formed by the transformation of a closed carbon network. The SWCNT growth continues by generating carbon precipitation from the catalyst below the SWCNT cap. The growth stops after the termination of the catalysts [24] by the complete sulfur coverage on the surface of the catalyst particle [22,23]. SWCNTs' diameter distribution is known to be affected by the composition of carrier gases. As an example, it was observed that increasing Ar as a carrier gas leads to a decrease in SWCNTs' diameter distributions [25]. A small diameter distribution is favorable for the reproductivity and consistency of SWCNT products.

Hydrogen is commonly used as a carrier gas in the FCCVD system. This gas is found to be very beneficial to the synthesis of SWCNTs. It can help the breakdown of the hydrocarbon precursors [26] and preserve the catalyst's lifetime [27]. Moreover, hydrogen induction can effectively remove amorphous carbon, leading to an improvement in CNT synthesis [28].

Controlling the electronic properties of SWCNTs can develop their technical applications in different fields. SWCNTs can show either semiconducting or metallic behaviors depending on their chiral vector. The electronic properties of SWCNTs can be improved by doping using acid treatments. Brønsted acids (HCL, $HNO_3$, $H_2SO_4$, ...) have been known to have electrochemical effects on graphite, leading to acceptor doping [29,30]. The use of SWCNT thin films, composed of either a random or oriented network, is an incredibly promising technology for the future of electronics. SWCNT films with both high electrical conductivity and optical transparency can serve as transparent electrodes in place of conventional indium tin oxide (ITO). Studies have demonstrated that SWCNTs possess high electronic mobilities ranging from 10,000 to 100,000 $cm^2/V$ s, and high conductivities up to 400,000 S/cm. However, it is still a challenge to extend the exceptional electrical properties of individual SWCNTs to two-dimensional networks. Without chemical doping, high-quality SWCNT films show sheet resistances of 300–1000 ohm/square at around 85% optical transmittance, which are significantly higher than ITO's 10–20 ohm/square at 90% transmittance [31]. Transferring charges (electrons or holes) to the nanotube through intercalation and/or functionalization processes is one way to change the carbon nanotube's electronic and vibrational characteristics [32–34]. Therefore, it is important to study the effect of doping on SWCNTs' electrical properties.

In this work, we report the production of clean SWCNT films using the FCCVD method. We followed a process similar to that used by Zhang et al. [35] to synthesize SWCNTs, with some changes in the type of carrier gases, their total flow rate, and the synthesis procedure. Our synthesis process, with a total carrier gas flow rate of 200 sccm, can be considered a low-cost procedure. In this research, thiophene ($C_4H_4S$) was used as a growth promoter, ferrocene as the catalyst, ethanol as the carbon precursor, and argon and hydrogen as the carrier gases. The morphology and characteristics of the thin films were examined through scanning electron microscopy (SEM) and transmission electron microscopy (TEM). Raman spectroscopy was used to show the quality and crystallinity of the SWCNT films. The transmittance of the synthesized SWCNT films was obtained for the thin film samples with collection times of 5, 10, 15, 20, and 25 min, and the electronic properties of the synthesized SWCNTs were investigated using Hall effect measurements. The SWCNT films showed a high carrier concentration of $10^{21}$–$10^{23}$ $cm^{-3}$, which is in good agreement with the theoretical prediction [36,37]. $H_2SO_4$, HCL, and $HNO_3$ were used for the treatment of the thin films, and the effect of different acid treatments on the mobility, conductivity, and sheet resistance of samples was studied. Fourier-Transform Infrared Spectroscopy (FTIR) was employed to determine the functional groups in the SWCNT films before and after the acid treatment.

Generally, SWCNTs are being actively researched and applied in various fields such as biomedicine, energy storage, and electronic devices. The p-type nature of our synthesized SWCNTs makes them suitable for use as a hole-transporting layer (HTL) in optoelectronic devices such as solar cells. The mobility of our pristine SWCNTs is in the order of $10^{-2}$ $cm^2/Vs$, and after the acid treatment, it reaches the highest value of 1.5 $cm^2/Vs$. This value is much higher than the reported value for the mobility of the spiro-OMeTAD (in the range of $8 \times 10^{-5}$ to $2 \times 10^{-4}$ $cm^2/Vs$) [38], which is one of the widely used HTLs in perovskite solar cells (PSCs). Moreover, employing SWCNTs as the hole-transporting layer in PSCs can dramatically increase the stability of the cells because they can act as a barrier to air and moisture. Transparent conducting films (TCFs) are another potential application for our synthesized SWCNTs, especially after the acid treatments; this is due to a lower sheet resistance being reached at a higher transmittance. SWCNT-based TCFs can be employed in the touch screen, organic light-emitting diode (OLED), flat panel display, and thin-film solar industries [39].

## 2. Experimental

### 2.1. Materials

Ferrocene (98%, Sigma-Aldrich, St. Louis, MO, USA) was used as a catalyst precursor. Thiophene (99+%, Acros Organics, Thermo Fisher Scientific, Fair Lawn, NJ, USA) and ethanol (89–91%, Fisher Chemical, Fisher Scientific, Fair Lawn, NJ, USA) were used as the growth promoter and the carbon source, respectively. Argon (AR UHP300, Airgas, Radnor, PA, USA) and hydrogen (HY UHP300, Airgas) were employed as carrier gases. Millipore Express membrane filters (0.45 μm PES Membrane) were used to collect the SWCNTs. The SWCNT thin films were doped using sulfuric acid ($H_2SO_4$, 96.1 $w/w$ %, Fisher Scientific, Fair Lawn, NJ, USA), nitric acid ($HNO_3$, 69.4 $w/w$ %, Fisher Scientific), and hydrochloric acid (HCL, 37.2 $w/w$ %, Fisher Scientific).

### 2.2. SWCNTs Synthesis

The precursor solution was prepared by dissolving ferrocene (0.4 wt %) and thiophene (molar ratio of S/Fe = 0.3) in 10 milliliters of ethanol. A syringe pump was used to inject the precursor solution into a heating line kept at 140 °C. The feeding rate of the syringe pump was adjusted at 6 μL/min. The precursor solution was evaporated in the heating line and carried into a vertical furnace by argon and hydrogen gases, each with a flow rate of 100 standard cubic centimeters (sccm). The temperature of the furnace was kept at 1000 °C during the experiment. During the growth time (30 min), valvel was kept open while valve 2 and 3 were closed, so residual air in the reaction chamber and any materials initially produced during this time were exhausted through the oil trap (Figure 1). After that, valve 2 was opened and valve 1 was simultaneously closed to collect the pure and clean SWCNTs on a membrane filter, which was kept in a collection tube at room temperature. To increase the mass production of the CNTs, we added an extra valve (valve 3) (Figure 1). The SWCNT films were removed from the collection tube after being purged with argon at 100 sccm, and a new membrane filter was installed and purged with argon to get rid of the air inside the collection tube. Then, valve 2 was opened and valves 1 and 3 closed to collect the SWCNTs again. This procedure was repeated several times without stopping the experiment to produce several CNT thin films in one experiment.

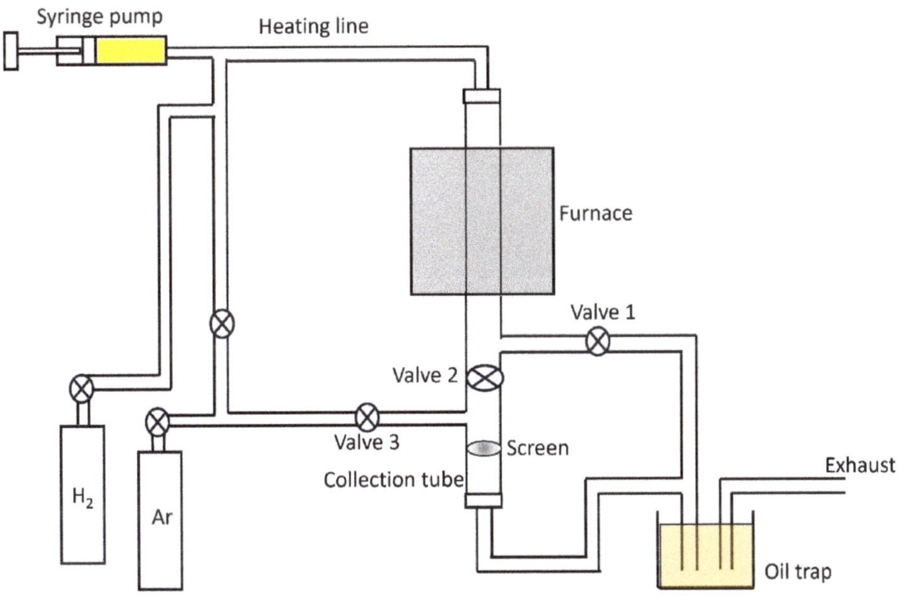


**Figure 1.** Schematic illustration of the FCCVD experimental setup for the SWCNT synthesis.

### 2.3. SWCNT Characterization

The overall morphology of the CNT networks and bundles was characterized using scanning electron microscopy (SEM FS100, JEOL-F100, JEOL Ltd., Tokyo, Japan) under 1 KV. The energy-dispersive X-ray spectroscopy (EDS) technique was used to determine the elemental composition of the samples. Transmission electron microscopy (TEM), performed on a monochromated and aberration-corrected FEI Titan operating at 300 keV, was used to analyze the crystallinity of the CNTs, and ImageJ software (free online version 1.53t) was employed to determine the distance between the fringes in the bundles. Raman spectroscopy (633.8 nm He-Ne Laser, 514 nm Argon ion laser, and 785 nm diode laser) was used to obtain the Raman spectra of the SWCNTs. A UV–visible spectroscopy system (Hewlett Packard 8453, Marshal Scientific, Hampton, NH, USA) was used to obtain optical absorption spectra from the SWCNT films on glass substrates. A Hall effect measurement system (Ecopia HMS-5300, Ecopia Corporation, Anyang-city, Gyeonggi-do, Republic of Korea) was employed to measure the carrier concentration, Hall coefficient, sheet resistance, and carrier mobility of the synthesized SWCNT films at a temperature of T = 300 K and a magnetic field of B = 0.518 T. The Cary 670 FTIR Spectrometer(Agilent Technologies, Miami, FL, USA) was used to identify the functional groups in the SWCNT thin films.

## 3. Results and Discussions

SWCNTs were synthesized using the FCCVD method, with ferrocene ($C_{10}H_{10}Fe$) as the catalyst precursor, thiophene ($C_4H_4S$) as the growth promoter, ethanol ($C_2H_5OH$) as the carbon source, and a mixture of argon (Ar) and hydrogen ($H_2$) as the carrier gas, with a total flow rate of 200 sccm. The growth temperature and the growth time were 1000 °C and 30 min, respectively. Figure 2a shows the optical images of samples with collection times of 5, 10, 15, and 20 min. As the collection time increases, the color of the deposited film changes from light gray to dark brown because of the increase in their thickness. Figure 2b illustrates a typical SEM image of SWCNT bundles that are connected and form a continuous 2D weblike structure [40,41]. Figure 2c shows that the deposited SWCNTs mainly contain C (carbon) and Fe (iron) elements. Fe is the catalyst nanoparticle coming from the catalyst precursor ferrocene and the carbon is from the SWCNTs.

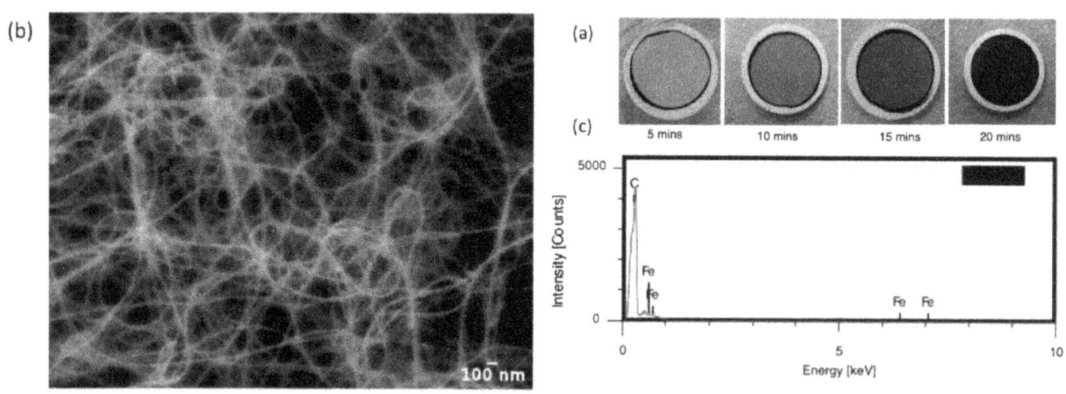


**Figure 2.** (**a**) Optical images of the samples with collection times of 5, 10, 15, and 20 min. (**b**) SEM image of SWCNT thin films. (**c**) EDS of the deposited SWCNTs.

Figure 3 shows the TEM images of SWCNT bundles. The morphology of the network structure of the SWCNTs and bundles can be seen in Figure 3a. Figure 3b–d are high-resolution TEM images of the SWCNTs. Using ImageJ software (free online version 1.53t), the average diameters of the SWCNTs in the bundles for the selected areas were measured as 1.54 nm, 1.45 nm, and 1.38 nm, respectively. The average diameter is 1.46 nm, comparable with the reported values for the SWCNTs' diameters, which were between 1 and 2 nm [42–44]. Hussain et al. [42] reported that the diameters of their SWCNTs synthesized by the FCCVD method were in the range of 1.3–1.5 nm. The diameter of the synthesized SWCNTs using the arc-discharge method was reported to be between 1.3 and 1.6 nm [45].

Figure 4a shows the Raman spectra for the three samples with collection times of 10, 15, and 20 min when employing a 633.8 nm He-Ne Laser, and Figure 4b shows a zoomed image for the Raman spectra of the sample with a collection time of 20 min when using a 514 nm argon ion laser. No peak appeared in the range of the D band when using the 633.8 nm He-Ne and 514 nm argon ion lasers. The emergence of a D band (usually around 1350 $cm^{-1}$) is related to the defects of crystalline $sp^2$ carbon structures [46,47]. Raman spectroscopy on graphene structures reveals that the D mode origin is from the edge defects [48]. The edge area in the well-crystallized SWCNTs is trivial because of the higher aspect ratio (>1000) and their atomic thickness. Therefore, they should not reveal the D band [49].

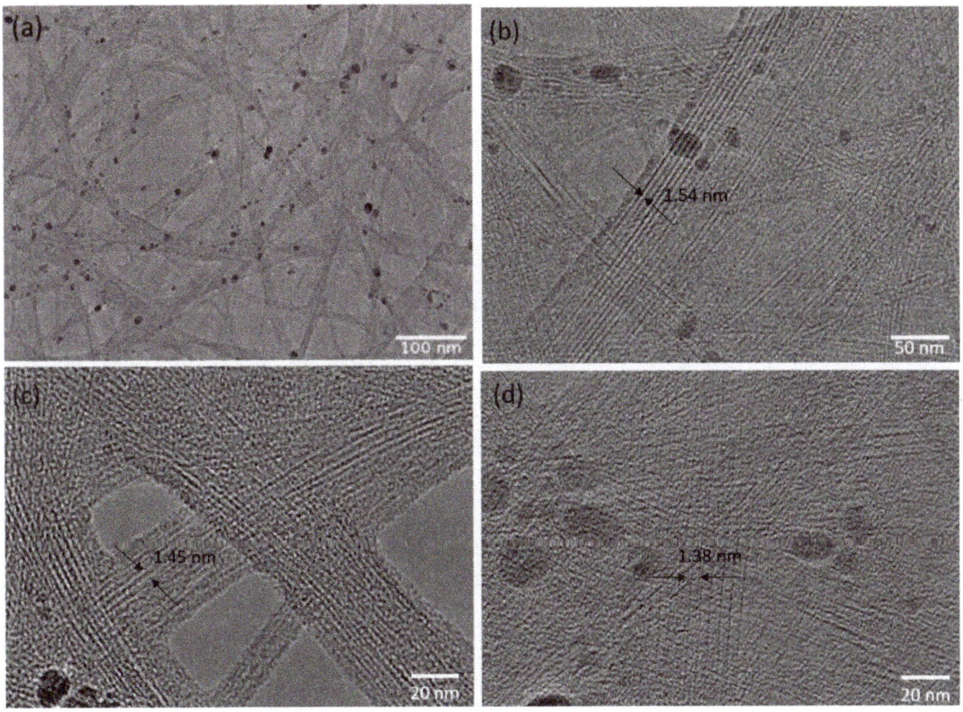


**Figure 3.** TEM images of deposited SWCNTs. (**a**) The overall morphology of the SWCNTs and bundles. (**b–d**) Different selected areas of the deposited film with the estimated average diameter of SWCNTs in the bundles, namely 1.54 nm, 1.45 nm, and 1.38 nm, respectively.

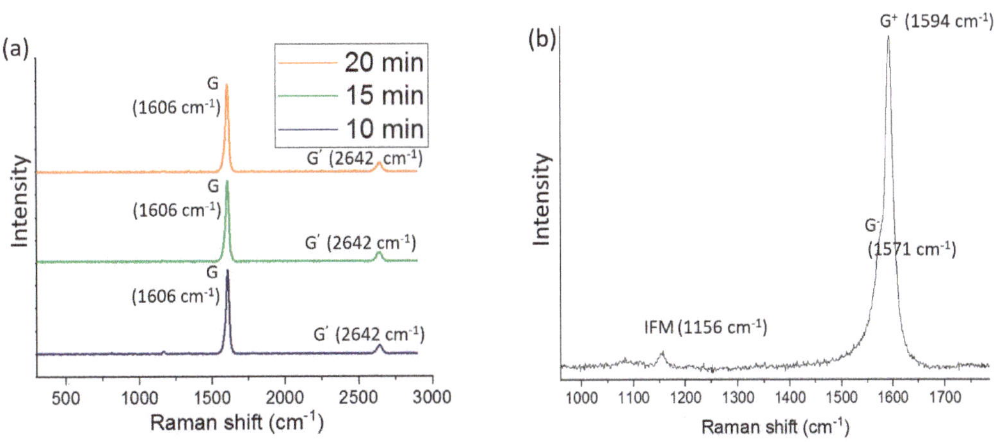


**Figure 4.** (**a**) Raman spectroscopy of the three samples with collection times of 10, 15, and 20 min using a 633.8 nm He-Ne Laser. (**b**) A zoomed image of the sample with a 20 min collection time using a 514 nm argon ion laser.

The G and G′ bands at 1606 $cm^{-1}$ and 2642 $cm^{-1}$, respectively, (Figure 4a), can be compared to the previously published values of 1591 $cm^{-1}$ and 2659 $cm^{-1}$ for the HiPCO SWCNTs (514 nm, Ar+ ion laser) [50]. The reason for the difference in the wavenumber can be attributed to the effect of the laser wavelength. The phonon frequencies of Raman signals can vary according to the change in the laser excitation energy [51]. The G band specified for all the $sp^2$ carbon materials is related to the in-plane bond stretching mode of the C-C bonds in the hexagonal lattice [52]. The G′ (or 2D) band is a peak in the spectra of most $sp^2$ carbon materials. The origin of this peak can be ascribed to a vibrational mode that was identified by the breathing of six carbons related to a hexagon in the graphene lattice. When increasing the collection time, no change was observed in the position of the peaks (Figure 4a). Because of the electron–phonon coupling or strain effect in SWCNTs, the G band splits into $G^+$ and $G^-$ peaks. The split between the $G^+$ (at 1594 $cm^{-1}$) and $G^-$ (at 1571 $cm^{-1}$) peaks can be seen in Figure 4b. The $G^-$ peak is usually unseparated from the $G^+$ peak; however, it can be varied in shape [53]. The signal related to the $G^-$ peak is attributed to the curvature of SWCNTs (i.e., diameter), which is specific to their electronic characteristics [54]. A small peak that appeared at 1156 $cm^{-1}$ (Figure 4b) is related to the intermediate frequency modes (IFM mode). These modes, which are usually reported around the range of 600–1200 $cm^{-1}$, are considered weak and insignificant features; they exist in all graphene-related materials [36] and are assigned to second-order, two-phonon, or one-phonon and one-elastic scattering double resonance Raman processes [55,56].

Figure 5 shows the Raman spectra of the pristine SWCNTs, nitric-acid-treated SWCNTs, and sulfuric-acid-treated SWCNTs using a 785 nm diode laser. Using this laser, a weak peak was observed at 1297 $cm^{-1}$ for all the samples that can be related to the D mode. The obtained $I_D/I_G$ values were 0.11, 0.22, and 0.14 for the pristine SWCNTs, nitric acid-treated SWCNTs, and sulfuric-acid-treated SWCNTs, respectively. The low $I_D/I_G$ values confirm that our synthesized SWCNTs are very well crystallized and have a high quality. The higher $I_D/I_G$ values for the acid-treated samples show that the acid treatment decreased the amount of graphenic $sp^2$-carbons in the pristine SWCNTs. The $I_D/I_G$ values we obtained can be compared to the values of 0.2 and 0.37 achieved by Wang et al. [57] for their pristine SWCNTs and sulfuric-acid-treated SWCNTs, respectively. An up-shift related to the G-frequency was observed after the acid treatments. This phenomenon was also detected by Tantang et al. [58] after the acid treatment. They related the change in the shift to the possibility of an increase in charge carrier-accepting defects after the acid treatment. The peak at 1762 $cm^{-1}$ can be related to the overtone of the out-of-plane infrared active mode in graphite, which is called the M band. The peak at 1886 $cm^{-1}$ can be assigned to the iTOLA band originating from a combination of a phonon from the in-plane transverse optical (iTO) or longitudinal optical (LO) branch and a longitudinal acoustic (LA) branch phonon [59]. The peaks at 112, 150, and 230 are related to the radial breathing modes (RBMs).

The structural characteristics of our synthesized SWCNTs showed the morphology of the SWCNTs in bundles, with an average diameter of 1.46 nm. The deposited SWCNTs mainly contain carbon and iron elements. The low value of $I_D/I_G$ (0.11) for the pristine SWCNTs in the Raman spectroscopy measurements indicates that our synthesized SWCNTs are very well crystallized and have a high quality. This ratio increased to 0.14 and 0.22 after the treatment with sulfuric acid and nitric acid, respectively.

We carried out the Hall effect measurements on the SWCNT films with different collection times (5, 10, 15, 20, and 25 min). Important electrical properties such as the carrier concentration (carrier density), mobility, Hall coefficient, and sheet resistance of the SWCNT films were explored through these measurements. The magnitudes of the carrier concentration were observed to be between $4.6 \times 10^{21}$ and $1.3 \times 10^{23}$ $cm^{-3}$. These values are comparable to the carrier density of $10^{21}$–$10^{22}$ $cm^{-3}$ obtained for the purified SWCNTs synthesized through a high-pressure CO conversion process (HiPCO CNTs) or laser ablation method (LA CNTs) and chemically treated with $SOCl_2$ [36]. Compared to the values of $10^{18}$–$10^{19}$ $cm^{-3}$, which were earlier reported for the CNT films [60–62] and bundles [63], our results are closer to the theoretically predicted value of ~$10^{22}$ $cm^{-3}$

that was calculated for the aligned metallic CNTs [37]. The higher carrier density of our SWCNTs can be attributed to the advances in the synthesis method used for SWCNT films and their purity [36]. For example, the dry FCCVD method allows for the production of films with longer CNTs and exceptional optoelectronic properties by resolving the tradeoff between the CNT length and solubility during film fabrication [20]. The mobility of the synthesized SWCNTs was in the same order of magnitude for all the collection times, with an average value of ~0.034 $cm^2/Vs$. This mobility is very close to the one observed for pristine HiPCO SWCNTs (0.04 $cm^2/Vs$) and reported by Lee et al. [36]. However, these Hall mobilities were much lower than the values for the field-effect mobility, namely 220 and 100,000 $cm^2/Vs$, reported for the individual semiconducting SWCNTs. The factors limiting the Hall mobility were thought to be random networks of CNTs and barriers at the inter-tube junctions of the CNT films [36,64,65].

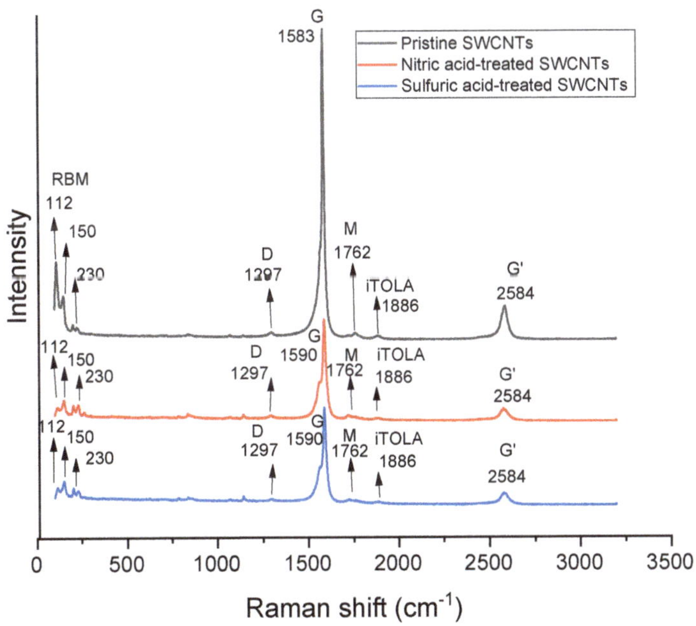


**Figure 5.** Raman spectra of the pristine SWCNTs, nitric-acid-treated SWCNTs, and sulfuric-acid-treated SWCNTs using a 785 nm diode laser.

In Figure 6a, the carrier concentration, Hall coefficient, and mobility are plotted versus the collection time of the SWCNT films. The carrier concentration increased from $4.6 \times 10^{21}$ to $1.3 \times 10^{23}$ $cm^{-3}$ by increasing the collection time from 5 to 25 min, while the Hall coefficient decreased from $1.35 \times 10^{-3}$ to $4.6 \times 10^{-5}$ (also see Table 1). However, the decrease in mobility was insignificant. Equation (1) shows that an increase in the carrier concentration will lead to a decrease in the Hall coefficient, as we can see in Figure 6b. Also, when increasing the carrier concentration, a slight decrease was observed in the mobility of the SWCNT thin films (Figure 6b). After reaching a certain thickness value (around 800 nm for 25 min of collection time), the further increase did not significantly change the carrier concentration of our thin films, meaning that the carrier concentration became independent of the thickness. The Hall effect measurements for the samples with collection times of 5, 10, 15, 20, and 25 min are summarized in Table 1. The sign of the Hall coefficients was positive for all samples, showing that the synthesized SWCNTs are p-type materials.

$$R_H = 1/ne \quad (1)$$

where $R_H$ is the Hall coefficient ($cm^3/C$), $n$ is the carrier concentration ($cm^{-3}$), and $e = 1.6 \times 10^{-19}$ (C).

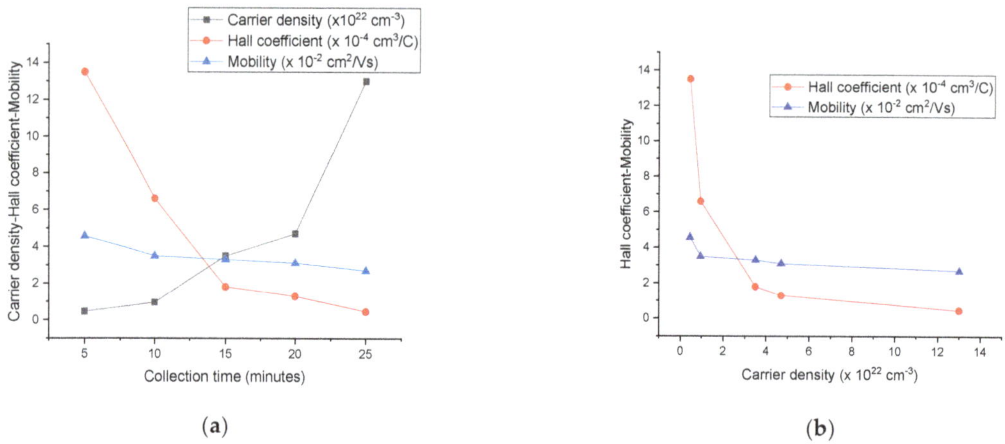


**Figure 6.** Graphs of (**a**) carrier density, Hall coefficient, and mobility versus collection time of SWCNT films, and (**b**) Hall coefficient and mobility versus carrier density.

**Table 1.** Hall effect measurements for the samples with collection times of 5, 10, 15, 20, and 25 min.

| Carrier Density ($cm^{-3}$) | Mobility ($cm^2/Vs$) | Hall Coefficient ($cm^3/C$) | Carrier Type | Sheet Resistance (ohm/Square) | Collection Time (Minutes) |
|---|---|---|---|---|---|
| $4.6 \times 10^{21}$ | $4.58 \times 10^{-2}$ | $1.35 \times 10^{-3}$ | P-type | 370 | 5 |
| $9.5 \times 10^{21}$ | $3.5 \times 10^{-2}$ | $6.6 \times 10^{-4}$ | P-type | 234 | 10 |
| $3.5 \times 10^{22}$ | $3.3 \times 10^{-2}$ | $1.8 \times 10^{-4}$ | P-type | 79 | 15 |
| $4.7 \times 10^{22}$ | $3.1 \times 10^{-2}$ | $1.3 \times 10^{-4}$ | P-type | 51 | 20 |
| $1.3 \times 10^{23}$ | $2.7 \times 10^{-2}$ | $4.6 \times 10^{-5}$ | P-type | 21 | 25 |

Figure 7a,b show the transmittance versus the collection time and sheet resistance. When increasing the collection time from 5 to 25 min, the transmittance and sheet resistance of the samples decreased from 95% to 43% and from 370 to 21 ohm/square, respectively. The best result for the sheet resistance and transmittance (T) achieved for the pristine Arc SWCNT TCFs was reported to be 390 ohm/square at 90% T for 550 nm light [66]. This can be compared to our SWCNT films with a sheet resistance of 370 ohm/square at 95% T. Increasing the collection time leads to an increase in the thickness of the thin films, which leads to a decrease in the sheet resistance of the materials, as indicated by Equation (2). The observed sheet resistance and transmittance magnitudes are comparable with the ones reported for the SWCNT thin films [35,39,42].

$$R_s = p/t \quad (2)$$

where $R_s$ is the sheet resistance (ohm/square), $p$ is the resistivity (ohm·cm), and t is the thickness of the thin film (cm).

Important electrical characteristics of the SWCNTs, including the carrier mobility, carrier density, Hall coefficient, and sheet resistance, were obtained using Hall effect measurements. The positive signs of the Hall coefficients showed that the majority of the carriers were holes in our synthesized SWCNT thin films. The samples with a transmittance between 95 and 43% showed a high carrier density of $10^{21}$–$10^{23}$ $cm^{-3}$, with a sheet resistance between 370 and 21 ohm/square.

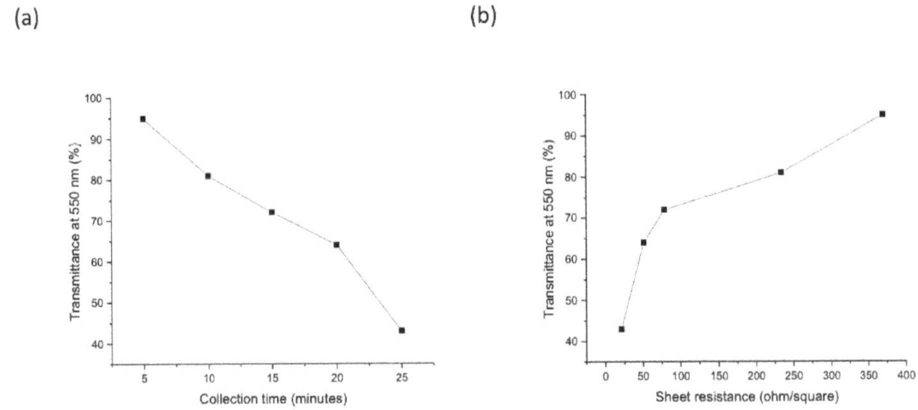


**Figure 7.** Graphs of the transmittance versus (**a**) collection time and (**b**) sheet resistance.

Different acid treatments were performed on the SWCNT films using the same collection time to investigate the effect of doping on their electrical properties. The conclusion obtained from treating one sample can be applied to the samples with different collection times. The samples with a 10 min collection time (transmittance of 80%) were treated with sulfuric, nitric, and hydrochloric acid at different acid concentrations (Table 2). All the samples were rinsed with deionized (DI) water after being treated with the acids; meanwhile, in some cases (Case C in Table 2), the samples were first rinsed with ethanol followed by DI water. In Case B, the samples were heated to 80 °C throughout the duration of the treatment (30 min). Hall effect measurements were applied to the samples after the treatments to study the change in their electrical characteristics.

**Table 2.** Hall effect measurements of the SWCNT film of collection time of 10 min after acid treatments.

| Case | Acid Treatment | Carrier Density ($cm^{-3}$) | Hole Mobility ($cm^2/Vs$) | Hall Coefficient ($cm^3/C$) | Carrier Type | Sheet Resistance (ohm/Square) | Conductivity (S/m) |
|---|---|---|---|---|---|---|---|
| A | 30 min $H_2SO_4$ (65 *w*/*w* %) treatment + DI water rinsing | $7.7 \times 10^{20}$ | 0.75 | $8 \times 10^{-3}$ | P-type | 139 | 9240 |
| B | 30 min $H_2SO_4$ (65 *w*/*w* %) treatment at 80 °C + DI water rinsing | $3.2 \times 10^{20}$ | 1.5 | $2 \times 0^{-2}$ | P-type | 163 | 7680 |
| C | 30 min $H_2SO_4$ (65 *w*/*w* %) treatment + DI water rinsing + ethanol rinsing | $8.4 \times 10^{21}$ | 0.09 | $7.7 \times 10^{-4}$ | P-type | 107 | 12,096 |
| D | 30 min $H_2SO_4$ treatment (4 M) + DI water rinsing | $5 \times 10^{21}$ | 0.08 | $1.4 \times 10^{-3}$ | P-type | 228 | 6400 |
| E | 30 min HCL treatment (4 M) + DI water rinsing | $8.8 \times 10^{21}$ | 0.1 | $8.8 \times 10^{-4}$ | P-type | 115 | 14,060 |
| F | 30 min $HNO_3$ (4 M) + DI water rinsing | $1.2 \times 10^{22}$ | 0.16 | $5.8 \times 10^{-4}$ | P-type | 43 | 30,720 |
| G | Without treatment | $9.5 \times 10^{21}$ | 0.035 | $6.6 \times 10^{-4}$ | P-type | 243 | 5320 |

All the samples maintained their p-type characteristic after being treated with the acids. Acceptor-type doping of the SWCNTs was observed by other researchers using Brønsted acids (like sulfuric, nitric, and hydrochloric acid) [29,66–68]. Compared to the sample without any treatment (Case G), the hole mobility increased in all acid-treated samples, the hole concentration of all sulfuric acid and hydrochloric-acid-treated samples decreased, but the hole concentration of the nitric-acid-treated sample increased. The experimental results showed that nitric acid treatment can improve the SWCNT film's mobility and carrier concentration (Case F). The corresponding conductivity was calculated for the samples using Equation (3).

$$\sigma = \mu n e \quad (3)$$

Here, $\sigma$ is the conductivity due to holes in a unit of Siemens per meter (S/m) or $\Omega^{-1}m^{-1}$, n and e stand for the electronic carrier concentration and electron charge, respectively, and $\mu$ is the hole mobility.

Despite a slight decrease in the hole concentration, the conductivity increased in all kinds of acid treatments while the sheet resistance decreased, as can be seen in Table 2. Although the sample treated with sulfuric acid at 80 °C (Case B) showed the highest hole mobility, the highest conductivity (30,720 S/m) was associated with the nitric acid treatment (Case F). Additionally, there was a significant decrease in the sheet resistance of the nitric-acid-treated sample (Case F), which decreased to 43 ohm/square, compared to the untreated sample, which had a sheet resistance of 243 ohm/square (Case G). This value of sheet resistance (43 ohm/square) is lower than the value of 59 ohm/square at 80% T that was achieved for the $HNO_3$-doped SWCNT films (synthesized via arc-discharge method) reported by Paul et al. [69]. The second and third highest conductivity are 14,060 S/m and 12,096 S/m for the films treated with hydrochloric acid and DI water rinsing (Case E) and those treated with the sulfuric acid treatment, with ethanol as well as DI water rinsing (Case C), respectively. All the calculated conductivities were in the range reported (in the order of $10^2$ to $10^6$ S/cm) for SWCNT [70]. The improvement in the conductivity of the SWCNT films can be related to the downshifts in the Fermi level toward the valence bands of the SWCNTs, leading to a reduction in the Schottky barrier height and increasing the conductivity of the films [35,71].

The bar graphs (Figure 8 ) show the comparison between the electrical properties of the SWCNT thin films under different acid treatments according to the data in Table 2. Overall, we can see the enhancement in the electrical properties of the SWCNT films after the acid treatments, with an increase in the hole mobility and conductivity, as well as a decrease in the sheet resistance in all cases. Out of all the samples, the ones treated with sulfuric acid at a temperature of 80 °C (Case B) showed the greatest hole mobility, at 1.5 $cm^2/Vs$ (Figure 8b). However, the sample treated with nitric acid (Case F) demonstrated the highest carrier conductivity, at 30,720 S/m (Figure 8e), the highest carrier density, at $1.2 \times 10^{22}$ $cm^{-3}$ (Figure 8a), and the lowest sheet resistance, at 43 ohm/square (Figure 8d).

Fourier-Transform Infrared Spectroscopy (FTIR) was utilized to analyze the chemical structure of the sulfuric-acid-treated SWCNTs at 80 °C (Case B) and the nitric-acid-treated SWCNTs (Case F) in comparison to the pristine SWCNTs (Case G). The analysis was conducted by identifying functional groups and comparing any of the changes observed in the three cases. Figure 9 shows the FTIR of the pristine SWCNTs (Case G), $HNO_3$-treated SWCNTs (Case F), and $H_2SO_4$-treated SWCNTs heated at 80 °C (Case B) in the wavelength ranges of 500–1600 $cm^{-1}$ (Figure 9a) and 1600–4000 $cm^{-1}$ (Figure 9b). The peaks at 630, 701, 719, 797, 837, and 1010 $cm^{-1}$ in all three samples are related to the C=C bending of the alkene functional group [72,73], while the peak at 872 $cm^{-1}$ comes from the bending of the C-H for the benzene derivative functional group [74]. Hydrogen exists in the precursor solution for the synthesis of SWCNTs and is also used as a carrier gas during the synthesis process. A new peak appeared at 1046 $cm^{-1}$ for the sulfuric-acid-treated sample, which represents the S=O stretching of sulfoxide [75] and is attributed to the sulfuric acid treatment. The peaks at 1073, 1106, 1153, 1244, and 1297 $cm^{-1}$, common in all three samples, are related to the C-O stretching mode of the primary alcohol, secondary alcohol, tertiary alcohol, and alkyl aryl ether functional groups [72,76]. Ethanol is a primary alcohol that is used in the precursor solution. Secondary and tertiary alcohols can form by attaching the hydroxyl group (-OH) to a carbon atom that is bonded to two and three alkyl groups (or hydrocarbon chains), respectively [77]. Ether is formed when an oxygen atom is attached to two alkyl or aryl groups (aromatic hydrocarbons) [78]. For all samples, the peak at 1322 cm$^{-1}$ represents the S=O stretching mode of the sulfone, and the peak at 1409 cm$^{-1}$ is related to the S=O stretching mode of the sulfate functional group [72,79]. Sulfur comes from thiophene, which is used in the precursor solution.

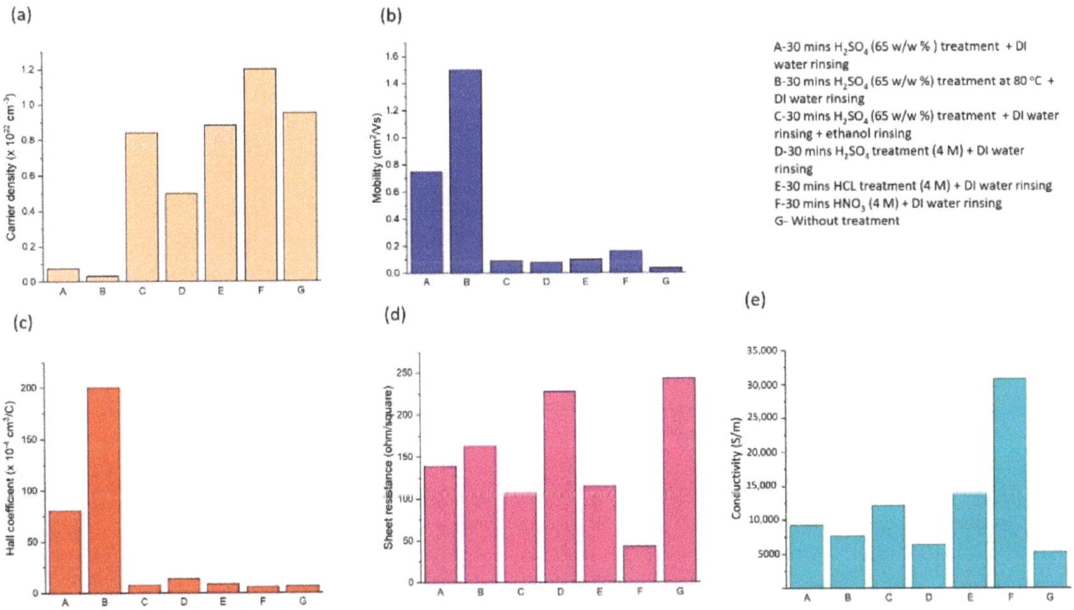


**Figure 8.** Bar graphs showing the effect of different acid treatments on the (**a**) carrier density, (**b**) hole mobility, (**c**) Hall coefficient, (**d**) sheet resistance, and (**e**) conductivity of the SWCNT thin films.

The 1488 and 1578 cm$^{-1}$ peaks appearing in all three samples originate from the C-H bending of alkane and the C=C stretching of cyclic alkane functional groups, respectively [80]. The peak at 1735 cm$^{-1}$ for pristine and nitric-acid-treated SWCNTs corresponds to the C=O stretching mode in the carbonyl or carboxylic acid groups [81]. This peak (1735 cm$^{-1}$) is downshifted to 1633 cm$^{-1}$ for the sulfuric-acid-treated sample, representing the C=C stretching of alkene [82]. The sulfuric-acid-treated sample also shows a new peak at 2113 cm$^{-1}$, which is related to the C≡C stretching of the alkyne functional group. This peak can be compared to the one at 1909 cm$^{-1}$ for the pristine SWCNT sample showing the C=C=C stretching of the allene functional group [72]. This change in the bond formation, from double-bond carbon in the pristine SWCNTs into triple-bond carbon in the sulfuric-acid-treated sample, can be attributed to the heating performed during the acid treatment. Wong et al. [83] conducted simulations on the distribution of chemical bonds in carbon chains. Their findings suggest that as the temperature rises, the carbon atoms in a compound may transition from a double bond to a single or triple bond. This transition is dependent on energy minimization and the Octet rule. The Octet rule dictates that an atom's valence shell can only accommodate a maximum of eight electrons, with no lone pair electrons are permitted.

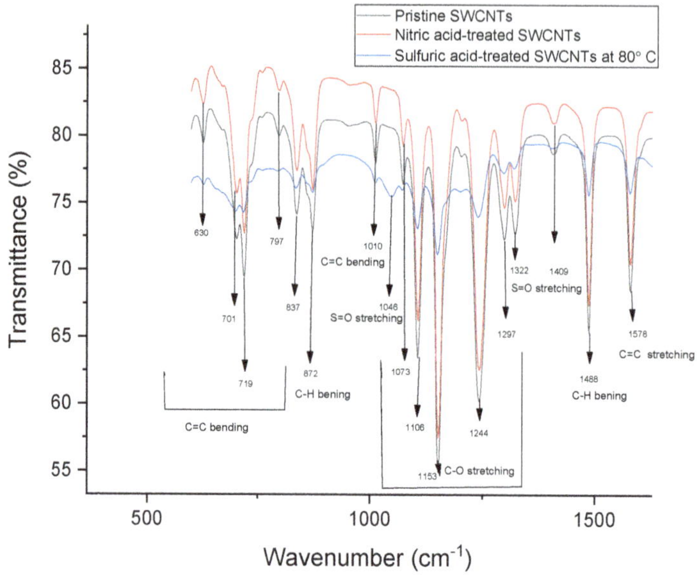


(**a**)

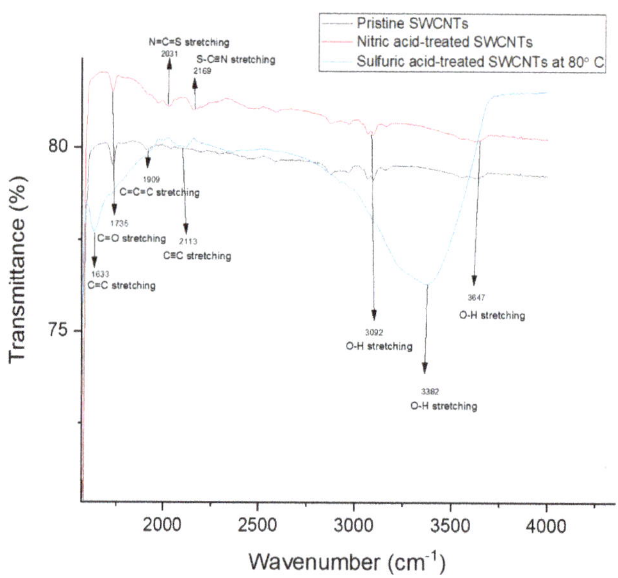


(**b**)

**Figure 9.** FTIR spectra of pristine SWCNTs, $HNO_3$-treated SWCNTs, and $H_2SO_4$-treated SWCNTs heated at 80 °C in the wavelength ranges of (**a**) 500–1600 cm$^{-1}$ and (**b**) 1600–4000 cm$^{-1}$.

In the range of 1600–4000 cm$^{-1}$, two new peaks appeared at 2031 and 2169 cm$^{-1}$ for the nitric-acid-treated sample. These peaks can be related to the N=C=S stretching of isothiocyanate and the S-C≡N stretching of thiocyanate, respectively [72]. Sulfur exists in the precursor solution created by thiophene. The peaks at 3092 and 3647 cm$^{-1}$ appeared for the pristine and nitric-acid-treated SWCNTs, corresponding to the O-H stretching of alcohol (hydroxyl functional group). The peak at 3382 $cm^{-1}$ for the sulfuric-acid-treated sample is also related to the O-H stretching of alcohol [84]. These peaks are usually broad and strong in the 3200–3550 $cm^{-1}$ range [72], as seen in Figure 9b. The disappearance of one of the peaks related to O-H stretching and also the peak related to C=O stretching (at 1735 $cm^{-1}$) for the sulfuric-acid-treated sample might be a reason for the increase in the hole mobility of this sample. Studies have shown that polar groups such as hydroxyl and carbonyl functional groups can form deep traps that can capture electric charge carriers [85,86]. The scattering and capture of free charges by deep traps can lead to energy loss in the charge carriers and a decrease in their mobility [86]. Therefore, the reduction in the number of polar groups such as carbonyl and hydroxyl functional groups in the sulfuric-acid-treated sample might play a role in improving its mobility. Compared to the sulfuric-acid-treated sample, the FTIR spectra of the nitric-acid-treated sample are more similar to the pristine SWCNTs' spectra. However, the improved electrical properties of the nitric-acid-treated SWCNTs might be related to the existence of two new peaks (2031 and 2169 $cm^{-1}$) in this sample, representing the isothiocyanate and thiocyanate functional groups. Polymers containing thiocyanate have been widely reported for use as hole-transporting layers in optoelectronic applications (such as solar cells) due to the improvement in the charge extraction efficiency associated with increased conductivity [87,88].

In summary, among the different acid treatments, the sample treated with sulfuric acid at a temperature of 80 °C (Case B) exhibited the highest hole mobility value, at 1.5 $cm^2/Vs$. In addition, the sample treated with nitric acid (Case F) demonstrated the highest carrier conductivity, at 30,720 S/m, the highest carrier density, at $1.2 \times 10^{22}$ $cm^{-3}$, and the lowest sheet resistance, at 43 ohm/square. These improvements in the electrical properties of both samples make them suitable for use in electronic applications such as thin film transistors or solar cells.

## 4. Conclusions

SWCNT films were synthesized through a floating catalyst chemical vapor deposition method using a total low flow rate of 200 sccm for the mixed argon and hydrogen gases. The structural properties of the thin films were investigated by SEM, TEM, FTIR, and Raman spectroscopy. The low value of $I_D/I_G$ (0.11) in the Raman spectroscopy measurements of our pristine synthesized SWCNTs confirms that the SWCNT films are highly crystallized. This value increased to 0.14 and 0.22 after the treatment with sulfuric acid and nitric acid, respectively. The significant electrical characteristics of the SWCNT films, such as the carrier concentration, carrier mobility, Hall coefficient, and sheet resistance, were investigated through Hall effect measurements. The samples showed a high carrier concentration of $10^{21}$–$10^{23}$ $cm^{-3}$, with a transmittance between 95 and 43%. The effect of different acid treatments was explored using nitric, sulfuric, and hydrochloric acids. In all the treatments, an increase was observed in their carrier mobility and conductivity, while their sheet resistance decreased. The samples retained their p-type nature after the acid treatments, showing that the treatments led to p doping. The samples treated with sulfuric acid at 80 °C showed the highest mobility, at 1.5 $cm^2/Vs$. On the other hand, the highest carrier conductivity (30,720 S/m) and the lowest sheet resistance (43 ohm/square) were obtained from the samples treated with nitric acid. The presence of multiple functional groups in our synthesized SWCNTs may increase their potential for various applications, since each functional group exhibits distinct characteristics and reactivity, enabling participation in various types of reactions and increasing functionality. Improving the electrical properties of SWCNTs can lead to a great enhancement in their electronic applications. Since the synthesized SWCNT films maintained their p-type nature after the acid treatments, the enhancement in their electrical characteristics makes them suitable candidates for hole-transporting layers in solar cells.

**Author Contributions:** Conceptualization, M.D.R. and W.L.; Methodology, M.D.R. and W.L.; Validation, M.D.R. and W.L.; Formal analysis, M.D.R. and W.L.; Investigation, M.D.R. and W.L.; Resources, J.W., D.R.H. and W.L.; Data curation, M.D.R., J.W. and W.L.; Writing—original draft, M.D.R.; Writing—review & editing, M.D.R., B.D., J.W., M.A., D.R.H. and W.L.; Supervision, W.L.; Project administration, W.L.; Funding acquisition, W.L. All authors have read and agreed to the published version of the manuscript.

**Funding:** This work was supported by the National Science Foundation under grants 1506640, 2134375, and 2213923. This work was performed, in part, at the Center for Integrated Nanotechnologies, an Office of Science User Facility operated for the U.S. Department of Energy (DOE) Office of Science. Los Alamos National Laboratory, an affirmative action equal opportunity employer, is managed by Triad National Security, LLC for the U.S. Department of Energy's NNSA, under contract 89233218CNA000001. The funder was not involved in the study design, collection, analysis, interpretation of data, the writing of this article or the decision to submit it for publication.

**Data Availability Statement:** The raw data required to reproduce these findings are available upon request from the corresponding author. The processed data required to reproduce these findings are available upon request from the corresponding author.

**Acknowledgments:** We thank Anamika Prasad, Alexi Switz, Jin He, Govinda Ghimire, and Vadym Drozd (FIU CESMEC) for their help with the Raman measurements, Indranil Chakraborty and the FIU Chemistry Department for the FTIR measurements, and Christopher Dares and Xiangyang Hou for the UV-Vis measurements. The authors would also like to acknowledge the support of the Advanced Materials Engineering Research Institutes (AMERI) at Florida International University.

**Conflicts of Interest:** The authors declare no conflict of interest.

## References

1. Maruyama, T. Carbon nanotubes. In *Handbook of Carbon-Based Nanomaterials*; Elsevier: Amsterdam, The Netherlands, 2021; pp. 299–319.
2. Dresselhaus, M.S.; Avouris, P. Introduction to carbon materials research. In *Carbon Nanotubes: Synthesis, Structure, Properties, and Applications*; Springer: Berlin/Heidelberg, Germany, 2001; pp. 1–9.
3. Iijima, S. Helical microtubules of graphitic carbon. *Nature* **1991**, *354*, 56–58. [CrossRef]
4. Iijima, S.; Ichihashi, T. Single-shell carbon nanotubes of 1-nm diameter. *Nature* **1993**, *363*, 603–605. [CrossRef]
5. Rezaee, M.D.; Dahal, B.; Li, W. *Single-Walled Carbon Nanotubes*; American Chemical Society: Washington, DC, USA, 2023.
6. Wu, K.; Niu, Y.; Zhang, Y.; Yong, Z.; Li, Q. Continuous growth of carbon nanotube films: From controllable synthesis to real applications. *Compos. Part A Appl. Sci. Manuf.* **2021**, *144*, 106359. [CrossRef]
7. Di, J.; Wang, X.; Xing, Y.; Zhang, Y.; Zhang, X.; Lu, W.; Zhu, Y.T. Dry-Processable Carbon Nanotubes for Functional Devices and Composites. *Small* **2014**, *10*, 4606–4625. [CrossRef] [PubMed]
8. Liu, L.; Ma, W.; Zhang, Z. Macroscopic carbon nanotube assemblies: Preparation, properties, and potential applications. *Small* **2011**, *7*, 1504–1520. [CrossRef] [PubMed]
9. Peng, B.; Locascio, M.; Zapol, P.; Li, S.; Mielke, S.L.; Schatz, G.C.; Espinosa, H.D. Measurements of near-ultimate strength for multiwalled carbon nanotubes and irradiation-induced crosslinking improvements. *Nat. Nanotechnol.* **2008**, *3*, 626–631. [CrossRef] [PubMed]
10. Bai, Y.; Zhang, R.; Ye, X.; Zhu, Z.; Xie, H.; Shen, B.; Wei, F. Carbon nanotube bundles with tensile strength over 80 GPa. *Nat. Nanotechnol.* **2018**, *13*, 589–595. [CrossRef] [PubMed]
11. Chen, H.; Zeng, S.; Chen, M.; Zhang, Y.; Li, Q. Fabrication and functionalization of carbon nanotube films for high-performance flexible supercapacitors. *Carbon* **2015**, *92*, 271–296. [CrossRef]
12. Zhou, Y.; Azumi, R. Carbon nanotube based transparent conductive films: Progress, challenges, and perspectives. *Sci. Technol. Adv. Mater.* **2016**, *17*, 493–516. [CrossRef]
13. Liao, Y.; Hussain, A.; Laiho, P.; Zhang, Q.; Tian, Y.; Wei, N.; Kauppinen, E.I. Tuning Geometry of SWCNTs by $CO_2$ in Floating Catalyst CVD for High-Performance Transparent Conductive Films. *Adv. Mater. Interfaces* **2018**, *5*, 1801209. [CrossRef]
14. Meysami, S.S.; Koos, A.A.; Dillon, F.; Grobert, N. Aerosol-assisted chemical vapour deposition synthesis of multi-wall carbon nanotubes: II. *Anal. Study Carbon* **2013**, *58*, 159–169. [CrossRef]
15. Choi, Y.C.; Bae, D.J.; Lee, Y.H.; Lee, B.S.; Park, G.S.; Choi, W.B.; Kim, J.M. Growth of carbon nanotubes by microwave plasma-enhanced chemical vapor deposition at low temperature. *J. Vac. Sci. Technol. A Vac. Surf. Film.* **2000**, *18*, 1864–1868. [CrossRef]
16. Makris, T.D.; Giorgi, R.; Lisi, N.; Pilloni, L.; Salernitano, E.; Sarto, F.; Alvisi, M. Carbon nanotubes growth by HFCVD: Effect of the process parameters and catalyst preparation. *Diam. Relat. Mater.* **2004**, *13*, 305–310. [CrossRef]
17. Byon, H.R.; Lim, H.S.; Song, H.J.; Choi, H.C. A synthesis of high purity single-walled carbon nanotubes from small diameters of cobalt nanoparticles by using oxygen-assisted chemical vapor deposition process. *Bull. Korean Chem. Soc.* **2007**, *28*, 2056–2060.

18. Feng, C.; Liu, K.; Wu, J.S.; Liu, L.; Cheng, J.S.; Zhang, Y.; Jiang, K. Flexible, stretchable, transparent conducting films made from superaligned carbon nanotubes. *Adv. Funct. Mater.* **2010**, *20*, 885–891. [CrossRef]
19. Bronikowski, M.J.; Willis, P.A.; Colbert, D.T.; Smith, K.A.; Smalley, R.E. Gas-phase production of carbon single-walled nanotubes from carbon monoxide via the HiPco process: A parametric study. *J. Vac. Sci. Technol. A Vac. Surf. Film.* **2001**, *19*, 1800–1805. [CrossRef]
20. Zhang, Q.; Wei, N.; Laiho, P.; Kauppinen, E.I. Recent developments in single-walled carbon nanotube thin films fabricated by dry floating catalyst chemical vapor deposition. In *Single-Walled Carbon Nanotubes: Preparation, Properties and Applications*; Springer: Berlin/Heidelberg, Germany, 2019; pp. 99–128.
21. Bogdanova, A.R.; Krasnikov, D.V.; Nasibulin, A.G. The role of sulfur in the CVD carbon nanotube synthesis. *Carbon* **2023**, *210*, 118051. [CrossRef]
22. Chen, D.R.; Chitranshi, M.; Schulz, M.; Shanov, V. A review of three major factors controlling carbon nanotubes synthesis from the floating catalyst chemical vapor deposition. *Nano Life* **2019**, *9*, 1930002. [CrossRef]
23. Yu, F.; Yang, M.; Li, F.; Su, C.; Ma, B.; Yuan, Z.; Ma, J. The growth mechanism of single-walled carbon nanotubes with a controlled diameter. *Phys. E Low-Dimens. Syst. Nanostruct.* **2012**, *44*, 2032–2040. [CrossRef]
24. Plata, D.L.; Meshot, E.R.; Reddy, C.M.; Hart, A.J.; Gschwend, P.M. Multiple alkynes react with ethylene to enhance carbon nanotube synthesis, suggesting a polymerization-like formation mechanism. *ACS Nano* **2010**, *4*, 7185–7192. [CrossRef]
25. Huang, S.; Cai, Q.; Chen, J.; Qian, Y.; Zhang, L. Metal-catalyst-free growth of single-walled carbon nanotubes on substrates. *J. Am. Chem. Soc.* **2009**, *131*, 2094–2095. [CrossRef] [PubMed]
26. Li, Y.; Ji, K.; Duan, Y.; Meng, G.; Dai, Z. Effect of hydrogen concentration on the growth of carbon nanotube arrays for gecko-inspired adhesive applications. *Coatings* **2017**, *7*, 221. [CrossRef]
27. Ma, Y.; Dichiara, A.B.; He, D.; Zimmer, L.; Bai, J. Control of product nature and morphology by adjusting the hydrogen content in a continuous chemical vapor deposition process for carbon nanotube synthesis. *Carbon* **2016**, *107*, 171–179. [CrossRef]
28. Kumar, A.; Kostikov, Y.; Zanatta, M.; Sorarù, G.D.; Orberger, B.; Nessim, G.D.; Mariotto, G. Carbon nanotubes synthesis using siliceous breccia as a catalyst source. *Diam. Relat. Mater.* **2019**, *97*, 107433. [CrossRef]
29. Graupner, R.; Abraham, J.; Vencelová, A.; Seyller, T.; Hennrich, F.; Kappes, M.M.; Ley, L. Doping of single-walled carbon nanotube bundles by Brønsted acids. *Phys. Chem. Chem. Phys.* **2003**, *5*, 5472–5476. [CrossRef]
30. Dresselhaus, M.S.; Dresselhaus, G. Intercalation compounds of graphite. *Adv. Phys.* **1981**, *30*, 139–326. [CrossRef]
31. Zhou, Y.; Shimada, S.; Saito, T.; Azumi, R. Understanding the doping effects on the structural and electrical properties of ultrathin carbon nanotube networks. *J. Appl. Phys.* **2015**, *118*, 215305. [CrossRef]
32. Barros, E.B.; Son, H.; Samsonidze, G.G.; Souza Filho, A.G.; Saito, R.; Kim, Y.A.; Dresselhaus, M.S. Raman spectroscopy of double-walled carbon nanotubes treated with $H_2SO_4$. *Phys. Rev. B* **2007**, *76*, 045425. [CrossRef]
33. Rao, A.M.; Eklund, P.C.; Bandow, S.; Thess, A.; Smalley, R.E. Evidence for charge transfer in doped carbon nanotube bundles from Raman scattering. *Nature* **1997**, *388*, 257–259. [CrossRef]
34. Fagan, S.B.; Souza Filho, A.G.D.; Mendes Filho, J.; Corio, P.; Dresselhaus, M.S. Electronic properties of Ag-and CrO3-filled single-wall carbon nanotubes. *Chem. Phys. Lett.* **2005**, *406*, 54–59. [CrossRef]
35. Ding, E.X.; Jiang, H.; Zhang, Q.; Tian, Y.; Laiho, P.; Hussain, A.; Kauppinen, E.I. Highly conductive and transparent single-walled carbon nanotube thin films from ethanol by floating catalyst chemical vapor deposition. *Nanoscale* **2017**, *9*, 17601–17609. [CrossRef]
36. Lee, S.H.; Uhm, T.W.; You, Y.G.; Kim, S.W.; Jhang, S.H.; Dettlaff-Weglikowska, U.; Park, Y.W. Hall effect in carbon nanotube thin films. *Synth. Met.* **2014**, *198*, 84–87. [CrossRef]
37. Mintmire, J.W.; Dunlap, B.I.; White, C.T. Are fullerene tubules metallic? *Phys. Rev. Lett.* **1992**, *68*, 631. [CrossRef] [PubMed]
38. Nakka, L.; Cheng, Y.; Aberle, A.G.; Lin, F. Analytical review of spiro-OMeTAD hole transport materials: Paths toward stable and efficient perovskite solar cells. *Adv. Energy Sustain. Res.* **2022**, *3*, 2200045. [CrossRef]
39. Kaskela, A.; Laiho, P.; Fukaya, N.; Mustonen, K.; Susi, T.; Jiang, H.; Kauppinen, E.I. Highly individual SWCNTs for high performance thin film electronics. *Carbon* **2016**, *103*, 228–234. [CrossRef]
40. Niu, Z.; Zhou, W.; Chen, J.; Feng, G.; Li, H.; Ma, W.; Xie, S. Compact-designed supercapacitors using free-standing single-walled carbon nanotube films. *Energy Environ. Sci.* **2011**, *4*, 1440–1446. [CrossRef]
41. Ma, W.; Song, L.; Yang, R.; Zhang, T.; Zhao, Y.; Sun, L.; Xie, S. Directly synthesized strong, highly conducting, transparent single-walled carbon nanotube films. *Nano Lett.* **2007**, *7*, 2307–2311. [CrossRef]
42. Hussain, A.; Liao, Y.; Zhang, Q.; Ding, E.X.; Laiho, P.; Ahmad, S.; Kauppinen, E.I. Floating catalyst CVD synthesis of single walled carbon nanotubes from ethylene for high performance transparent electrodes. *Nanoscale* **2018**, *10*, 9752–9759. [CrossRef]
43. A Shi, Z.; Lian, Y.; Liao, F.H.; Zhou, X.; Gu, Z.; Zhang, Y.; Zhang, S.L. Large scale synthesis of single-wall carbon nanotubes by arc-discharge method. *J. Phys. Chem. Solids* **2000**, *61*, 1031–1036. [CrossRef]
44. Ajayan, P.M.; Ebbesen, T.W. Nanometre-size tubes of carbon. *Rep. Prog. Phys.* **1997**, *60*, 1025. [CrossRef]
45. Ruoff, R.S.; Qian, D.; Liu, W.K. Mechanical properties of carbon nanotubes: Theoretical predictions and experimental measurements. *Comptes Rendus Phys.* **2003**, *4*, 993–1008. [CrossRef]
46. Dresselhaus, M.S.; Dresselhaus, G.; Saito, R.; Jorio, A. Raman spectroscopy of carbon nanotubes. *Phys. Rep.* **2005**, *409*, 47–99. [CrossRef]
47. Pimenta, M.A.; Dresselhaus, G.; Dresselhaus, M.S.; Cancado, L.G.; Jorio, A.; Saito, R. Studying disorder in graphite-based systems by Raman spectroscopy. *Phys. Chem. Chem. Phys.* **2007**, *9*, 1276–1290. [CrossRef] [PubMed]

48. Ferrari, A.C.; Meyer, J.C.; Scardaci, V.; Casiraghi, C.; Lazzeri, M.; Mauri, F.; Geim, A.K. Raman spectrum of graphene and graphene layers. *Phys. Rev. Lett.* **2006**, *97*, 187401. [CrossRef] [PubMed]
49. Park, Y.; Hembram, K.P.S.S.; Yoo, R.; Jang, B.; Lee, W.; Lee, S.G.; Lee, J.K. Reinterpretation of single-wall carbon nanotubes by Raman spectroscopy. *J. Phys. Chem. C* **2019**, *123*, 14003–14009. [CrossRef]
50. Kim, K.K.; Park, J.S.; Kim, S.J.; Geng, H.Z.; An, K.H.; Yang, C.M.; Lee, Y.H. Dependence of Raman spectra G′ band intensity on metallicity of single-wall carbon nanotubes. *Phys. Rev. B* **2007**, *76*, 205426. [CrossRef]
51. Lopez-Lorente, A.I.; Simonet, B.M.; Valcarcel, M. Qualitative detection and quantitative determination of single-walled carbon nanotubes in mixtures of carbon nanotubes with a portable Raman spectrometer. *Analyst* **2013**, *138*, 2378–2385. [CrossRef]
52. Jorio, A.; Saito, R. Raman spectroscopy for carbon nanotube applications. *J. Appl. Phys.* **2021**, *129*, 021102. [CrossRef]
53. Jorio, A.; Souza Filho, A.G.; Dresselhaus, G.; Dresselhaus, M.S.; Swan, A.K.; Ünlü, M.S.; Saito, R. G-band resonant Raman study of 62 isolated single-wall carbon nanotubes. *Phys. Rev. B* **2002**, *65*, 155412. [CrossRef]
54. Kataura, H.; Kumazawa, Y.; Maniwa, Y.; Umezu, I.; Suzuki, S.; Ohtsuka, Y.; Achiba, Y. Optical properties of single-wall carbon nanotubes. *Synth. Met.* **1999**, *103*, 2555–2558. [CrossRef]
55. Saito, R.; Jorio, A.; Souza Filho, A.G.; Dresselhaus, G.; Dresselhaus, M.S.; Pimenta, M.A. Probing phonon dispersion relations of graphite by double resonance Raman scattering. *Phys. Rev. Lett.* **2001**, *88*, 027401. [CrossRef] [PubMed]
56. Saito, R.; Grüneis, A.; Samsonidze, G.G.; Brar, V.W.; Dresselhaus, G.; Dresselhaus, M.S.; Souza Filho, A.G. Double resonance Raman spectroscopy of single-wall carbon nanotubes. *New J. Phys.* **2003**, *5*, 157. [CrossRef]
57. Wang, P.C.; Liao, Y.C.; Lai, Y.L.; Lin, Y.C.; Su, C.Y.; Tsai, C.H.; Hsu, Y.J. Conversion of pristine and p-doped sulfuric-acid-treated single-walled carbon nanotubes to n-type materials by a facile hydrazine vapor exposure process. *Mater. Chem. Phys.* **2012**, *134*, 325–332. [CrossRef]
58. Tantang, H.; Ong, J.Y.; Loh, C.L.; Dong, X.; Chen, P.; Chen, Y.; Li, L.J. Using oxidation to increase the electrical conductivity of carbon nanotube electrodes. *Carbon* **2009**, *47*, 1867–1870. [CrossRef]
59. Rao, R.; Reppert, J.; Podila, R.; Zhang, X.; Rao, A.M.; Talapatra, S.; Maruyama, B. Double resonance Raman study of disorder in CVD-grown single-walled carbon nanotubes. *Carbon* **2011**, *49*, 1318–1325. [CrossRef]
60. Yomogida, Y.; Horiuchi, K.; Okada, R.; Kawai, H.; Ichinose, Y.; Nishidome, H.; Yanagi, K. Hall effect in gated single-wall carbon nanotube films. *Sci. Rep.* **2022**, *12*, 101. [CrossRef] [PubMed]
61. Hall, E.H. On a new action of the magnet on electric currents. *Am. J. Math.* **1879**, *2*, 287–292. [CrossRef]
62. Baumgartner, G.; Carrard, M.; Zuppiroli, L.; Bacsa, W.; de Heer, W.A.; Forró, L. Hall effect and magnetoresistance of carbon nanotube films. *Phys. Rev. B* **1997**, *55*, 6704. [CrossRef]
63. Song, S.N.; Wang, X.K.; Chang, R.P.H.; Ketterson, J.B. Electronic properties of graphite nanotubules from galvanomagnetic effects. *Phys. Rev. Lett.* **1994**, *72*, 697. [CrossRef]
64. Martel, R.; Schmidt, T.; Shea, H.R.; Hertel, T.; Avouris, P. Single-and multi-wall carbon nanotube field-effect transistors. *Appl. Phys. Lett.* **1998**, *73*, 2447–2449. [CrossRef]
65. Dürkop, T.; Getty, S.A.; Cobas, E.; Fuhrer, M.S. Extraordinary mobility in semiconducting carbon nanotubes. *Nano Lett.* **2004**, *4*, 35–39. [CrossRef]
66. Jiang, S.; Hou, P.X.; Liu, C.; Cheng, H.M. High-performance single-wall carbon nanotube transparent conductive films. *J. Mater. Sci. Technol.* **2019**, *35*, 2447–2462. [CrossRef]
67. Hennrich, F.; Wellmann, R.; Malik, S.; Lebedkin, S.; Kappes, M.M. Reversible modification of the absorption properties of single-walled carbon nanotube thin films via nitric acid exposure. *Phys. Chem. Chem. Phys.* **2003**, *5*, 178–183. [CrossRef]
68. Sumanasekera, G.U.; Allen, J.L.; Fang, S.L.; Loper, A.L.; Rao, A.M.; Eklund, P.C. Electrochemical oxidation of single wall carbon nanotube bundles in sulfuric acid. *J. Phys. Chem. B* **1999**, *103*, 4292–4297. [CrossRef]
69. Paul, S.; Kang, Y.S.; Sun, Y.K.; Kim, D.W. Highly conductive and transparent thin films fabricated with predominantly semiconducting single-walled carbon nanotubes. *Carbon* **2010**, *48*, 2646–2649. [CrossRef]
70. Earp, B.; Dunn, D.; Phillips, J.; Agrawal, R.; Ansell, T.; Aceves, P.; Luhrs, C. Enhancement of electrical conductivity of carbon nanotube sheets through copper addition using reduction expansion synthesis. *Mater. Res. Bull.* **2020**, *131*, 110969. [CrossRef]
71. Kim, K.K.; Bae, J.J.; Park, H.K.; Kim, S.M.; Geng, H.Z.; Park, K.A.; Lee, Y.H. Fermi level engineering of single-walled carbon nanotubes by $AuCl_3$ doping. *J. Am. Chem. Soc.* **2008**, *130*, 12757–12761. [CrossRef] [PubMed]
72. Coates, J. Interpretation of infrared spectra, a practical approach. *Encycl. Anal. Chem.* **2000**, *12*, 10815–10837.
73. Kamble, R.V.; Bhinge, S.D.; Mohite, S.K.; Randive, D.S.; Bhutkar, M.A. In vitro targeting and selective killing of mcf-7 and colo320dm cells by 5-fluorouracil anchored to carboxylated SWCNTs and MWCNTs. *J. Mater. Sci. Mater. Med.* **2021**, *32*, 71. [CrossRef]
74. Klein, D.R. *Organic Chemistry*; John Wiley & Sons: Hoboken, NJ, USA, 2022.
75. Poudel, Y.R.; Zhao, X.; Jungjohann, K.L.; Thapa, A.; Guo, R.; Li, W. $Ni_3S_2$ nanowires filled carbon nanotubes of ultra-high quality: Synthesis methods, structure, and electrical properties. *Diam. Relat. Mater.* **2022**, *127*, 109156. [CrossRef]
76. Youn, H.C.; Bak, S.M.; Park, S.H.; Yoon, S.B.; Roh, K.C.; Kim, K.B. One-step preparation of reduced graphene oxide/carbon nanotube hybrid thin film by electrostatic spray deposition for supercapacitor applications. *Met. Mater. Int.* **2014**, *20*, 975–981. [CrossRef]
77. Smith, B. The CO bond, Part I: Introduction and the infrared spectroscopy of alcohols. *Spectroscopy* **2017**, *32*, 14–21.

78. Mandal, S.; Mandal, S.; Ghosh, S.K.; Sar, P.; Ghosh, A.; Saha, R.; Saha, B. A review on the advancement of ether synthesis from organic solvent to water. *RSC Adv.* **2016**, *6*, 69605–69614. [CrossRef]
79. Lin, W.; Xiu, Y.; Zhu, L.; Moon, K.S.; Wong, C.P. Assembling of carbon nanotube structures by chemical anchoring for packaging applications. In Proceedings of the 2008 58th Electronic Components and Technology Conference, Lake Buena Vista, FL, USA, 27–30 May 2008; IEEE: Piscataway, NJ, USA, 2008; pp. 421–426.
80. Rahmam, S.; Mohamed, N.M.; Sufian, S. Effect of acid treatment on the multiwalled carbon nanotubes. *Mater. Res. Innov.* **2014**, *18* (Suppl. 6), S6–S196. [CrossRef]
81. De Menezes, B.R.C.; Ferreira, F.V.; Silva, B.C.; Simonetti, E.A.N.; Bastos, T.M.; Cividanes, L.S.; Thim, G.P. Effects of octadecylamine functionalization of carbon nanotubes on dispersion, polarity, and mechanical properties of CNT/HDPE nanocomposites. *J. Mater. Sci.* **2018**, *53*, 14311–14327. [CrossRef]
82. Kim, S.D.; Kim, J.W.; Im, J.S.; Kim, Y.H.; Lee, Y.S. A comparative study on properties of multi-walled carbon nanotubes (MWCNTs) modified with acids and oxyfluorination. *J. Fluor. Chem.* **2007**, *128*, 60–64. [CrossRef]
83. Wong, C.H.; Buntov, E.A.; Rychkov, V.N.; Guseva, M.B.; Zatsepin, A.F. Simulation of chemical bond distributions and phase transformation in carbon chains. *Carbon* **2017**, *114*, 106–110. [CrossRef]
84. Ţucureanu, V.; Matei, A.; Avram, A.M. FTIR spectroscopy for carbon family study. *Crit. Rev. Anal. Chem.* **2016**, *46*, 502–520. [CrossRef] [PubMed]
85. Zhang, C.; Chang, J.; Zhang, H.; Li, C.; Zhao, H. Improved direct current electrical properties of crosslinked polyethylene modified with the polar group compound. *Polymers* **2019**, *11*, 1624. [CrossRef]
86. Teyssedre, G.; Laurent, C. Charge transport modeling in insulating polymers: From molecular to macroscopic scale. *IEEE Trans. Dielectr. Electr. Insul.* **2005**, *12*, 857–875. [CrossRef]
87. Wijeyasinghe, N.; Eisner, F.; Tsetseris, L.; Lin, Y.H.; Seitkhan, A.; Li, J.; Anthopoulos, T.D. p-Doping of Copper (I) Thiocyanate (CuSCN) Hole-Transport Layers for High-Performance Transistors and Organic Solar Cells. *Adv. Funct. Mater.* **2018**, *28*, 1802055. [CrossRef]
88. Wijeyasinghe, N.; Anthopoulos, T.D. Copper (I) thiocyanate (CuSCN) as a hole-transport material for large-area opto/electronics. *Semicond. Sci. Technol.* **2015**, *30*, 104002. [CrossRef]

nanomaterials

MDPI

*Article*

# Actuator-Driven, Purge-Free Formaldehyde Gas Sensor Based on Single-Walled Carbon Nanotubes

**Shinsuke Ishihara [1,*], Mandeep K. Chahal [1,2], Jan Labuta [1,3], Takeshi Tanaka [4], Hiromichi Kataura [5], Jonathan P. Hill [1] and Takashi Nakanishi [1]**

1 Research Center for Materials Nanoarchitectonics (MANA), National Institute for Materials Science (NIMS), 1-1 Namiki, Tsukuba 305-0044, Ibaraki, Japan
2 School of Chemistry and Forensic Science, University of Kent, Canterbury CT2 7NH, UK
3 NMR Spectroscopy Group, Institute of Organic Chemistry and Biochemistry (IOCB), Czech Academy of Sciences (CAS), Flemingovo nám. 2, 160 00 Prague, Czech Republic
4 Research Institute of Core Technology for Materials Innovation, National Institute of Advanced Industrial Science and Technology (AIST), Tsukuba 305-8565, Ibaraki, Japan
5 Nanomaterials Research Institute, National Institute of Advanced Industrial Science and Technology (AIST), Tsukuba 305-8565, Ibaraki, Japan
* Correspondence: ishihara.shinsuke@nims.go.jp

**Abstract**

Formaldehyde vapor (HCHO) is a harmful chemical substance and a potential air contaminant, with a permissible level in indoor spaces below 0.08 ppm (80 ppb). Thus, highly sensitive gas sensors for the continuous monitoring of HCHO are in demand. The electrical conductivity of semiconducting nanomaterials (e.g., single-walled carbon nanotubes (SWCNTs)) makes them sensitive to chemical substances adsorbed on their surfaces, and a variety of portable and highly sensitive chemiresistive gas sensors, including those capable of detecting HCHO, have been developed. However, when monitoring low levels of vapors (<1 ppm) found in ambient air, most chemiresistive sensors face practical issues, including false responses to interfering effects (e.g., fluctuations in room temperature and humidity), baseline drift, and the need to apply a purge gas. Here, we report an actuator-driven, purge-free chemiresistive gas sensor that is capable of reliably detecting 0.05 ppm of HCHO in the air. This sensor is composed of an HCHO→HCl converter (powdery hydroxylamine salt, HA), an HCl detector (a SWCNT-based chemiresistor), and an HCl blocker (a thin plastic plate). Upon exposure to HCHO, the HA emits HCl vapor, which diffuses onto the adjacent SWCNTs, increasing their electrical conductivity through *p*-doping. Meanwhile, inserting a plastic plate between HA and SWCNTs makes the conductivity of SWCNTs insensitive to HCHO. Thus, via periodic actuation (insertion and removal) of the plastic plate, HCHO can be detected reliably over a wide concentration range (0.05–15 ppm) with excellent selectivity over other volatile organic compounds. This actuator-driven system is beneficial because it does not require a purge gas for sensor recovery or baseline correction. Moreover, since the response to HCHO is synchronized with the actuation timing of the plate, even small (~0.8%) responses to 0.05 ppm of HCHO can be clearly separated from larger noise responses (>1%) caused by interfering effects and baseline drift. We believe that this work provides substantial insights into the practical implementation of nanomaterial-based chemiresistive gas sensors.

**Keywords:** gas sensor; semiconductor; chemiresistor; formaldehyde; carbon nanotube

## 1. Introduction

Formaldehyde vapor (HCHO) is a harmful pollutant that continuously evaporates from construction materials (e.g., adhesives, resins, paints, and preservatives) in the interiors of buildings and constitutes a health hazard leading to severe conditions, including dermatitis, asthma, and cancer [1]. For this reason, the World Health Organization (WHO) has set the permissible level of HCHO indoors below 0.08 ppm (80 ppb) [2]. Although conventional analytical methods such as hydrazine-labeled liquid chromatography and detector tubes can be used for the quantification of HCHO [3,4], its continuous monitoring is made inconvenient by requirements including the frequent exchange of consumable items, individual manual sampling, or cumbersome instruments. Thus, a demand for advanced gas sensors with excellent sensitivity, selectivity, portability, cost-effectiveness, low power consumption, and continuous operability has emerged [5–8].

Of the various potential detection mechanisms, chemiresistive gas sensors that monitor the variation in electrical conductivity in semiconducting materials are ideal for integration into portable devices (e.g., smartphones and IoT devices) due to their simple configurations [9–13]. The electrical conductivity of semiconducting nanomaterials (e.g., sorted single-walled carbon nanotubes (SWCNTs) [14,15]) can be affected by chemical substances adsorbed onto their surfaces, and highly sensitive and selective chemiresistive gas sensors (including for HCHO sensing [16–24]) have been developed by using post-synthesis chemical functionalization [25–30]. However, when monitoring very low levels of trace gases (<1 ppm) in ambient air, most chemiresistive sensors face practical issues, including false responses to sensing conditions (e.g., fluctuations in room temperature and humidity), baseline drift during long-term operation [31], or the requirement for a purge gas for sensor recovery and baseline correction. These issues arise largely a consequence of a material's weak sensing response to low concentrations of gas-phase analytes, making it difficult to establish an applicable signal-to-noise ratio.

We recently reported an HCHO sensor comprising a non-deliquescent hydroxylamine salt sandwiched between membrane filters (HA) and an SWCNT network embedded in a comb-shaped gold electrode (SWCNT-based chemiresistor) [32]. HA and SWCNTs are spatially isolated (ca. 0.4 mm), and SWCNT-based chemiresistors monitor the diffusion of HCl vapor resulting from the condensation of HCHO and HA. Since HA is present in excess compared to the ppm-level HCHO analyte, the sensor can be used to continuously and repetitively monitor HCHO with confirmed operational stability for 6 months. However, the sensor continuously reacts with HCHO, so purging air (i.e., with HCHO-free air) is required for baseline correction and recovery. Although high-pressure gas cylinders or purified air can be used for purging, this requirement limits continuous operation and negatively affects the dimensions of the sensor system as a whole.

In this work, we demonstrate an actuator-driven, purge-free chemiresistive HCHO sensor realized by the introduction of a thin plastic plate into our abovementioned sensor [32] (Figure 1a). When a plastic plate is inserted between HA and SWCNT, the SWCNT-based chemiresistor is insensitive to HCHO since HCl vapor is no longer able to access the chemiresistive region, thus making sensor recovery and baseline setting possible even in the presence of HCHO in the surrounding air. Moreover, since the response to HCHO is synchronized with the actuation timing of the plate's removal, small responses (~0.8%) to 0.05 ppm of HCHO can be separated from larger noise responses (>1%) caused by interfering effects and baseline drift. Thus, through periodic actuation (insertion and removal) of the plate, HCHO can be reliably detected over a wide concentration range (0.05–15 ppm), with excellent selectivity over other volatile organic compounds (VOCs). As a general concept, an actuator-driven chemiresistive sensor consists of a selector component (in this case, HA), a detector component (in this case, a SWCNT-based chemiresistor), and an actuator component (in this case, the plastic plate). The selector and detector components are spatially isolated. The actuator enables temporary cessation of selector–detector interactions, so sensor recovery and baseline corrections are possible without the need to apply a purge gas. We believe that this work provides substantial insights into the practical implementation of nanomaterial-based chemiresistive gas sensors, particularly for detecting low-level trace gases in real-world environments.

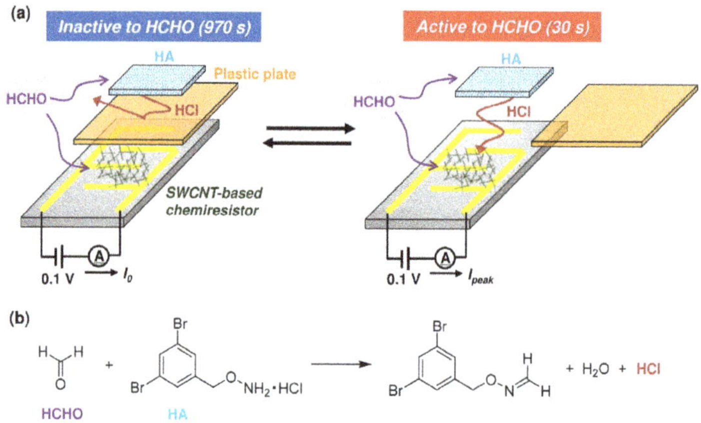


**Figure 1.** Actuator-driven, purge-free chemiresistive HCHO sensor used in this work. (**a**) HCHO sensor composed of a powdered hydroxylamine salt sandwiched between hydrophobic polytetrafluoroethylene (PTFE) membrane filters (HA), an SWCNT network embedded on a comb-shaped gold electrode (SWCNT-based chemiresistor), and a thin plastic plate. The HA and SWCNT-based chemiresistor are spatially isolated, and the intervening plastic plate is periodically removed to monitor HCHO in the air. Note that the sensor was only exposed to analyte gases without switching to purge gas. (**b**) Condensation reaction between HCHO and HA, generating HCl vapor.

## 2. Experimental Section

### 2.1. Materials

Chemicals and solvents were purchased from commercial sources and used as received unless otherwise stated. Purified water (18.2 MΩ cm) was obtained from a Milli-Q system (Direct-Q 3 UV, Merck (Tokyo, Japan)). Formalin solution (37 wt%) and 2-butanone were obtained from TCI Chemical Co., Ltd. (Tokyo, Japan). Acetaldehyde was obtained as a permeation tube (P-92-1) from GASTEC (Kanagawa, Japan). Methanol, ethanol, tetrahydrofuran (THF), ethyl acetate, toluene, and hexane were obtained from Fujifilm Wako Pure Chemicals Co., Ltd. (Osaka, Japan). Acetone and chloroform-d were obtained from Kanto Chemical Co., Ltd. (Tokyo, Japan).

### 2.2. Preparation of the HCHO Sensor

SWCNT network chemiresistor was prepared by using semiconducting SWCNTs (sorted using the column chromatography technique) (Figure S4) [33], a dispersing agent based on metallo-supramolecular polymer (Figure S5) [34,35], and interdigitated gold electrode with a gap of 200 μm (CC1.W1, BVT Technologies (Strážek, Czech Republic)) (see our previous report for details [32]). As the hydroxylamine salt (HA), we employed *o*-3,5-dibromobenzylhydroxylamine hydrochloride (see Figure 1b) due to its excellent correlation between sensing response and HCHO concentration as well as its confirmed stability for 6 months [32]. HA solid was ground using a mortar and pestle, and the powdered material (ca. 1 mg) was sandwiched between hydrophobic PTFE membrane filters (pore size: 3 μm, T300A047A, Advantec (Tokyo, Japan)). The HA patch was affixed to the SWCNT-based chemiresistor. A thin plastic plate (polypropylene, thickness = 0.20 mm) was inserted between the HA patch and SWCNT-based chemiresistor to block HCl vapor generated by the condensation reaction between HCHO and HA.

### 2.3. Generation of HCHO and Acetaldehyde Vapors

HCHO vapor was generated according to our previously reported methods [22,32]. Briefly, formalin solution (37%) was diluted with water; then, formalin solution (1 mL; individually for 37%, 7.4%, 1.85%, 0.46%, 0.123%, and 0% (wt%)), in a glass test tube, was placed in a two-necked flask (200 mL) maintained at 40 ± 1 °C. Relative humidity (RH) was controlled by blending dry and water-bubbled air by using a dry-air generator (P4-EFA, IAC (Kanagawa, Japan)), mass-flow controllers (CUBE MFC-1005-4S2-2L-AIR, FCON (Kochi, Japan)), and a humidity sensor (HMI41, VAISALA (Tokyo, Japan)). Humidity-adjusted air (50% RH, 0.3 L/min) was continuously injected into the flask as a carrier gas. The HCHO concentration was determined using a formaldehyde analyzer (FP-31, RIKEN KEIKI (Tokyo, Japan)). The HCHO concentrations obtained using this method were 14.6, 3.8, 1.2, 0.27, 0.05, and 0 ppm for each formalin solution. Acetaldehyde vapor was generated as follows. Permeation tubes (GASTEC, P-92-1) containing neat acetaldehyde liquid were placed in a two-necked flask (200 mL) maintained at room temperature (20 ± 2 °C). Humidity-adjusted air (50% RH, 0.3 L/min) was injected into the flask as a carrier gas. Acetaldehyde concentration was determined by using a detector tube (GASTEC, 92L/92M). A steady concentration of 2 ppm of acetaldehyde vapor was obtained using this method. The 2 ppm acetaldehyde vapor was diluted with humidity-adjusted air to control its concentration.

### 2.4. Generation of VOC Vapors

When generating VOC vapors, the same apparatus we used for HCHO vapor generation (diffusion tube method) was used, but this time, we used neat VOC liquid (1 mL) in a glass test tube located in the flask (except for acetone and 2-butanone). At constant flow rate and temperature (typically 0.3 L/min and 40 °C), the mass of analyte liquid in the test tube was measured, typically at 1, 2, and 3 h, and the average evaporation rate of the analyte was calibrated based on the decrease in mass. When the vapor pressure of the analyte was low, the measurement interval of mass was extended (e.g., 24 h). When detector tubes (methanol, ethanol, acetone, and 2-butanone) were available, the concentration of analyte vapors was measured to confirm that mass changes were consistent. In the cases of acetone (10, 50 ppm) and 2-butanone (8.4, 42 ppm), neat liquids were diluted with water 150 or 30 times. VOC vapors in air (50% RH) were continuously injected into the sensor chamber.

### 2.5. Evaluation of Sensors

The SWCNT-based chemiresistor was connected to a potentiostat (EmStat, PalmSens (Houten, The Netherlands)), and the electric current under constant voltage (0.10 V) was recorded using PSTrace software (v. 4.8). This low operation voltage (0.10 V) is energetically economical. Analyte gas (20 ± 2 °C, 50% RH, 0.3 L/min) was continuously injected into the sensor chamber. In sensing experiments, the thin plastic plate between the HA and SWCNT-based chemiresistor was periodically actuated by using a programmable syringe pump (YSP-201, YMC Co., Ltd. (Kyoto, Japan)) (see Figures 1a and S1 for details). Typically, the plastic plate was removed for 30 s every 1000 s.

## 3. Results and Discussion

Under continuous exposure to HCHO in air (50% RH), the thin plastic plate located between the HA and SWCNT-based chemiresistor was periodically removed for 30 s every 1000 s (Figures 1a and S1). As the HCHO concentration increases, the electric current at the SWCNT-based chemiresistor also increases due to *p*-doping with HCl (Figure 2a–f) [36–38].

Insertion of the plastic plate leads to the recovery of the original electric current even in the presence of HCHO because of the desorption of HCl from the SWCNT surface. Thus, regeneration of the sensor is possible without using a purge (i.e., HCHO-free) gas. The normalized response (%) can be calculated from Equation (1).

$$\text{Normalized response (\%)} = (I_{\text{peak}} - I_0)/I_0 \times 100, \quad (1)$$

where $I_0$ denotes the electric current before the removal of the plate, and $I_{\text{peak}}$ denotes the peak electric current after the removal of the plate (see Figure 2f for details). Generally, $I_0$ refers to the baseline electric current obtained under purge air conditions (i.e., in the absence of the target analyte). However, our actuator-driven system offers $I_0$ values even in the presence of HCHO in the surrounding air. Thus, the normalized response and the concentration of HCHO correlate well, allowing a calibration curve for the quantification of HCHO concentration from the sensing response to be constructed (Figures 2g and S2).

The limit of detection (LoD) of our sensor is estimated to be 0.020 ppm based on the calibration curve in the minimum concentration range (Figure 2h) and on Equation (2) [39].

$$\text{LoD} = \text{mean}_{\text{blank}} + 1.645 \times \sigma_{\text{blank}} + 1.645 \times \sigma_{\text{lowest conc.}}, \quad (2)$$

where $\text{mean}_{\text{blank}}$, $\sigma_{\text{blank}}$, and $\sigma_{\text{lowest conc.}}$ represent the mean response for the blank, the standard deviation of the response for the blank, and the lowest gas concentration (in this case, HCHO = 0.05 ppm), respectively. The LoD of the response is estimated to be 0.46%, which corresponds to 0.020 ppm of HCHO based on the initial response ($y = 9.56\,x - 0.276$).

We also tested a longer removal time (180 s) of the plastic plate for 0.05 ppm of HCHO, although the signal ratio of 0.05 ppm of HCHO and blank air did not improve (Figure S3). Longer removal of the plate probably leads to accumulation of the sensing response (cumulative adsorption of HCl on the SWCNT surface) and incomplete recovery (desorption of HCl from the SWCNT surface) within a reasonable period of time. Our previous study showed that it takes ca. 12 h for complete recovery following high-level HCHO exposure [22]. In addition, the risk of environmental fluctuations (e.g., variation in temperature and RH) during the absence of the plate is increased. Therefore, a short removal time is preferred for our actuator-driven system (as far as the sensing response for HCHO is concerned).

There is a subtle response to blank air (0 ppm), with the possible origins of this being (i) contamination of trace HCHO in the blank air (i.e., humidity-adjusted ambient air) or (ii) proton exchange equilibrium between HA and acidic gases in air ($NO_2$, $SO_2$, $CO_2$, etc.) leading to evaporation of trace HCl vapor. Further investigations are necessary to elucidate the blank response mechanism.

Figure 3 indicates that our sensor exhibits excellent selectivity for HCHO, with negligible responses to non-carbonyl VOC vapors (methanol, ethanol, ethyl acetate, tetrahydrofuran, chloroform, toluene, and hexane at 100–1000 ppm). These compounds are unreactive with HA, so the SWCNT resistor was not doped with HCl vapor.

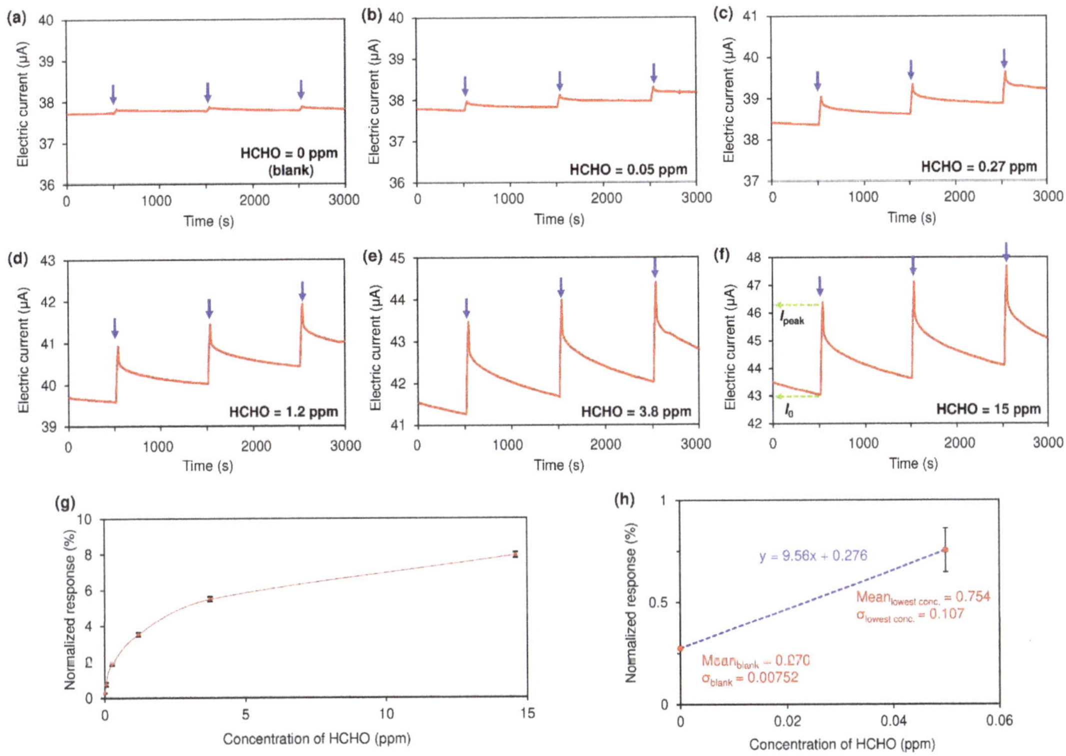


**Figure 2.** Sensing responses for HCHO at different concentrations: (**a**) blank air (50% RH), (**b**) 0.05 ppm, (**c**) 0.27 ppm, (**d**) 1.2 ppm, (**e**) 3.8 ppm, and (**f**) 15 ppm. Temporary removal (30 s) of the thin plastic plate was repeated every 1000 s (three times per run as indicated by the blue arrows). (**g**) Calibration curve: HCHO concentration versus normalized sensing response (mean of three repeated measurements with standard deviations). See Figure S2 and the corresponding test in SI for details on the calibration curve's construction. (**h**) Normalized responses for blank air and 0.05 ppm of HCHO used for LoD estimation.

In contrast, carbonyl-containing VOC vapors such as acetaldehyde, acetone, and 2-butanone are reactive with HA, and our sensor exhibits responses when concentrations of these vapors are high (Figure 4a–g). However, the responses to 0.088 ppm acetaldehyde and 10 ppm acetone are close to those for blank air and appreciably smaller than those for 0.05 ppm of HCHO. The responses to 0.05 ppm of HCHO and 0.67 ppm acetaldehyde are similar, so the selectivity to HCHO over acetaldehyde is approximately 10-fold greater. Moreover, the responses to 0.05 ppm of HCHO and 42–50 ppm ketones (acetone and 2-butanone) are similar, so the selectivity for HCHO over ketones is ca. $10^3$-fold greater. The order of selectivity (HCHO > acetaldehyde >> ketones) is in good agreement with the electrophilic reactivity of aldehydes and ketones, which is weakened by the presence of $\alpha$-alkyl groups due to steric and electron-donating effects at the carbonyl group [40].

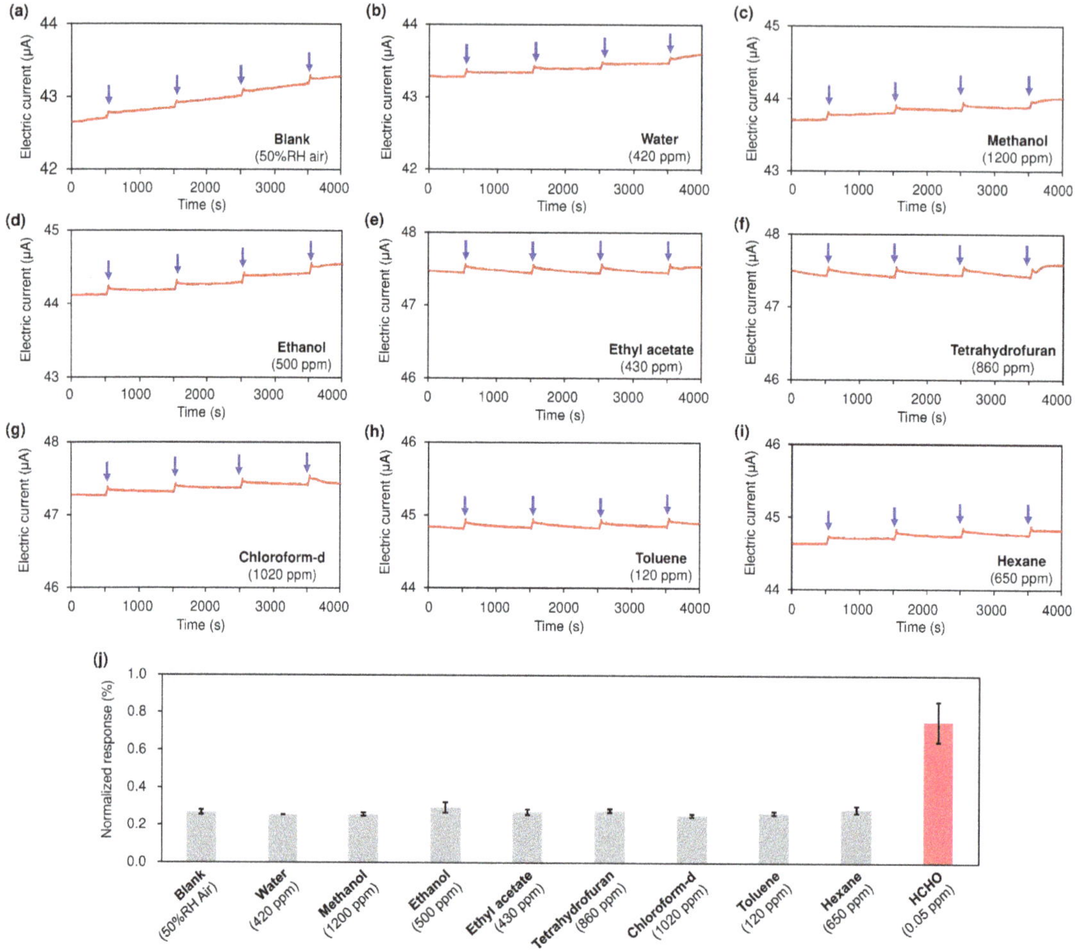

**Figure 3.** Sensing responses for non-carbonyl VOC vapors: (**a**) blank air (50% RH), (**b**) water, (**c**) methanol, (**d**) ethanol, (**e**) ethyl acetate, (**f**) tetrahydrofuran, (**g**) chloroform-d, (**h**) toluene, and (**i**) hexane. Temporary removal (30 s) of the thin plastic plate was repeated every 1000 s (four times per run as indicated by the blue arrows). Values (in ppm) quoted at the bottom right in each panel indicate the concentration of water or each VOC added into the airflow (50% RH). (**j**) Normalized sensing responses for each vapor. The average of four repeated measurements, along with their standard deviation, is shown. Since conventional chloroform ($CHCl_3$) contains ethanol as a stabilizer, chloroform-d ($CDCl_3$, NMR grade) was used in this study.

In our actuator-driven sensor, a sensing response to HCHO should occur only when the plastic plate is intentionally removed at a known time, and the actual response to HCHO can be easily distinguished from interfering effects and baseline drift. Figure 5 shows the long-term monitoring of HCHO in air. In blank air (green region; 50% RH air), baseline drift was observed, presumably due to subtle changes in ambient temperature, but sensing responses upon periodic removal of the plastic plate (30 s removal every 1000 s) were negligibly low, indicating the absence of HCHO. When trace amounts of HCHO (0.05 ppm) were present, the actuator-driven sensor exhibited periodic spike signals synchronizing with the timing of plate removal. Even after 60 repeats, our sensor demonstrated stable performance because the amount of solid HA was much greater relative to HCHO. As soon as HCHO was cleared, the spike signals disappeared. Thus, even small responses (~0.8%) to 0.05 ppm of HCHO can be clearly isolated from larger noise responses (>1%) caused by interfering effects and unpredictable baseline drift.

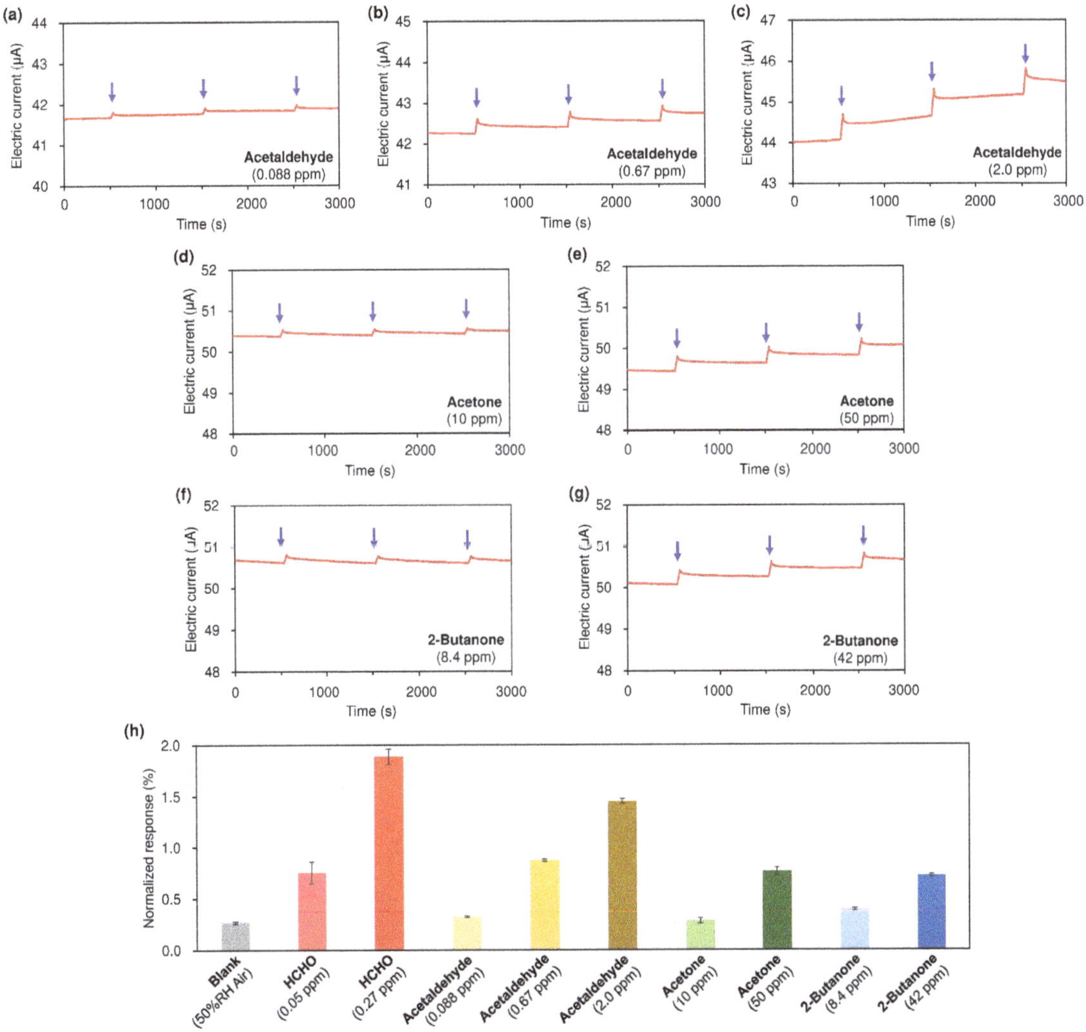


**Figure 4.** Sensing responses for carbonyl-containing VOC vapors. (**a**–**c**) acetaldehyde, (**d**,**e**) acetone, and (**f**,**g**) 2-butanone. (**h**) Normalized sensing responses for carbonyl vapors. The average of three repeated measurements, along with their standard deviation, is shown.

It should be noted that our sensing mechanism is based on the condensation reaction between HCHO and HA and the vaporization of HCl (Figure 1), whose reaction kinetics and equilibria should be affected by the temperature and RH of the environment [22]. Thus, in practical situations, these parameters should be separately monitored for proper quantification of HCHO.

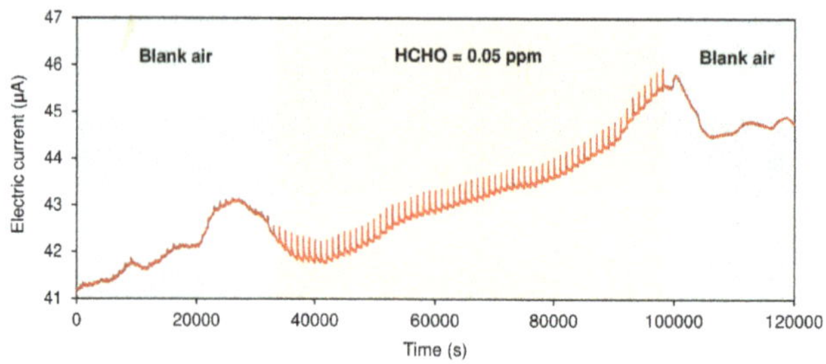


**Figure 5.** Continuous monitoring of HCHO in the air. Temporal removal (30 s) of the thin plastic plate was repeated every 1000 s in the presence and absence of HCHO.

## 4. Conclusions

In conclusion, our actuator-driven, purge-free chemiresistive sensor enables the continuous monitoring of sub-ppm level HCHO, with a clear separation of noise signals due to interfering effects and baseline drift. The sensing response to HCHO is based on the nucleophilic condensation reactivity of HA to carbonyl compounds and shows excellent selectivity to HCHO over other VOC vapors. With periodic actuation (pushing/pulling motion) of the plate, sensor recovery and baseline correction are possible without requiring purging with clean air. This feature is advantageous for the miniaturization of our sensory system. This HCHO sensor is unique in that a selector component (in this case, HA) and a detector component (in this case, an SWCNT-based chemiresistor) are spatially isolated, which enables temporary cessation of selector–detector interactions via the physical insertion of the plate between the two components. Most chemiresistive sensors comprise an inseparable mixture of selector and detector components, allowing the sensor to be persistently sensitive to the target analyte. A possible solution for these sensors might involve the introduction of a filter or adsorbent (selective to the target analyte) to render the sensor temporarily insensitive to the target, leading to the construction of a different type of actuator-driven, purge-free chemiresistive gas sensor. Thus, the concept of this work can be applied generally to address some of the fundamental issues in the practical implementation of nanomaterial-based chemiresistive gas sensors. We believe this work provides a practical demonstration of the implementation of nanomaterial-based chemiresistive gas sensors, particularly useful for targeting low levels of analytes present in real environments.

**Supplementary Materials:** The following supporting information can be downloaded at https://www.mdpi.com/article/10.3390/nano15130962/s1. Figure S1: (a) Photograph of HA patch (left) and SWCNT-based chemiresistive sensor (right). (b) Experimental set-up used for periodic actuation of the thin plastic plate. (c) Actuation of the thin plastic plate. Left: The thin plastic plate is inserted between HA and SWCNT-based chemiresistor, making the system insensitive to HCHO. Right: The thin plastic plate is removed from between the HA and SWCNT-based chemiresistor, making the system sensitive to HCHO. Typically, the plate insertion time is 970 s, and the plate removal time is 30 s. (d) Actuation of the thin plastic plate when HA patch is absent (provided for clarity). Left: SWCNT is covered by the plate. Right: SWCNT is not covered by the plate.; Figure S2: Fitted $y = F(x)$ and inverse $x = G(y)$ calibration curves. (a) HCHO concentration ($x$) to sensor response ($y$) calibration curve $y = F(x)$. (b) Sensor response ($y$) to HCHO concentration ($x$) calibration curve $x = G(y)$. Note that the concentration range (in ppm) for the $F(x)$ calibration curve is from 0 to 15 ppm of HCHO, and the response range (in %) for the $G(y)$ calibration curve is from 0 to 8%.; Figure S3: (a) Sensing responses for blank and 0.05 ppm HCHO in air (50% RH). Temporary removal (180 s) of the thin plastic plate was repeated every 1000 s (three times per run as indicated by the blue arrows). (c) Normalized sensing responses for blank and 0.05 ppm HCHO upon 30 s and 180 s removal of the plate. The average of three repeated measurements, along with its standard deviation, is shown.; Figure S4: Characterization of the SWCNT used. (a) Photograph of interdigitated electrode comprised of ceramic substrate and gold lines with a gap distance of 200 μm. (b) Photograph of the interdigitated electrode with and without SWCNT. (c–f) SEM image of interdigitated electrode embedded with SWCNT. (g) Scanning probe microscopy images of SWCNT spin-coated (3000 rpm) on flat $SiO_2$/Si substrate. (h) I-V curve (0 V → 3 V → −3V → 0 V) for interdigitated electrode embedded with SWCNT (scan rate = 0.1 V $s^{-1}$) measured under ambient air. (i) Raman spectra of SWCNTs (metallic and semiconducting) separated by column chromatography. RBM and BWF denote radial breathing mode and Breit-Wigner-Fano band, respectively. Reprinted with permission from reference [28]. Copyright 2020 American Chemical Society; Figure S5: The metallo-supramolecular polymer used as a dispersing agent of SWCNT.

**Author Contributions:** Conceptualization, S.I.; synthesis of HA, M.K.C.; purification of SWCNTs, T.T. and H.K.; sensing experiments, S.I. and M.K.C.; formal analysis, S.I. and J.L.; writing—original draft preparation, S.I.; writing—review and editing, M.K.C., J.L., J.P.H. and T.N. All authors have read and agreed to the published version of the manuscript.

**Funding:** This work was supported by JSPS KAKENHI (JP18H02016, JP24K01282).

**Data Availability Statement:** The data presented in this study are available upon reasonable request from the corresponding author.

**Acknowledgments:** This work was supported by the World Premier International Research Center Initiative (WPI), MEXT, Japan. Shihori Shimizu and Hiroyasu Ichikawa are acknowledged for their assistance during this work.

**Conflicts of Interest:** The authors declare no conflicts of interest.

## References

1. Salthammer, T.; Mentese, S.; Marutzky, R. Formaldehyde in the Indoor Environment. *Chem. Rev.* **2010**, *100*, 2536–2572. [CrossRef]
2. WHO Regional Office for Europe. *Air Quality Guidelines for Europe*, 2nd ed.; World Health Organization, Regional Office for Europe: Copenhagen, Denmark, 2000.
3. Formaldehyde (Diffusive Samplers). Available online: https://www.osha.gov/sites/default/files/methods/osha-1007.pdf (accessed on 11 April 2025).
4. Formaldehyde HCHO. Available online: https://www.gastec.co.jp/files/user/asset/pdf/en/detector_tube/91LL.pdf (accessed on 11 April 2025).
5. Chung, P.-R.; Tzeng, C.-T.; Ke, M.-T.; Lee, C.-Y. Formaldehyde Gas Sensors: A Review. *Sensors* **2013**, *13*, 4468–4484. [CrossRef] [PubMed]
6. Chen, D.; Yuan, Y.J. Thin-Film Sensors for Detection of Formaldehyde: A Review. *IEEE Sens. J.* **2015**, *15*, 6749–6760. [CrossRef]
7. Kukkar, D.; Vellingiri, K.; Kaur, R.; Bhardwaj, S.K.; Deep, A.; Kim, K.-H. Nanomaterials for Sensing of Formaldehyde in Air: Principles, Applications, and Performance Evaluation. *Nano Res.* **2019**, *12*, 225–246. [CrossRef]
8. Roy, S.; Pan, S.; De, P. Recent Progress on Polymeric Probes for Formaldehyde Sensing: A Comprehensive Review. *Sci. Technol. Adv. Mater.* **2024**, *25*, 2423597. [CrossRef]
9. Neri, G. First Fifty Years of Chemoresistive Gas Sensors. *Chemosensors* **2015**, *3*, 1–20. [CrossRef]
10. Kaushik, A.; Kumar, R.; Arya, S.K.; Nair, M.; Malhotra, B.D. Organic-Inorganic Hybrid Nanocomposite-Based Gas Sensors for Environmental Monitoring. *Chem. Rev.* **2015**, *115*, 4571–4606. [CrossRef]
11. Bandodkar, A.J.; Jeerapan, I.; Wang, J. Wearable Chemical Sensors: Present Challenges and Future Prospects. *ACS Sens.* **2016**, *1*, 464–482. [CrossRef]
12. Lim, H.-R.; Kim, H.S.; Qazi, R.; Kwon, Y.-T.; Jeong, J.-W.; Yeo, W.-H. Advanced Soft Materials, Sensor Integrations, and Applications of Wearable Flexible Hybrid Electronics in Healthcare, Energy, and Environment. *Adv. Mater.* **2020**, *32*, 190192. [CrossRef]
13. Ates, H.C.; Nguyen, P.Q.; Gonzalez-Macia, L.; Morales-Narváez, E.; Güder, F.; Collins, J.J.; Dincer, C. End-to-End Design of Wearable Sensors. *Nat. Rev. Mater.* **2022**, *7*, 887–907. [CrossRef]
14. Bati, A.S.R.; Yu, L.; Batmunkh, M.; Shapter, J.G. Recent Advances in Applications of Sorted Single-Walled Carbon Nanotubes. *Adv. Funct. Mater.* **2019**, *29*, 1902273. [CrossRef]
15. Tang, D.-M.; Cretu, O.; Ishihara, S.; Zheng, Y.; Otsuka, K.; Xiang, R.; Maruyama, S.; Cheng, H.-M.; Liu, C.; Golberg, D. Chirality Engineering for Carbon Nanotube Electronics. *Nat. Rev. Electr. Eng.* **2024**, *1*, 149–162. [CrossRef]

16. Itoh, T.; Matsubara, I.; Shin, W.; Izu, N.; Nishibori, M. Preparation of Layered Organic-Inorganic Nanohybrid Thin Films of Molybdenum Trioxide with Polyaniline Derivatives for Aldehyde Gases Sensors of Several Tens ppb Level. *Sens. Actuators B* **2008**, *128*, 512–520. [CrossRef]
17. Zhang, L.; Zhao, J.; Lu, H.; Gong, L.; Li, L. High Sensitive and Selective Formaldehyde Sensors Based on Nanoparticle-assembled ZnO Micro-Octahedrons Synthesized by Homogeneous Precipitation Method. *Sens. Actuators B* **2011**, *160*, 364–370. [CrossRef]
18. Lu, Y.; Meyyappan, M.; Li, J. A Carbon-Nanotube-Based Sensor Array for Formaldehyde Detection. *Nanotechnology* **2011**, *22*, 055502. [CrossRef]
19. Kim, J.Y.; Lee, J.; Hong, S.; Chung, T.D. Formaldehyde Gas Sensing Chip Based on Single-Walled Carbon Nanotubes and Thin Water Layer. *Chem. Commun.* **2011**, *47*, 2892–2894. [CrossRef]
20. Xie, H.; Sheng, C.; Chen, X.; Wang, X.; Li, Z.; Zhou, J. Multi-Wall Carbon Nanotube Gas Sensors Modified with Amino-Group to Detect Low Concentration of Formaldehyde. *Sens. Actuator B-Chem.* **2012**, *168*, 34–38. [CrossRef]
21. Shi, D.; Wei, L.; Wang, J.; Zhao, J.; Chen, C.; Xu, D.; Geng, H.; Zhang, Y. Solid Organic Acid Tetrafluorohydroquinone Functionalized Signle-Walled Carbon Nanotube Chemiresistive Sensors for Highly Sensitive and Selective Formaldehyde Detection. *Sens. Actuator B Chem.* **2013**, *177*, 370–375. [CrossRef]
22. Ishihara, S.; Labuta, J.; Nakanishi, T.; Tanaka, T.; Kataura, H. Amperometric Detection of Sub-ppm Formaldehyde Using SWCNTs and Hydroxylamines: A Referenced Chemiresistive System. *ACS Sens.* **2017**, *2*, 1405–1409. [CrossRef]
23. Gu, F.; Di, M.; Han, D.; Hong, S.; Wang, Z. Atomically Dispersed Au on $In_2O_3$ Nanosheets for Highly Sensitive and Selective Detection of Formaldehyde. *ACS Sens.* **2020**, *5*, 2611–2619. [CrossRef]
24. Deng, Z.; Zhang, Y.; Xu, D.; Zi, B.; Zeng, J.; Lu, Q.; Xiong, K.; Zhang, J.; Zhao, J.; Liu, Q. Ultrasensitive Formaldehyde Sensor Based on $SnO_2$ with Rich Adsorbed Oxygen Derived from a Metal Organic Framework. *ACS Sens.* **2022**, *7*, 2577–2588. [CrossRef] [PubMed]
25. Snow, E.S.; Perkins, F.K.; Robinson, J.A. Chemical Vapor Detection Using Single-Walled Carbon Nanotubes. *Chem. Soc. Rev.* **2006**, *35*, 790–798. [CrossRef] [PubMed]
26. Kauffman, D.R.; Star, A. Carbon Nanotube Gas and Vapor Sensors. *Angew. Chem. Int. Ed.* **2008**, *47*, 6550–6570. [CrossRef]
27. Schroeder, V.; Savagatrup, S.; He, M.; Lin, S.; Swager, T.M. Carbon Nanotube Chemical Sensors. *Chem. Rev.* **2019**, *119*, 599–663. [CrossRef]
28. Ishihara, S.; Bahuguna, A.; Kumar, S.; Krishnan, V.; Labuta, J.; Nakanishi, T.; Tanaka, T.; Kataura, H.; Kon, Y.; Hong, D. Cascade Reaction-Based Chemiresistive Array for Ethylene Sensing. *ACS Sens.* **2020**, *5*, 1405–1410. [CrossRef]
29. Luo, S.-X.L.; Swager, T.M. Chemiresistive Sensing with Functionalized Carbon Nanotubes. *Nat. Rev. Methods Primers* **2023**, *3*, 73. [CrossRef]
30. Luo, K.; Peng, H.; Zhang, B.; Chen, L.; Zhang, P.; Peng, Z.; Fu, X. Advances in Carbon Nanotube-Based Gas Sensors: Exploring the Path to the Future. *Coord. Chem. Rev.* **2024**, *518*, 216049. [CrossRef]
31. Fennell, J.F.; Liu, S.F.; Azzarelli, J.M.; Weis, J.G.; Rochat, S.; Mirica, K.A.; Ravnsbæk, J.B.; Swager, T.M. Nanowire Chemical/Biological Sensors: Status and a Roadmap for the Future. *Angew. Chem. Int. Ed.* **2016**, *55*, 1266–1281. [CrossRef]
32. Chahal, M.K.; Sumita, M.; Labuta, J.; Payne, D.T.; Hill, J.P.; Yamauchi, Y.; Nakanishi, T.; Tanaka, T.; Kataura, H.; Koga, K.; et al. Selective Detection of Toxic C1 Chemicals Using a Hydroxylamine-Based Chemiresistive Sensor Array. *ACS Sens.* **2023**, *8*, 1585–1592. [CrossRef]
33. Yomogida, Y.; Tanaka, T.; Zhang, M.; Yudasaka, M.; Wei, X.; Kataura, H. Industrial-Scale Separation of High-Purity Single-Chirality Single-Wall Carbon Nanotubes for Biological Imaging. *Nat. Commun.* **2016**, *7*, 12056. [CrossRef]
34. Ishihara, S.; Azzarelli, J.M.; Krikorian, M.; Swager, T.M. Ultratrace Detection of Toxic Chemicals: Triggered Disassembly of Supramolecular Nanotube Wrappers. *J. Am. Chem. Soc.* **2016**, *138*, 8221–8227. [CrossRef] [PubMed]
35. Ishihara, S.; O'Kelly, C.J.; Tanaka, T.; Kataura, H.; Labuta, J.; Shingaya, Y.; Nakayama, T.; Ohsawa, T.; Nakanishi, T.; Swager, T.M. Metallic Versus Semiconducting SWCNT Chemiresistors: A Case for Separated SWCNTs Wrapped by a Metallosupramolecular Polymer. *ACS Appl. Mater. Interface* **2017**, *9*, 38062–38067. [CrossRef] [PubMed]
36. Graupner, R.; Abraham, J.; Vencelová, A.; Seyller, T.; Hennrich, F.; Kappes, M.M.; Hirsch, A.; Ley, L. Doping of Single-Walled Carbon Nanotube Bundles by Brønsted Acids. *Phys. Chem. Chem. Phys.* **2003**, *5*, 5472–5476. [CrossRef]
37. Zhang, Z.-B.; Li, J.; Cabezas, A.L.; Zhang, S.-L. Characterization of Acid-Treated Carbon Nanotube Thin Films by Means of Raman Spectroscopy and Field-Effect Response. *Chem. Phys. Lett.* **2009**, *476*, 258–261. [CrossRef]
38. Tey, J.N.; Ho, X.; Wei, J. Effect of Doping on Single-Walled Carbon Nanotubes Network of Different Metallicity. *Nanoscale Res. Lett.* **2012**, *7*, 548. [CrossRef]
39. Armbruster, D.A.; Pry, T. Limit of Blank, Limit of Detection and Limit of Quantitation. *Clin. Biochem. Rev.* **2008**, *29*, S49–S52.
40. McMurry, J. *Organic Chemistry*, 4th ed.; Brooks/Cole Publishing Company: Pacific Grove, CA, USA, 1996; pp. 726–727.

*nanomaterials*

*Review*

# Advances, Challenges, and Applications of Graphene and Carbon Nanotube-Reinforced Engineering Ceramics

**Alaa Almansoori [1,2,*], Katalin Balázsi [1] and Csaba Balázsi [1,*]**

1 Institute for Technical Physics and Materials Science, HUN-REN Centre for Energy Research, Konkoly-Thege Miklós Str. 29-33, 1121 Budapest, Hungary; balazsi.katalin@ek.hun-ren.hu
2 Technical Institute of Basra, Southern Technical University, AlZubair Str., Basra 42001, Iraq
* Correspondence: alaa.almansoori@ek.hun-ren.hu (A.A.); balazsi.csaba@ek.hun-ren.hu (C.B.)

**Abstract:** Engineering ceramics and their composites are widely used owing to their excellent properties, including high wear, corrosion and heat resistance, low friction coefficient, and low thermal conductivity; thus, the current paper presents a comprehensive review of the most common types of engineering ceramics, demonstrating their key properties, advantages, potential applications, and challenges. This paper also provides prevailing methods for tackling the engineering ceramic challenges and maximizing their applicability. This review paper focuses on alumina ($Al_2O_3$), silicon carbide (SiC), zirconia ($ZrO_2$), aluminum nitride (AlN), and silicon nitride ($Si_3N_4$), and explores their usability in automotive, aerospace, and tribological applications. Additionally, the incorporation of reinforcing nanomaterials, i.e., graphene and carbon nanotubes or their combination with second-phase reinforcing nanomaterials in these types of ceramics to improve their physico mechanical properties is also discussed. By strategically adding these reinforcing materials, the brittleness of ceramics can be mitigated, leading to materials that are more suitable for demanding applications in various high-performance industries.

**Keywords:** ceramics; graphene; carbon nanotubes; functional properties; spark plasma sintering

## 1. Introduction

Engineering ceramic materials are well known for their superior properties, such as mechanical properties, including high stiffness, hardness, wear resistance, and thermal properties, for example, their capacity to withstand high temperatures without degrading. Furthermore, engineering ceramics offer preferred properties such as being lightweight and highly resistant to corrosion in harsh environments [1,2]. Thus, ceramics offer a unique combination of diverse properties, making them a suitable class of materials for a broad range of demanding applications. Engineering ceramics, therefore, are extensively used in fields such as manufacturing of batteries [3,4], electronics [5–7], and thermal and mechanical parts [8]. Engineering ceramics play a pivotal role in the automotive industry, such as in oxygen sensors, exhaust gas catalysts, and brake systems due to their durability and reliability [9,10]. Moreover, engineering ceramics are capable of withstanding high temperatures and destructive action in an aggressive environment, making them crucial for applications like aerospace engineering, especially in hot structures and thermal protection systems [11,12].

Despite the numerous advantages offered by engineering ceramics, there are still several challenges that need to be addressed, for example, high brittleness, machining difficulties, and costly production [13–15]. These drawbacks can limit the range of applications in various industries. However, the use of ceramic composites by incorporating small amounts of nano- or micro-sized materials provides a promising approach to tackling these challenges. Graphene and carbon nanotubes (CNTs) have demonstrated great potential for enhancing fracture and toughness resistance of engineering ceramics as well as reducing their brittleness [16–19]. Furthermore, the machinability and thermal stability of ceramic composites have been remarkably improved, making them ideal for complex applications [20–23]. The development of ceramic composites not only eliminates inherent weaknesses but also offers a combination of superior mechanical, thermal, and electrical properties, thereby expanding the potential uses of ceramics in various fields [24–26].

Hence, ceramics and their composites have become an outstanding class of materials and are progressively replacing metals and other materials in various fields like grey cast iron in brake systems. This is due to ceramic composites having multifunctional and physiomechanical exceptional properties, for example, high wear resistance and low friction coefficient, which are pivotal for reducing losses in components during rotation and enhancing the durability of components [1,27]. Thus, ceramic composites are continuously used in the production of brake system components like brake discs and pads, which last longer and have less wear compared to other traditional materials [10,28].

Ceramic composites are renowned for their ability to tolerate high wear and corrosion, which highlights their significance in improving the performance and efficiency of mechanical systems in high tribological applications [28–32]. Hence, the use of ceramic composites in these fields helps to reduce energy losses and processes waste and maintenance costs while increasing the overall performance and service life of the equipment. In brief conclusion, the unique and diverse properties of engineering ceramics and their composites make them an ideal choice in modern technologies and industrial applications, enabling innovation and sustainability and improving performance in several sectors [10,15].

The aim of this review paper is to present an overview of recent progress in engineering ceramics and their composites, highlighting their superior mechanical properties, including high wear and corrosion resistance, low friction coefficient, and robustness. In addition, the current paper aims to demonstrate the different kinds of engineering ceramics, especially concentrating on alumina ($Al_2O_3$), silicon carbide (SiC), zirconia ($ZrO_2$), aluminum nitride (AlN), and silicon nitride ($Si_3N_4$), and to discuss their advantages and potential applications in fields such as automotive, aerospace, and tribology. In addition, this review undertakes a comprehensive investigation of the role of graphene and carbon nanotubes (CNTs) in ceramic composites.

## 2. Engineering Ceramics

The development of engineering ceramics can be traced back several decades to the need for materials that can withstand high temperatures and harsh environmental conditions. Until the late 20th century, when their potential for engineering purposes was recognized, engineering ceramics were gradually developed, and their applications were limited to traditional applications [33]. Over time, significant progress has been made in the development of advanced technical ceramics, making the promise of structural ceramics much more likely to be fulfilled, except for applications where brittleness and strength are essential. Additionally, technological challenges may exist in the production of objects with complex-shaped and microsized features [34]. However, engineering ceramics are diverse and offer superior properties that make them ideal for a variety of demanding applications. These ceramic materials, such as alumina, silicon carbide, zirconia, and silicon nitride, are known for their exceptional properties, including thermal, mechanical, and chemical properties. For example, engineering ceramics possess high-temperature stability, high wear resistance, lightweight, and ability to withstand harsh environments, making them suitable for a wide array of applications in industries, including aerospace, biomedical, and automotive sectors [14,35,36]. Due to their excellent insulating properties and ability to dissipate energy as heat, as well as their favorable dielectric properties, engineering ceramics are also suited for electronic applications [6,10]. Although ceramics are brittle in nature, and machining and joining them together or to other materials is challenging, advances in processing techniques and a better understanding of their microstructures are driving their expanded uses in various high-performance fields [13,37,38].

### 2.1. Typical Engineering Ceramics and Their Respective Applications

The following are some key types of engineering ceramics:

Silicon Carbide (SiC): Silicon carbide is composed of silicon and carbon atoms joined together by a strong covalent bond. The covalently attached SiC atoms are arranged in a tetrahedral crystal structure to form a very durable ceramic. Although SiC occurs naturally, synthetic SiC can also be obtained when silicon and carbon grains are bonded together using sintering, deposition, reduction, or sol-gel methods [39,40]. Figure 1 displays selected SEM images of synthesized SiC particles in sphere-shaped (a–c) and needle-shaped particles (d–f). The excellent structure of SiC particles and their shapes, shown in Figure 1, play a crucial role in developing nanomaterials used in various engineering fields, making them ideal for high-performance applications.

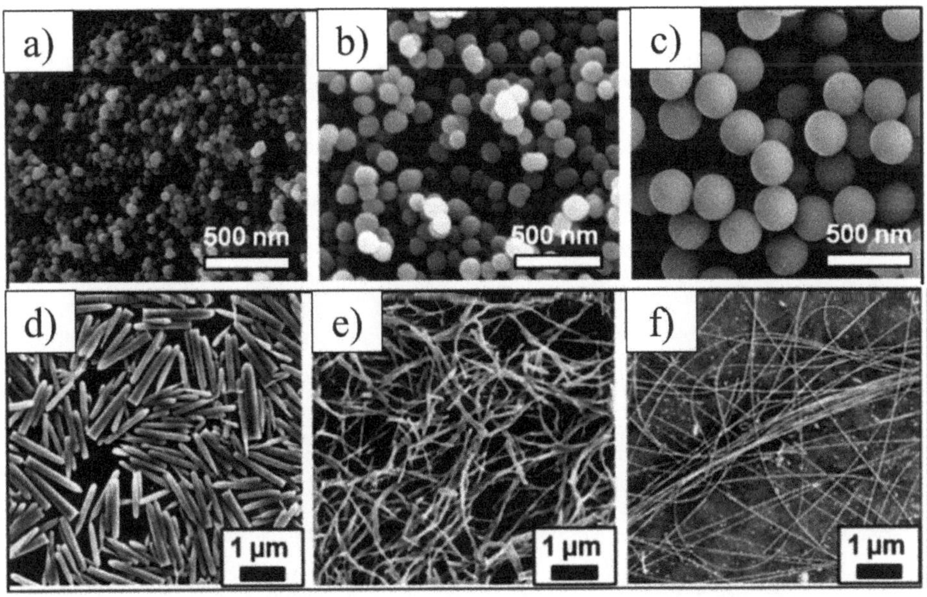


**Figure 1.** Scanning electron microscopy (SEM) images of SiC nanoparticles with diameters equal to (**a**) 51.3 ± 5.5 nm, (**b**) 92.8 ± 6.6 nm, and (**c**) 278.3 ± 8.2 nm, and rods (**d**) with length of 1.5 ± 0.2 μm, and diameter of 0.20 ± 0.05 μm, (**e**) fibers with length > 10 μm and diameter of 0.18 ± 0.04 μm, and (**f**) fibers with length > 25 μm and diameter = 0.050 ± 0.001 μm [41]. Adapted with permission from [41], published by [RSC Publishing], (2013).

SiC, therefore, is known for its exceptional hardness and thermal conductivity, making it ideal for high-temperature applications such as components in gas turbines and automotive [39,42,43]. SiC also has excellent chemical resistance and mechanical strength, which are crucial in abrasive environments [9,10].

Alumina ($Al_2O_3$): Aluminum oxide, commonly known as alumina, is a crystalline material produced from aluminum hydroxide, which is extracted from naturally occurring bauxite ores in the Bayer process. Although alumina exists in various crystalline phases, known as transition alumina phases, i.e., γ-, θ-, δ-, η-, and α-alumina, the latter is considered the most thermodynamically stable phase [44]. α-$Al_2O_3$ crystallizes in a corundum structure in which oxygen ions are arranged in an approximate close-packed hexagonal structure surrounding aluminum cations occupying two-thirds of the octahedral sites [45,46]. Thus, α-$Al_2O_3$ has exceptional properties, making it ideal in applications where high-temperature resistance, toughness, and hardness are required, such as cutting tools, grinding media, and high-temperature bearings [47]. Table 1 shows the typical properties of high-density alumina α-$Al_2O_3$.

**Table 1.** Typical properties of high-density α-alumina with 99.99 wt% content [48].

| Property | Alumina (Content: 99.99 wt%) |
|---|---|
| Density, $g/cm^3$ | 3.97–3.99 |
| Melting point, °C | 2054 |
| Ultimate strength, (flexure (σf)) MPa | 282 |
| Compression, MPa | 2550–3100 |
| Modulus of elasticity, GPa | 366–410 |
| Crack resistance ($K_{1c}$), $MPa{\cdot}m^{0.5}$ | 2.8–4.5 |
| Hardness: Vickers (HV), GPa | 19.3 |
| Thermal conductivity at room temperature, W/(m·K) | 38.9 |
| Thermal expansion coefficient ($10^{-6}$/K) at 200–1200 °C | 6.5–8.9 |
| Specific volumetric electrical resistivity (ρ), Ohm·m | $2 \times 10^{14}$ |

In addition, Alumina is produced from a relatively low-cost process characterized by its excellent thermal stability and resistance to chemical corrosion, making it widely used in components exposed to aggressive chemical environments [47]. Alumina provides a safe option for biomedical implants due to its biocompatibility and high wear resistance. In addition, alumina is a good choice for the production of dielectric substrates [49]. Moreover, Alumina-based ceramics with improved mechanical properties and controllable pore structures have been fabricated using the green biocompatible foaming method [50]. This method is suitable for applications that require highly porous structures such as oil absorption, drug delivery, and tissue engineering. Figure 2 shows an example of porous alumina ceramics with good permeability to water [50].

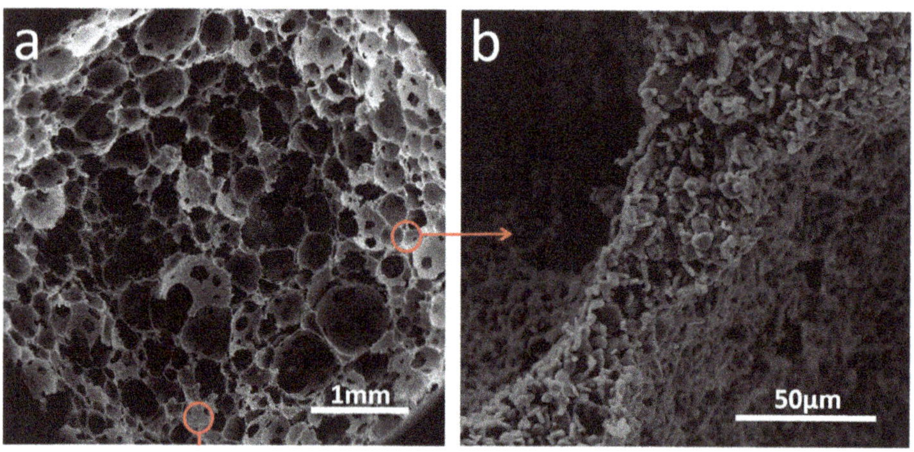


**Figure 2.** (**a**) SEM images of the pore structure of alumina ceramics, and (**b**) magnified image of pore ridge [50]. Adapted with permission from [50], published by [Elsevier], (2016).

The ultra-fine α-alumina powder can be synthesized by several methods, including the hydrothermal treatment method, precipitation method, sol-gel method, as well as the Bayer process [51]. In the hydrothermal method, acidic or basic environments can be used to produce alumina ground powder from aluminum salts like aluminum nitrate or aluminum chloride and ammonia solution for the basic route and Tetraethylammonium hydroxide (TEAH) for the acidic route [52], whereas in the precipitation method, the α-alumina powder is synthesized from the aluminum salt (i.e., aluminum nitrate or aluminum chloride) in ammonia solution [51,53]. However, the α-alumina powders synthesized using the hydrothermal approach do not require high temperature or extensive milling [51]. Furthermore, powder made by hydrothermal processes shows less agglomeration, high phase purity, and better chemical purity [54].

In addition, the high-temperature plasma treatment followed by the desired cooling rate can optimize the phase structure, resulting in improved end-use properties and expanding the usability of the final products [44]. However, plasma spheroidization can be customized to maintain the most stable α-phase for applications in highly corrosive environments [44]. Figure 3 shows SEM images of untreated and plasma-treated α-alumina (a,b) and EBSD maps of as-received α-alumina (c) and high-temperature plasma-treated α-alumina (d).

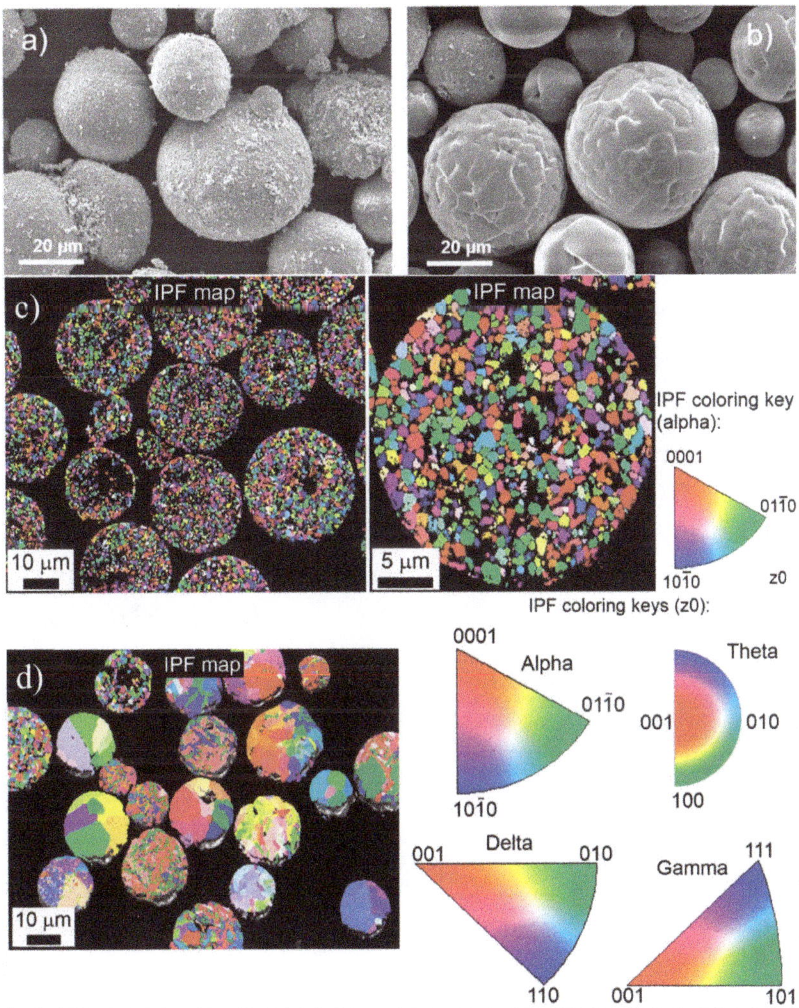


**Figure 3.** SEM images of typical α-alumina particles before treatment (**a**) and after treatment (**b**). EBSD maps (**c**), inverse pole figure (IPF) map superimposed on the band contrast map of the as-received α-alumina, and (**d**) high-temperature plasma-treated α-alumina [44]. Adapted with copyright from Elsevier under the CC BY.

Zirconia ($ZrO_2$), also known as Zirconium dioxide, is a white crystalline ceramic material that occurs naturally as the mineral baddeleyite with a monoclinic crystalline structure. The crystalline content of Zirconia is about 96%–99%, with no glassy phase, exhibiting high fracture toughness and hardness in addition to its superior biocompatibility and aesthetic quality. This makes it suitable for dental implantology and structural applications where durability and precision are required [55,56]. Thus, zirconia is valued in dental applications for the construction of crowns, bridges, and implants, replacing titanium implants [57–59]. Moreover, Zirconia nanopowder can be employed as a filling material [60] or nanocoating [61,62] in dentistry and tissue engineering. Owing to zirconia's excellent strength, hardness, and biocompatibility, it can be employed for the fabrication of porous scaffolds for potential use in bone tissue engineering [63–65]. Figure 4a–d show SEM images of fabricated porous scaffolds from zirconia at different sintering cycles: one, three, five, and seven sintering cycles, respectively.

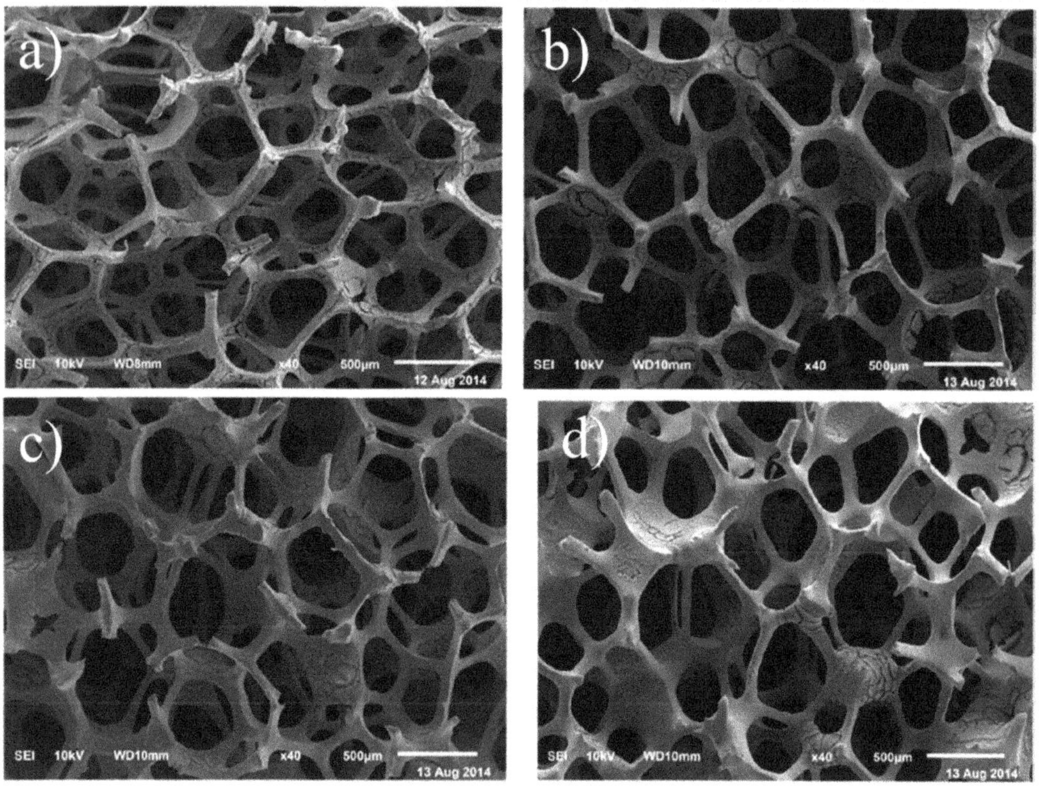


**Figure 4.** SEM images of porous zirconia scaffold at different sintering cycles (**a**–**d**): one, three, five, and seven sintering cycles, respectively [63]. Reproduced with permission from [63]; published by [IOP Science], (2015).

Besides its biocompatibility, Zirconia is renowned for its high strength, hardness, wear resistance, and low coefficient of friction properties, making it an excellent option for applications like cutting tools, refractories, and abrasives [66]. Its excellent resistance to crack propagation further enhances its reliability in mechanical applications [67].

Silicon Nitride ($Si_3N_4$): Silicon nitride is a structural ceramic material crystallized in two major phases, i.e., α- and β-$Si_3N_4$ phases [68,69]. The α-phase and β-phase are the most stable and common phases of $Si_3N_4$ that appear with a trigonal and hexagonal structure, respectively [70]. Moreover, the basic structural units of α- and β-$Si_3N_4$ phases are the same, with tetrahedral bonding (Si atoms) and polyhedral bonding (N atoms) [71]. However, the trigonal α-$Si_3N_4$ may convert to hexagonal β-$Si_3N_4$ at high temperatures of around 1400 °C [70,72]. The third phase, γ-$Si_3N_4$, appears only at high temperatures and pressure [73].

$Si_3N_4$ is one of the most attractive ceramic materials used for applications requiring high temperature and high stress such as gas turbines, diesel engines, automotive, and industrial heat exchangers [74]. $Si_3N_4$ ceramics have a diverse group of superior mechanical properties at high temperatures, including high strength, toughness, and resistance to thermal shock and corrosion at ambient and high temperatures. These properties make it suitable for demanding applications such as engine components, turbochargers, bearings, and cutting tools [75,76]. $Si_3N_4$ also exhibits good wear resistance and low friction, which are advantageous in tribological applications [69,75]. In addition, $Si_3N_4$ offers high biocompatibility, fracture toughness, and hardness properties, making it a good choice for hip and knee joint replacement, dental plants, bone grafts, and scaffolds [70,77].

Aluminum Nitride (AlN): Aluminum nitride is one of the ceramic compounds found in the form of nanoparticles with remarkable physiochemical properties, making it the ideal choice in various sectors. It exhibits high thermal conductivity, good thermal stability, and excellent electrical insulation, making it a suitable choice for electronic substrates and heat sinks [78,79]. AlN can also efficiently dissipate heat from electronic equipment, thereby enhancing their performance and reliability [80–82].

AlN powder can be synthesized using various methods, such as direct nitridation, carbothermal reduction, and chemical vapor deposition [79,83–85]. Direct nitridation is a good choice for synthesizing AlN powder with high purity and better sinterability, while the carbothermal method is recommended as a low-temperature synthesis method [79,84,85].

In summary, each of these selected ceramics has unique thermal, mechanical, and chemical properties that contribute to their widespread use in various high-performance applications. A comparison of the properties of the selected ceramics in this review paper is shown in Figure 5. These examples highlight the significant role of engineering ceramics in advancing technology and improving the performance of various industrial applications. Table 2 briefly demonstrates that engineering ceramics have a wide range of important applications owing to their unique properties. Although engineering ceramics, as discussed earlier in this paper, have a variety of applications, the challenges and limitations associated with their use in engineering applications need to be addressed. For example, their brittleness, high cost, and difficulty of processing make ceramics difficult to use in structural applications, which limits their widespread use; therefore, investigation and testing, including wear and hardness, are essential in determining the properties of ceramics. Additionally, processing methods, such as ball milling and spark plasma sintering, can be analyzed to offer insights into the optimization of ceramic properties for specific applications.

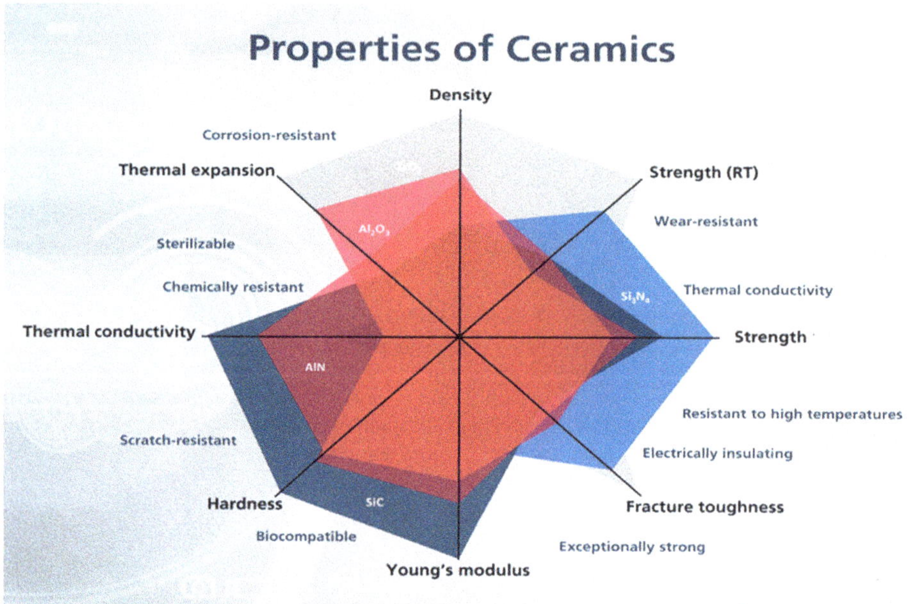


**Figure 5.** A comparison of the properties of selected ceramics [86].

**Table 2.** Summary of selected ceramics and their properties and applications in industry.

| Ceramics | Key Properties | Applications |
|---|---|---|
| Silicon Carbide (SiC) | Exceptional hardness, high thermal conductivity, chemical resistance, mechanical strength | Gas turbines and automotive [42,43], abrasive materials [9,10], biomedical devices [87]. |
| Alumina ($Al_2O_3$) | High wear resistance, temperature resistance, toughness, hardness, thermal stability, chemical corrosion resistance, biocompatibility. | Cutting tools, grinding media, high-temperature bearings [47], aggressive chemical environment [47], biomedical implants [88], dielectric substrates [49]. |
| Zirconia ($ZrO_2$) | High fracture toughness, hardness, thermal expansion, resistance to crack propagation, biocompatibility. | Dental implants, structural applications, bone tissue engineering [57,58,63], nanocoating [61,62]. |
| Silicon Nitride ($Si_3N_4$) | Superior mechanical properties at high temperatures, high strength, toughness, resistance to thermal shock, wear resistance, low friction | Engine components, turbochargers, bearings, and cutting tools [75,76], tribological applications [69,75], joint replacement [70,77]. |
| Aluminum Nitride (AlN) | High thermal conductivity, thermal stability, electrical insulation, heat dissipation performance | Electronic substrates and heat sinks [78,79] |

### *2.2. Investigation and Testing of Engineering Ceramics*

Hardness Testing: Hardness testing is essential for determining the wear resistance and durability of ceramic materials and their composites [89]. Hardness is essential for applications where ceramic materials are subjected to mechanical stress and surface interactions, such as wear-resistant coatings [90], linings, and components; thus, hardness behavior is crucial for optimizing material performance in environments demanding hard materials such as aerospace, automotive, and biomedical applications [91]. Graphene has shown significant potential for improving the hardness of ceramic materials, such as SiC, for coating purposes by enabling epitaxial growth of graphene on SiC [42]. However, it should be noted that the hardness of the ceramic material itself without a protective coating does not improve with the addition of graphene, as shown in Tables 3 and 4. Similarly, the addition of CNT could reduce the hardness of ceramic materials such as alumina, attributed to the reduction in density and increase in porosity [92].

### 2.3. Wear Test

With the development of ceramic and composite materials, it is crucial to evaluate their surface properties and substrate-coating combinations, particularly for applications demanding high surface performance. Wear is a major concern in the tribological use of even the hardest materials, including ceramics, during contact with rotating parts or abrasive media, which involves removing material and occurs mostly in the outer surfaces. Therefore, researchers are particularly interested in wear resistance tests to investigate the effects of contact abrasion or erosion on coated and uncoated ceramic components. Owing to their good wear properties, ceramics such as silicon nitride, alumina, and zirconia have been proposed for applications where wear performance is critical; thus, they are preferred for use in industrial or biomedical applications such as cutting tools, turbine blades, and bearings, as well as in hip and knee replacements [47,63,75,93,94]. However, the wear behavior of the main ceramic bulk material can be further improved to increase their application by combining them with secondary-phase nanomaterials to create hybrid materials with improved performance [95]. For example, Yttria-stabilized zirconia (YSZ), zirconia-toughened alumina (ZTA), alumina-toughened zirconia (ATZ), and others are ceramic–ceramic systems that exhibit improved wear performance [95]. Silicon carbide (SiC)/graphene ceramic composites have also been investigated for their potential use in face seals. These composites containing thermally reduced graphene oxide (TRGO), graphene-enhanced silicon carbide (GSiC), and benchmark materials with carbon-reinforcing materials were produced and tested tribologically under water-lubricated conditions. Notably, the results showed that all the produced composites, except graphite, exhibited improved friction and wear performance in an aqueous environment compared to the material without graphene [30]. Thus, it can be concluded that ceramics and ceramic composites show promise in terms of wear performance, making them useful in a wide range of applications that require high wear resistance and biocompatibility for specific uses.

Graphene and CNT integration into ceramic matrices is accomplished using advanced processing techniques. Hot pressing applies heat and pressure at the same time, resulting in densification and reduced chances of agglomeration. Nanomaterials are uniformly distributed in the ceramic matrix through this method, therefore improving the composites' mechanical properties. For example, a study conducted by Yazdani et al. demonstrated that the tribological performance of graphene/CNT-reinforced $Al_2O_3$ composites produced by hot pressing showed remarkable improvements in friction coefficients along with a promising wear rate reduction [96]. However, hot-pressed graphene/$Si_3N_4$ composites did not improve or improved wear rates only slightly, while the coefficient of friction results showed improvement, especially when 3% graphene was added, as shown in a study by Maria et al. [97].

To produce ceramic nanocomposites, spark plasma sintering (SPS) is widely used. SPS has become a key technology for producing a new generation of advanced materials. Utilizing a pulsed electric current, SPS heats up the material, making it densify quickly at low temperatures compared to other sintering methods. This technique helps to maintain the nano-structure of GNPs and CNTs, which is necessary for maintaining their reinforcement effects [98]. In fact, SPS-processed ceramics possess higher hardness, toughness, and improved electrical conductivity, plus thermal conductivities [99,100]. SPS is particularly effective in enhancing the properties of zirconia ($ZrO_2$) and alumina ($Al_2O_3$) composites. Liu et al. (2012) demonstrated that adding graphene platelets to these composites significantly improves their toughness and strength. The rapid sintering process of SPS helps with preserving the integrity of the graphene platelets, thereby enhancing the overall composite performance [101]. Anselmi-Tamburini et al. (2004) conducted a study on the densification of fully stabilized zirconia using SPS. Based on their results, densification is highly influenced by temperature and pressure, but not heating rates and holding time. Thus, they found that SPS could achieve near-theoretical density at a temperature of about 1200 °C. This is attributed to the simultaneous application of pressure and electric current, which enhances the diffusion processes and promotes particle bonding [102]. Table 3 compares the results of the density and hardness of $Si_3N_4$ composites fabricated by hot press and SPS techniques performed by Maria et al. [97].

**Table 3.** Properties of $Si_3N_4$ and graphene/$Si_3N_4$ composite produced by HIP and SPS: apparent density, relative density, Vickers hardness, and VIF toughness [97]. Reproduced with permission from [97]; published by [Elsevier], (2016).

| Sintering Method | MLG (wt%) | Apparent Density (g/cm$^3$) | Relative Density (%) | HVM0.5 (GPa) | VIF Toughness (MPa·m$^{1/2}$) |
|---|---|---|---|---|---|
| HIP | 0 | 3.23 | 96.52 | 13.3 ± 0.48 | 4.7 ± 0.38 |
| | 1 | 3.27 | 97.94 | 11.8 ± 0.68 | 4.2 ± 0.14 |
| | 3 | 2.80 | 84.66 | 5.8 ± 0.65 | 5.7 ± 0.28 |
| SPS | 0 | 3.23 | 96.54 | 15.8 ± 0.84 | 5.1 ± 0.47 |
| | 1 | 3.29 | 98.52 | 15.4 ± 0.70 | 4.9 ± 0.72 |
| | 3 | 3.11 | 93.84 | 13.7 ± 0.82 | 2.7 ± 0.37 |

## 3. Engineering Ceramics Challenges

Despite the numerous advantages of engineering ceramics, they have significant drawbacks that can limit their applications. One of the primary disadvantages is their intrinsic brittleness, which leads to weak mechanical properties, particularly in terms of toughness. Ceramics are generally hard and resistant to wear but tend to fracture easily under tensile stress or impact, which can compromise their reliability in applications requiring high mechanical strength and toughness [103]. The ceramics' brittleness originates from their atomic structures, which lack the ability to deform plastically. This means that, unlike metals, ceramics cannot absorb and dissipate energy through plastic deformation, making them prone to catastrophic failure when subjected to stress [38]. This property is particularly detrimental in structural applications where mechanical reliability is critical. In medical applications, brittle failure of ceramic materials can lead to severe clinical complications associated with ceramic crowns and hip replacements, which highlights the importance of fracture toughness of ceramics [14,104].

To overcome this limitation, various strategies have been developed to enhance ceramic toughness and increase their wearability. These strategies involve the addition of a second-phase ceramic material or reinforced materials, such as carbon nanotubes and graphene, to create ceramic composites. These composites can significantly improve fracture toughness by mechanisms such as crack deflection, bridging, and pull-out [14]. In addition, advanced processing techniques, most importantly hot pressing and spark plasma sintering, can improve the microstructural properties of ceramics, resulting in improved mechanical performance [7,26,39]. However, other physio-chemical properties can also be improved to expand applications in various sectors. Hence, improving the toughness, strength, thermal, wear, and corrosion properties of ceramics is pivotal for expanding their uses in various sectors where mechanical reliability is required, like in the aerospace, automotive, and biomedical fields [9,31]. By addressing these challenges, the full potential of engineering ceramics can be realized, thus enabling their applications in more demanding environments, including automotive, aerospace, and medical industries. It could also reduce energy consumption while protecting the environment.

## 4. Improving the Properties of Engineering Ceramics

Improving the properties of engineering ceramics is commonly achieved by developing ceramic composites. These composites improve the mechanical behavior of traditional polycrystalline ceramics, which are typically characterized by high brittleness and low fracture toughness [105,106]. By incorporating small quantities of materials, such as reinforcing phases, the overall properties of ceramics can be significantly enhanced. Ceramic composites are generally created by embedding at least one micro- or nano-sized material and/or reinforcing secondary phase into a ceramic matrix:

(1) The secondary phase can include ceramic micro- or nanoparticles, such as silicon carbide (SiC) and silicon nitride ($Si_3N_4$), which enhance toughness and wear resistance.
(2) The inclusion of carbon nanotubes (CNTs) as a reinforcing material for enhancing toughness and functional properties, such as electrical and thermal conductivity, of ceramic matrices.
(3) Graphene nanoplatelets are another effective reinforcing material that can be used to improve the fracture toughness of ceramics through mechanisms like crack deflection, bridging, and pull-out, leading to increased mechanical performance.
(4) Hybrid composites can be obtained by combining two or more reinforcing materials. For example, CNTs and graphene, or the combination of second-phase nano/micro materials with graphene and CNTs can further enhance the properties of ceramics, resulting in a balance of strength, toughness, and other functional properties.

The mechanical properties of ceramic composites are essentially influenced by the interphase between the ceramic matrix and the reinforcing material. In the case of weak interphase, initiated cracks are more likely to deflect along the interface between the ceramic matrix and the reinforcing phase. This deflection, however, prevents the crack from propagating through the reinforcing phase, maintaining the integrity of the reinforcement and improving the toughness of the composite. Hence, these composites can effectively resist fracture and perform well under mechanical loads [100,107].

In contrast, if the interphase and associated interfaces between the two material phases are too strong, initiated cracks penetrate the reinforcing phase, leading to transgranular fracture. This penetration, as a result, makes the composite brittle, similar to the behavior of pure ceramics, which lack the ability to absorb and dissipate energy during deformation [1]. Therefore, achieving the optimal balance in the matrix–reinforcement interaction plays a key role in enhancing the mechanical performance of ceramic composites [24].

By selecting the right combination of materials (matrices and reinforcements) and optimizing the interactions between them, ceramic composites can achieve excellent mechanical behavior, making them an ideal choice for a wide range of high-performance applications.

### 4.1. *Using Nanosized Ceramic Particles as Secondary Phases*

The incorporation of second-phase ceramic nanoparticles has remarkably improved the mechanical and functional properties of ceramics. Various nanoparticles, including silicon carbide (SiC), silicon nitride ($Si_3N_4$), titanium carbide (TiC), titanium dioxide ($TiO_2$), and zirconium dioxide ($ZrO_2$), have been successfully embedded in ceramic matrices such as alumina ($Al_2O_3$), silicon nitride ($Si_3N_4$), and magnesium oxide (MgO) to produce an important class of ceramic nanocomposites [7]. Some of these composite materials include $Si_3N_4$–SiC, $Al_2O_3$–SiC, $Al_2O_3$–$Si_3N_4$, $Al_2O_3$–TiC, $Al_2O_3$–$ZrO_2$, and MgO–SiC, which have been extensively studied [108].

Zirconia ($ZrO_2$), for example, has been widely investigated as a ceramic composite material, either as a secondary phase or matrix, owing to its exceptional properties. $ZrO_2$ exhibits excellent ionic conductivity and superior mechanical properties such as excellent wear resistance, high mechanical strength, and stability at high temperatures [7]. This makes these materials ideal for structural, tribological, and multifunctional applications. For instance, the inclusion of $ZrO_2$ particles in $Al_2O_3$ as a second phase, the so-called zirconia-toughened alumina (ZTA), enables ZTA to exhibit higher fracture toughness than an alumina ceramic matrix, but with moderate Vickers hardness [67,109]. Additionally, alumina-toughened zirconia (ATZ) exhibits enhanced fracture toughness when $Al_2O_3$ acts as a secondary phase [14].

Silicon carbide (SiC) nanoparticles have been incorporated as a semiconductive second-phase material in alumina matrices to enhance thermal and electrical conductivities. This combination is beneficial for applications requiring high thermal conductivity as well as durability and resistance to abrasion, such as cutting tools, wear-resistant tools, and gas radiant burners. Considering electrical conductivity, an $Al_2O_3$-SiC composite with moderate electrical conductivity is suited to DC electronic applications [110]. Similarly, silicon nitride ($Si_3N_4$) reinforced with titanium nitride (TiN) nanoparticles improves toughness and thermal shock resistance. This enhancement is critical for applications in high-temperature environments, such as engine components and turbine impellers [111]. Titanium dioxide ($TiO_2$) nanoparticles embedded in magnesium oxide (MgO) matrices improve the optical and photocatalytic properties of ceramics, making them suitable for photocatalysis (such as reduction of $CO_2$, degradation of $CH_3CHO$, and dye degradation) or other applications like dye-sensitized solar cells or nanofluids with high thermal conductivity [112].

These examples demonstrate how incorporating different nanoparticles into ceramic matrices can tailor the properties of the resulting nanocomposites for specific applications. The enhanced properties, including mechanical strength, thermal stability, and functional capabilities, have expanded the utility of ceramics in various high-performance fields.

### *4.2. Using Graphene Nanomaterials to Increase Mechanical and Functional Properties of Ceramics*

Graphene is increasingly being used to improve the mechanical properties of ceramic composites due to its exceptional strength and flexibility. In particular, adding small amounts of graphene to ceramics such as silicon nitride ($Si_3N_4$) can significantly enhance fracture resistance [22] or improve tribological behavior [31]. Typically, a graphene content of 2 wt% improves the fracture resistance of $Si_3N_4$ ceramic composites through three primary toughening mechanisms: crack bridging, crack deflection, and pull-out [107,113]. These mechanisms help to prevent crack propagation and enhance the material's toughness [113].

However, increasing the graphene content beyond this range can sometimes lead to a decrease in mechanical properties. This decline is often attributed to the inhomogeneous distribution of graphene within the matrix or reduced surface interaction between graphene and the ceramic matrix. Ensuring uniform dispersion and optimal interaction is crucial for maximizing the benefits of graphene reinforcement [107,113].

Extensive research has focused on optimizing the composition, formation, and sintering techniques for graphene-reinforced ceramics, including zirconia ($ZrO_2$) [114], alumina ($Al_2O_3$) [115], silicon carbide (SiC) [116], zirconium diboride ($ZrB_2$) [117], and boron carbide ($B_4C$) [118]. In the case of silicon nitride ($Si_3N_4$), studies have also demonstrated the beneficial toughening effects of graphene [1,107]. These studies, as well as others, highlight that the strengthening and toughening performance of graphene depends on its dispersion, volume fraction, type of ceramic matrix, and processing parameters of the composite [119,120]. Figure 6 represents a typical example of using graphene with alumina.

A variety of applications benefit from the enhanced mechanical and functional properties provided by the addition of graphene [121]. Applications of ceramic/graphene composites are diverse and include energy production and storage, sensors, tissue engineering, electromagnetic interference shielding, thermal management, protection against wear and corrosion, and catalysis [121]. Table 4 lists some of the ceramic/graphene applications.

**Table 4.** Summary of selected ceramics reinforced with graphene, including some examples of their applications in different fields.

| Applications | Type of Ceramics and References |
|---|---|
| Energy production and storage | Li-ion battery cathodes: vanadium pentoxide ($V_2O_5$) [122], cobalt oxide ($Co_3O_4$) [123]<br>Anodes: void-containing $Al_2O_3$/coated porous Si [124]<br>$Si_3N_4$-coated Si core [125], |
| Piezoelectric energy harvesting | Lead zirconate titanate (PZT) [126], Barium titanate [127] |
| Sensors | Tin oxide ($SnO_2$) [128,129]<br>$\gamma$-alumina [130]<br>Zin oxide (ZnO) [131,132] |
| Electromagnetic interference shielding | Boron carbide ($B_4C$) [133]<br>Magnetic iron oxide ($Fe_3O_4$) [134]<br>Nickel cobalt sulfide (NiCoS) [135] |

**Table 4.** *Cont.*

| Applications | Type of Ceramics and References |
|---|---|
| Catalytic applications | Titanium dioxide ($TiO_2$) [136]<br>Bismuth vanadate–silicon dioxide ($BiVO_4/SiO_2$) [137]<br>Iron oxide/Nickle oxide ($Fe_3O_4/NiO$) [138] |
| Heat sinks and thermal energy storage | Alumina ($Al_2O_3$) [139,140]<br>Dual silicon oxycarbide (SiOC) [141,142] |

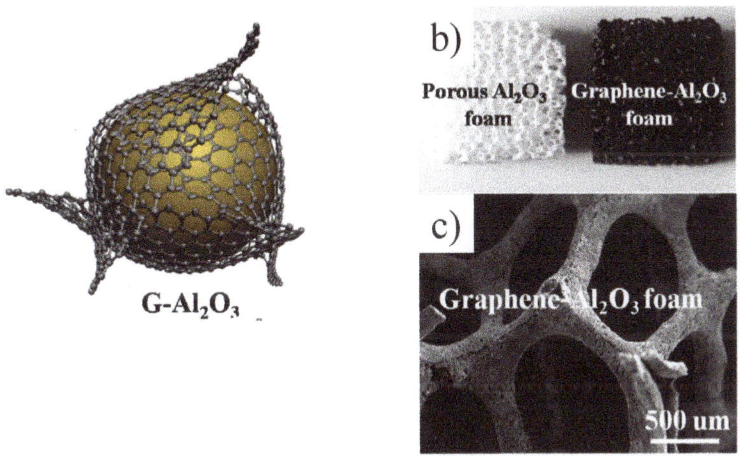


**Figure 6.** A schematic representation of graphene on $Al_2O_3$ particle (**a**), porous $Al_2O_3$ foam with and without graphene (**b**), and high-resolution SEM image of graphene/$Al_2O_3$ foam (**c**) [140]. Adapted with permission from [140]; published by [Elsevier], (2014).

### *4.3. Using Carbon Nanotubes to Increase the Properties of Ceramics*

Reinforcing ceramic matrices with carbon nanotubes (CNTs) has gained considerable interest due to the significant improvements CNTs can bring to both the mechanical and functional properties of ceramics. The incorporation of CNTs into engineering ceramic matrices improves the toughness of ceramics, addressing their intrinsic brittleness and improving their fracture resistance [143]. CNTs contribute to toughness enhancement through mechanisms such as crack bridging, crack deflection, and pull-out, which help to prevent crack propagation within the ceramic matrix [144]. These mechanisms significantly improve the durability and mechanical performance of composite materials with different ceramic matrices such as $Al_2O_3$ [144], B4C [145], and $ZrB_2$–SiC [146,147]. The toughness of ceramic/carbon nanotube composites can be calculated using two different methods: single-edge notched beam (SENB) method and direct crack measurement (DCM) method [144]. A comparison between the two methods is illustrated in the inset table within the schematic diagram in Figure 7. The inset graphs obtained by Shah et al. represent the fracture toughness of alumina ceramic composites with the addition of 1 wt% carbon nanotubes and 0–1.2 wt% graphene. The results showed that the fracture toughness increased by 49.5% with 1 wt% carbon nanotubes and 0.4 wt% graphene [99].

Beyond mechanical properties, CNTs also improve the functional properties of ceramics, particularly their electrical and thermal conductivities [148,149]. The excellent electrical conductivity of CNTs allows for the creation of ceramic composites with enhanced electrical properties, which are useful in various electronic applications [143]. Similarly, the high thermal conductivity of CNTs enables the development of ceramic composites that can efficiently dissipate heat, making them suitable for applications in thermal management [150]. However, achieving a uniform dispersion of CNTs within the ceramic matrix is crucial for maximizing these benefits. Homogeneous distribution ensures that the reinforcing effects of CNTs are evenly applied throughout the material, preventing areas of weakness and enhancing overall performance [150]. Advanced processing techniques, such as spark plasma sintering and hot pressing, have been employed to achieve better dispersion and bonding of CNTs within ceramic matrices [118].

Applications of CNT-reinforced ceramics are diverse, including structural components, electronic devices, thermal management systems, and more. The combination of improved mechanical toughness and enhanced functional properties makes CNT-reinforced ceramics a promising material for high-performance applications [143].

| **Method** | **(1) Direct crack measurement (DCM) method** | **(2) Single-edge notched beam (SENB) method** |
| --- | --- | --- |
| **Test** | Vickers hardness test | 3-point bending test |
| **Fracture Toughness Equation** | $K_{IC} = 0.016\left(\frac{E}{H}\right)^{0.5}.\left(\frac{P}{c^{2/3}}\right)$ | 3-point bending test: (ASTM E399-90) $K_{IC} = \frac{3.P_{max}.L}{2.b.h^{3/2}}.\alpha^{0.5}.Y$ Where: $Y = \frac{1.99-\alpha(1-\alpha).(2.15-3.93\alpha+2.7\alpha^2)}{(1+2\alpha).(1-\alpha)^{3/2}}$ |
| **Results** | (i) 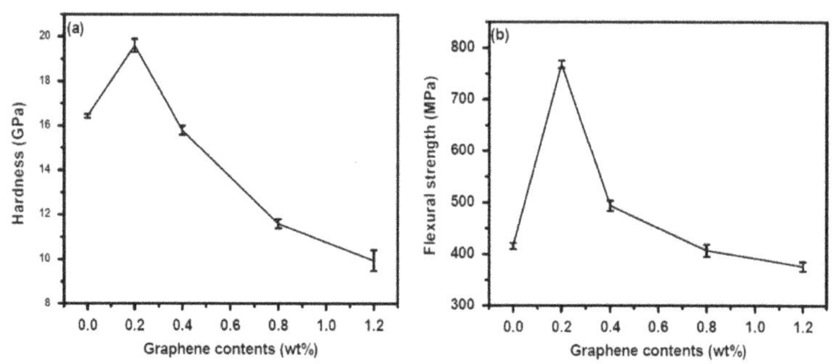  | |
| | (ii) 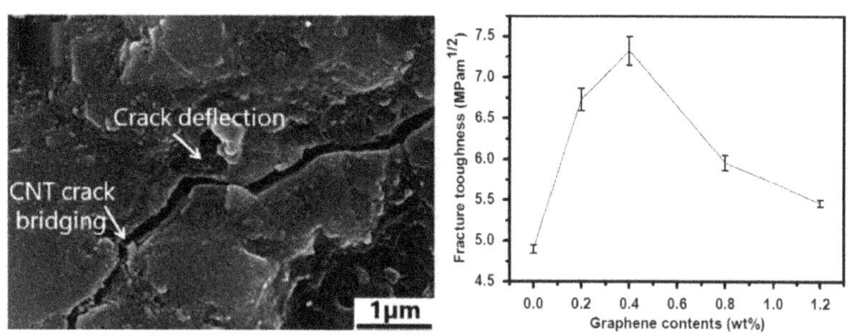  | |

**Figure 7.** A schematic diagram of the fracture toughness calculation methods. In method one, fracture toughness is calculated by the equation proposed by Anstis et al. [151,152], while equations in method two are based on ASTM E399 [153]. The inset graphs (**i**,**ii**) obtained by Shah et al. [99] represent the results of hardness and flexural strength ((**a**,**b**) in graph (**i**)) and calculated fracture toughness and SEM

image of alumina ceramic composites with the addition of 1 wt% carbon nanotubes and 0–1.2 wt% graphene (graph (**ii**)). Inset graphs are reproduced with permission from [99] published by [Elsevier], (2021).

### 4.4. Combination of Graphene and Carbon Nanotubes

Combining graphene and carbon nanotubes (CNTs) as reinforcing materials in ceramic matrices has been shown to significantly enhance the properties and performance of ceramics. The synergistic effects of these two materials (mostly in nanosize) enhance both mechanical and functional properties, making them suited to advanced applications in various sectors [154]. Figure 8 displays a schematic diagram of a hybrid ceramic composite consisting of carbon nanotube and graphene and the ceramic matrix represented by a Voronoi element [155].

**Figure 8.** A schematic graphic of random introduction of CNT and graphene into a ceramic matrix. (**a**) Carbon nanotube, (**b**) graphene, and (**c**,**d**) hybrid CNT and graphene [155]. Reproduced with permission from [155]. Published by [Elsevier], (2021).

The combination of these two reinforcing materials, i.e., graphene and CNTs, in ceramic matrices improves fracture toughness through mechanisms such as crack bridging, crack deflection, and necking [99,156,157]. These mechanisms are effective because graphene provides high tensile strength and flexibility, and CNTs contribute to a high aspect ratio and superior mechanical properties [99,158]. This combination influence arises from preventing crack propagation and distributing stress more evenly within the ceramic matrix, thus significantly improving toughness and durability [154,156]. Moreover, the functional properties of ceramics, such as electrical and thermal conductivity, are also enhanced by incorporating both graphene and CNTs. Graphene's high electrical conductivity and CNTs' superior thermal conductivity make composites ideal for applications that need efficient heat dissipation and electrical performance [156]. Thus, GNT-reinforced ceramics are ideal for electronic equipment, thermal management systems, and structural parts subjected to high thermal and electrical loads [159].

Various ratios of graphene nanoplatelets to CNTs were used to prepare these composites, providing insights into the optimal composition required for maximum enhancement. For example, it was found that the mechanical properties of $Si_3N_4$ were improved with the addition of up to 5% CNT/graphene content, whereas the ceramic/CNT/graphene composite materials exhibited excellent improvement in electrical and wear properties with a very small quantity (only 1%) of CNT/graphene content [154]. The optimal reinforcement for $Al_2O_3$–TiC composites was obtained at 0.8 wt% MWCNTs and 0.2 wt% GNPs, according to a study performed by Daming Sun et al. [158]. Other properties of ceramics, like the thermal conductivity of monolithic alumina ($Al_2O_3$), were optimized to best values at low CNT and graphene concentrations, which were 0.4 wt% graphene and 1 wt% CNTs [160].

Achieving uniform dispersion of graphene and CNTs within the ceramic matrix is crucial for maximizing these benefits [155]. This is due to the fact that increasing the amount of reinforcing materials, i.e., graphene and CNT, leads to agglomeration, which results in the deterioration of the mechanical properties of the composite and limits their usability [155,158]. However, the dispersion of graphene is easier than that of CNT due to the fiber morphology of CNT and the van der Waals interactions between them, which result in nonuniform distribution [99,158]. Furthermore, advanced processing techniques like HP and SPS are employed to ensure homogeneous distribution and strong interfacial bonding. These methods not only enhance the mechanical properties but also preserve the inherent advantages of the nanomaterials [154,158].

In summary, the combined use of graphene and CNTs in ceramic composites results in materials with superior mechanical strength, toughness, and functional properties, paving the way for their application in a variety of high-performance fields. However, achieving uniform dispersion and strong interfacial bonding remains a key challenge that must be addressed through innovative processing techniques and careful material design.

## 5. Challenges of Using Carbon Nanotubes in Ceramics:

(i) Homogeneous Dispersion

Hybrid ceramic nanocomposites represent an advanced class of materials that incorporate two or more different nanoscale reinforcements, mainly CNTs and graphene, to improve the mechanical and functional properties of ceramics such as fracture toughness, thermal stability, and electrical conductivity [89,140]. The inclusion of CNTs helps mitigate the brittleness of ceramics by providing mechanisms such as crack bridging, pull-out, and crack deflection, which prevent crack propagation and improve toughness [160]. SEM images in Figure 9 display the dispersion of the hybrid ceramic composite of CNTs, graphene, and alumina in which the pull-out and bridging mechanisms were observed [161].

However, to create ceramic/CNT composites with improved properties, it is crucial to achieve proper interfacial bonding between the ceramic matrix and CNTs and ensure uniform dispersion of CNTs along the grain boundaries. Silvestre et al. [162] reviewed the enhancement of mechanical properties in ceramic composites, emphasizing the role of homogeneous dispersion of CNTs within the ceramic matrix. This is because CNTs often agglomerate due to their high aspect ratio, resulting in stress concentration and subsequent reduction in the overall properties of the composites. This agglomeration is the most significant challenge when preparing CNT–ceramic matrix composites (CMCs). When CNTs bundle or agglomerate, they cause non-uniform distribution within the composites, leading to degradation of fracture toughness, particularly noticeable during indentation analysis. Large amounts of CNT bundles also cause porosity, reducing the contribution of CNTs to the reinforcement mechanism. Uniform dispersion using surfactants and sonication ensures the reinforcing effects of CNTs are optimized, leading to enhanced mechanical strength and toughness [163]. However, surfactantless CNTs were also successfully prepared and homogeneously incorporated within α-alumina ceramics, providing promising reinforcing efficiency [164]. Rubel et al. [165] reviewed the challenges associated with CNT agglomeration in reinforced composites. They noted that achieving a homogeneous distribution of CNTs is essential for optimizing the mechanical properties of CMNCs. Advanced processing techniques, such as sol-gel, hydrothermal/solvothermal, and colloidal processing, are critical for achieving this dispersion [165,166]. Additionally, the in situ growth of CNTs by chemical vapor deposition (CVD) can achieve uniform CNT distribution by growing CNTs at the designed growth sites (metal catalysts) within the ceramic precursor [167].

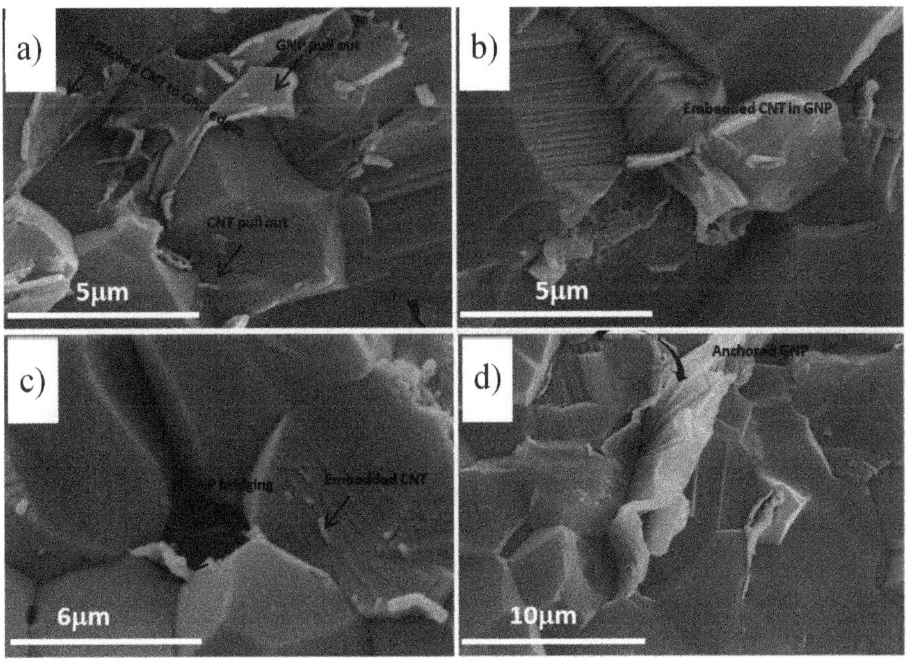


**Figure 9.** SEM images of the fractured surfaces of hybrid CNT/graphene alumina nanocomposite showing the graphene and CNT pull-out mechanism (**a**), CNTs embedded in graphene flake surfaces (**b**), CNTs and graphene bridging phenomena (**c**), and large graphene rolled along the alumina grain (**d**) [161]. Adapted with permission from [161] published by [Elsevier], (2015).

(ii) Suitable Interfacial Adhesion between CNTs and Ceramics

The functional properties of ceramic composites, including their mechanical and thermal characteristics, are highly dependent on the interfacial adhesion between carbon nanotubes (CNTs) and the ceramic matrix. This is because graphene is easier to disperse in ceramic matrices than CNTs [155]; thus, strong interfacial adhesion ensures efficient load transfer from the ceramic matrix to the CNTs, enhancing the composite's overall strength and toughness. For instance, when the CNT–ceramic interaction is weak, strong phonon scattering occurs, resulting in high thermal resistance, which diminishes the composite's thermal conductivity [143]. Moreover, weak interfacial adhesion between CNTs and the ceramic matrix can exacerbate crack propagation. This weak bonding fails to effectively utilize toughening mechanisms such as crack bridging, where the CNTs bridge the crack surfaces and impede crack growth, thus improving toughness [113]. Additionally, poor interfacial adhesion undermines the pull-out mechanism, where CNTs are pulled out of the matrix during fracture, dissipating energy and enhancing toughness [156]. Evidence of

CNTs sticking to alumina at the interface is shown in the high-resolution TEM images in Figure 10a,b and illustrated in the schematic model in Figure 10c–e [150]. The unknown interface layer shown in Figure 10a,b is suggested to be primarily composed of $Al_2OC$ after the reaction during the high-temperature and high-pressure sintering process, as illustrated in Figure 10c–e [150].

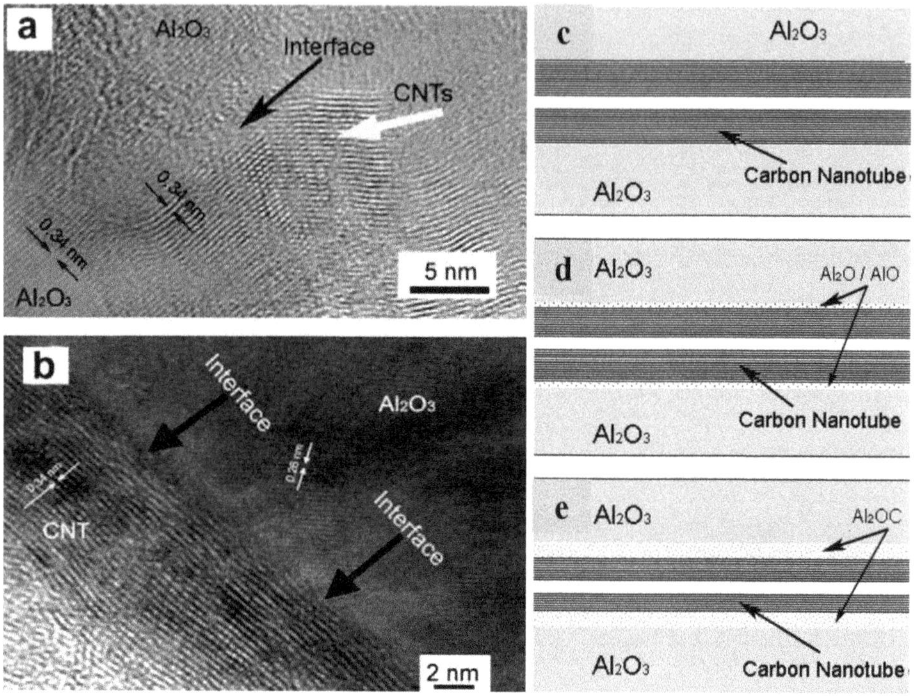


**Figure 10.** CNT/ceramic interfaces are observed in high-resolution TEM images (**a**,**b**) and illustrated by a schematic model (**c**–**e**) [150]. Adapted with permission from [150]; published by [Elsevier], (2010).

To overcome these challenges, surface functionalization of CNTs can be employed to enhance interfacial bonding. Techniques such as sol-gel processing and chemical vapor deposition (CVD) can modify the surface of CNTs, increasing their compatibility with the ceramic matrix and improving the overall composite properties [166]. Properly engineered interfacial adhesion not only enhances mechanical properties like fracture toughness but also improves thermal conductivity, making CNT-reinforced ceramic composites suitable for high-performance applications.

(iii) Thermal Degradation of Carbon Nanotubes in Ceramics

Thermal degradation is considered a significant concern in the fabrication of CNT-reinforced ceramic composites. Previous studies have demonstrated that at high pressures, elevated sintering temperatures, and prolonged sintering durations, the structural integrity of CNTs can be influenced [168]. It was found that spark plasma sintering (SPS) at high temperatures can induce phase transformations in CNTs, resulting in a reduction in their mechanical and thermal behaviors [169]. Additionally, the synthesis of ceramic–CNT nanocomposites at high temperatures leads to degradation of CNTs, affecting the overall performance of ceramic composites [170]. To eliminate these effects, optimizing sintering parameters such as temperature, pressure, and duration is essential. Thus, this approach helps preserve the structural integrity of CNTs, ensuring that the ceramic composite material retains its improved mechanical and thermal properties. Thermal degradation of CNTs in ceramic composites by SPS as can be observed by SEM images and Raman spectra [168]. According to a study conducted by F. Inam et al., the SEM images showed that the multi-walled carbon nanotubes in boron carbide composite were damaged and became fibrous, while the Raman spectra revealed that there was severe structural degradation [168]. The decomposition stages of the CNTs under different SPS conditions can be explained by the thermal gravimetric analysis (TGA) test, as shown in Figure 11, based on a study conducted by E. Suslova et al. [169].

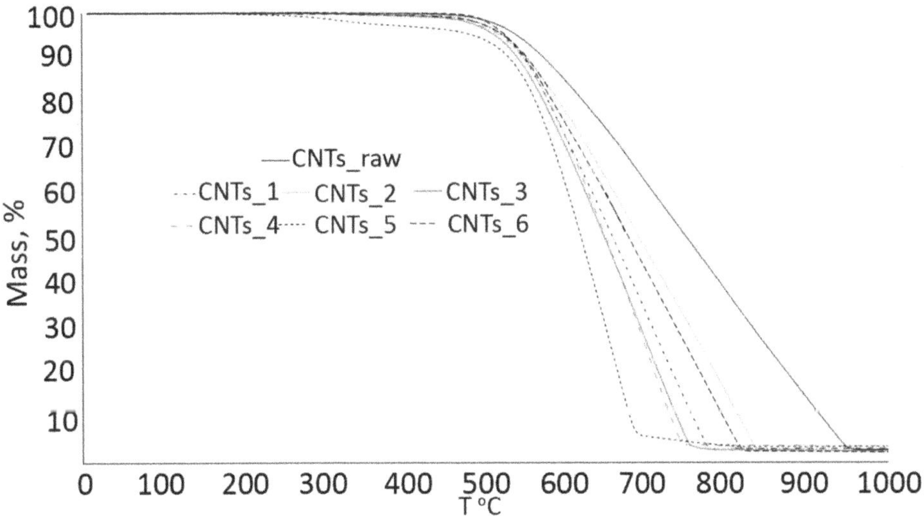


**Figure 11.** TGA curves of CNTs after SPS processing [169]. Reproduced with permission from [169]; published by [Elsevier], (2020).

## 6. Conclusions

This comprehensive review highlights that engineering ceramics and their composites play a crucial role in modern industrial applications in various areas, thanks to their remarkable properties like high wear and corrosion resistance, low friction coefficient, and structural robustness. By focusing on key materials such as silicon carbide, alumina, zirconia, silicon nitride, and aluminum nitride, the article underscores their versatile applications across industries such as automotive, aerospace, medical, and tribological sectors.

Furthermore, the incorporation of advanced materials like graphene and carbon nanotubes is shown to significantly enhance the prevailing properties of engineering ceramics, including mechanical, thermal, and electrical properties. Hence, graphene and CNT-based ceramic composites are ideal for demanding applications in various high-performance industries.

Below are routes for improving the key properties of engineering ceramics by the inclusion of nano or micro-sized second-phase materials:

(1) The inclusion of carbon nanotubes (CNTs) as a reinforcing phase has been shown to enhance both the toughness and functional properties, such as electrical and thermal conductivity, of ceramic matrices.
(2) Graphene nanoplatelets are another effective reinforcing material. They improve the fracture toughness of ceramics through mechanisms like crack deflection, bridging, and pull-out, leading to increased mechanical performance.
(3) Hybrid composites, combining different reinforcing materials such as CNTs and graphene, can further enhance the properties of ceramics, providing a balance of strength, toughness, and other functional attributes.

Additionally, various preparation methods, including ball milling and spark plasma sintering, are critically analyzed to offer insights into optimizing ceramic properties for specific applications. This comprehensive review not only consolidates current knowledge but also paves the way for future research and development aimed at harnessing the full potential of engineering ceramics in cutting-edge technological advancements.

**Author Contributions:** Conceptualization, K.B., A.A. and C.B.; methodology, A.A. and K.B.; validation, C.B. and K.B.; formal analysis, A.A. and K.B.; investigation, A.A., K.B. and C.B.; resources, K.B. and C.B.; data curation, A.A.; writing—original draft preparation, A.A.; writing—review and editing, A.A., K.B. and C.B.; visualization, A.A. and K.B.; supervision, K.B. and C.B.; project administration, C.B.; funding acquisition, K.B. and C.B. All authors have read and agreed to the published version of the manuscript.

**Funding:** This work was supported by the National Research, Development, and Innovation Office–OTKA NKFI 146076.

**Conflicts of Interest:** The authors declare no conflict of interest.

## References

1. Balázsi, K.; Furkó, M.; Liao, Z.; Fogarassy, Z.; Medved, D.; Zschech, E.; Dusza, J.; Balázsi, C. Graphene added multilayer ceramic sandwich (GMCS) composites: Structure, preparation and properties. *J. Eur. Ceram. Soc.* **2020**, *40*, 4792–4798. [CrossRef]
2. Xia, Y.; Zeng, Y.P.; Jiang, D. Microstructure and mechanical properties of porous $Si_3N_4$ ceramics prepared by freeze-casting. *Mater. Des.* **2012**, *33*, 98–103. [CrossRef]
3. Palakkathodi Kammampata, S.; Thangadurai, V. Cruising in ceramics—Discovering new structures for all-solid-state batteries—Fundamentals, materials, and performances. *Ionics* **2018**, *24*, 639–660. [CrossRef]
4. Takada, K. Secondary batteries—Lithium rechargeable systems—Lithium-ion | Electrolytes: Solid Oxide. In *Encyclopedia of Electrochemical Power Sources*; Elsevier: Amsterdam, The Netherlands, 2009; pp. 328–336. ISBN 9780444527455.
5. Vinila, V.S.; Isac, J. Synthesis and structural studies of superconducting perovskite $GdBa_2Ca_3Cu_4O_{10.5+\delta}$ nanosystems. In *Design, Fabrication, and Characterization of Multifunctional Nanomaterials*; Elsevier: Amsterdam, The Netherlands, 2022; pp. 319–341.
6. Eda, G.; Chhowalla, M. Graphene-based composite thin films for electronics. *Nano Lett.* **2009**, *9*, 814–818. [CrossRef] [PubMed]
7. Lamnini, S.; Pugliese, D.; Baino, F. Zirconia-Based Ceramics Reinforced by Carbon Nanotubes: A Review with Emphasis on Mechanical Properties. *Ceramics* **2023**, *6*, 1705–1734. [CrossRef]
8. Yin, S.; Wan, W.; Fang, X.; Ma, H.; Xie, X.; Zhou, C.; Li, T.; Zuo, R. Mechanical and thermal properties of $Si_3N_4$ ceramics prepared by gelcasting using high-solid-loading slurries. *Ceram. Int.* **2023**, *49*, 40930–40941. [CrossRef]
9. Krupka, M.; Kienzle, A. *Fiber Reinforced Ceramic Composite for Brake Discs*; SAE Technical Paper; SAE International: Warrendale, PA, USA, 2000. [CrossRef]
10. Hardwicke, C.U.; Lau, Y.-C. Advances in Thermal Spray Coatings for Gas Turbines and Energy Generation: A Review. *J. Therm. Spray Technol.* **2013**, *22*, 564–576. [CrossRef]
11. Cheng, Z.; Liu, M. Characterization of sulfur poisoning of Ni–YSZ anodes for solid oxide fuel cells using in situ Raman microspectroscopy. *Solid State Ion.* **2007**, *178*, 925–935. [CrossRef]
12. Valenzuela-Gutiérrez, A.; López-Cuevas, J.; González-Ángeles, A.; Pilalua-Díaz, N. Addition of ceramics materials to improve the corrosion resistance of alumina refractories. *SN Appl. Sci.* **2019**, *1*, 784. [CrossRef]
13. Kong, N.; Chen, A.; Yan, W.; Zhang, H. Ceramic implant fracture: A clinical report. *J. Prosthet. Dent.* **2019**, *122*, 425–429. [CrossRef]
14. Bai, R.; Sun, Q.; He, Y.; Peng, L.; Zhang, Y.; Zhang, L.; Lu, W.; Deng, J.; Zhuang, Z.; Yu, T.; et al. Ceramic Toughening Strategies for Biomedical Applications. *Front. Bioeng. Biotechnol.* **2022**, *10*, 840372. [CrossRef] [PubMed]
15. Zoli, L.; Servadei, F.; Bassi, G.; Rossi, A.; Montesi, M.; Vinci, A.; Sciti, D.; Panseri, S. From outer space to inside the body: Ultra-high temperature ceramic matrix composites for biomedical applications. *J. Eur. Ceram. Soc.* **2024**, *44*, 729–737. [CrossRef]
16. Rincón, A.; Chinelatto, A.S.A.; Moreno, R. Tape casting of alumina/zirconia suspensions containing graphene oxide. *J. Eur. Ceram. Soc.* **2014**, *34*, 1819–1827. [CrossRef]
17. Balázsi, C.; Kónya, Z.; Wéber, F.; Biró, L.P.; Arató, P. Preparation and characterization of carbon nanotube reinforced silicon nitride composites. *Mater. Sci. Eng. C* **2003**, *23*, 1133–1137. [CrossRef]
18. Sayyadi-Shahraki, A.; Rafiaei, S.M.; Ghadami, S.; Nekouee, K.A. Densification and mechanical properties of spark plasma sintered $Si_3N_4/ZrO_2$ nano-composites. *J. Alloys Compd.* **2019**, *776*, 798–806. [CrossRef]
19. Balázsi, K.; Furkó, M.; Liao, Z.; Gluch, J.; Medved, D.; Sedlák, R.; Dusza, J.; Zschech, E.; Balázsi, C. Porous sandwich ceramic of layered silicon nitride-zirconia composite with various multilayered graphene content. *J. Alloys Compd.* **2020**, *832*, 154984. [CrossRef]
20. Liao, Z.; Standke, Y.; Gluch, J.; Balázsi, K.; Pathak, O.; Höhn, S.; Herrmann, M.; Werner, S.; Dusza, J.; Balázsi, C.; et al. Microstructure and fracture mechanism investigation of porous silicon nitride–zirconia–graphene composite using multi-scale and in-situ microscopy. *Nanomaterials* **2021**, *11*, 285. [CrossRef]

21. Rutkowski, P.; Stobierski, L.; Górny, G. Thermal stability and conductivity of hot-pressed $Si_3N_4$-graphene composites. *J. Therm. Anal. Calorim.* **2014**, *116*, 321–328. [CrossRef]
22. Cygan, T.; Wozniak, J.; Kostecki, M.; Adamczyk-Cieslak, B.; Olszyna, A. Influence of graphene addition and sintering temperature on physical properties of $Si_3N_4$ matrix composites. *Int. J. Refract. Met. Hard Mater.* **2016**, *57*, 19–23. [CrossRef]
23. Zhu, Y.; Murali, S.; Cai, W.; Li, X.; Suk, J.W.; Potts, J.R.; Ruoff, R.S. Graphene and graphene oxide: Synthesis, properties, and applications. *Adv. Mater.* **2010**, *22*, 3906–3924. [CrossRef]
24. Qadir, A.; Balazsi, K.; Balazsi, C.; Ivor, M.; Dusza, J. Properties of MWCNTs added $Si_3N_4$ composites processed from oxidized silicon nitride powders. *Process. Appl. Ceram.* **2020**, *14*, 25–31. [CrossRef]
25. Kovalčikova, A.; Balázsi, C.; Dusza, J.; Tapasztó, O. Mechanical properties and electrical conductivity in a carbon nanotube reinforced silicon nitride composite. *Ceram. Int.* **2012**, *38*, 527–533. [CrossRef]
26. Ge, P.; Sun, K.; Li, A.; Pingji, G. Improving the electrical and microwave absorbing properties of $Si_3N_4$ ceramics with carbon nanotube fibers. *Ceram. Int.* **2018**, *44*, 2727–2731. [CrossRef]
27. Tapasztó, O.; Balko, J.; Puchy, V.; Kun, P.; Dobrik, G.; Fogarassy, Z.; Horváth, Z.E.; Dusza, J.; Balázsi, K.; Balázsi, C.; et al. Highly wear-resistant and low-friction $Si_3N_4$ composites by addition of graphene nanoplatelets approaching the 2D limit. *Sci. Rep.* **2017**, *7*, 1–8. [CrossRef] [PubMed]
28. Amutha Rani, D.; Yoshizawa, Y.; Hyuga, H.; Hirao, K.; Yamauchi, Y. Tribological behavior of ceramic materials ($Si_3N_4$, SiC and $Al_2O_3$) in aqueous medium. *J. Eur. Ceram. Soc.* **2004**, *24*, 3279–3284. [CrossRef]
29. Subramaniam, S.; Nithyaprakash, R.; Abbas, G.; Pramanik, A.; Basak, A.K. Tribological behavior of silicon nitride-based ceramics-a review. *J. Tribol.* **2021**, *29*, 57–71.
30. Schlüter, B.; Schröder, C.; Zhang, W.; Mülhaupt, R.; Degenhardt, U.; Sedlák, R.; Dusza, J.; Balázsi, K.; Balázsi, C.; Kailer, A. Influence of Graphene Type and Content on Friction and Wear of Silicon Carbide/Graphene Nanocomposites in Aqueous Environment. *Materials* **2022**, *15*, 7755. [CrossRef]
31. Llorente, J.; Ramírez, C.; Belmonte, M. High graphene fillers content for improving the tribological performance of silicon nitride-based ceramics. *Wear* **2019**, *430–431*, 183–190. [CrossRef]
32. Shrivastava, S.; Rajak, D.K.; Joshi, T.; Singh, D.K.; Mondal, D.P. Ceramic Matrix Composites: Classifications, Manufacturing, Properties, and Applications. *Ceramics* **2024**, *7*, 652–679. [CrossRef]
33. Ferraris, E.; Vleugels, J.; Guo, Y.; Bourell, D.; Kruth, J.P.; Lauwers, B. Shaping of engineering ceramics by electro, chemical and physical processes. *CIRP Ann. Manuf. Technol.* **2016**, *65*, 761–784. [CrossRef]
34. Franks, G.V.; Tallon, C.; Studart, A.R.; Sesso, M.L.; Leo, S. Colloidal processing: Enabling complex shaped ceramics with unique multiscale structures. *J. Am. Ceram. Soc.* **2017**, *100*, 458–490. [CrossRef]
35. Terrani, K.A.; Kiggans, J.O.; Silva, C.M.; Shih, C.; Katoh, Y.; Snead, L.L. Progress on matrix SiC processing and properties for fully ceramic microencapsulated fuel form. *J. Nucl. Mater.* **2015**, *457*, 9–17. [CrossRef]
36. Ramachandran, K.; Boopalan, V.; Bear, J.C.; Subramani, R. Multi-walled carbon nanotubes (MWCNTs)-reinforced ceramic nanocomposites for aerospace applications: A review. *J. Mater. Sci.* **2022**, *57*, 3923–3953. [CrossRef]
37. Fernie, J.A.; Drew, R.A.L.; Knowles, K.M. Joining of engineering ceramics. *Int. Mater. Rev.* **2009**, *54*, 283–331. [CrossRef]
38. Katz, R.N. Overview of Ceramic Materials, Design, and Application. *Mech. Eng. Handb. Mater. Mech. Des. Third Ed.* **2006**, *1*, 433–449. [CrossRef]
39. Xu, M.; Girish, Y.R.; Rakesh, K.P.; Wu, P.; Manukumar, H.M.; Byrappa, S.M.; Udayabhanu; Byrappa, K. Recent advances and challenges in silicon carbide (SiC) ceramic nanoarchitectures and their applications. *Mater. Today Commun.* **2021**, *28*, 102533. [CrossRef]
40. Abderrazak, H.; Hadj Hmi, E.S.B. Silicon Carbide: Synthesis and Properties. In *Properties and Applications of Silicon Carbide*; IntechOpen: London, UK, 2011; pp. 361–388. [CrossRef]
41. Dasog, M.; Smith, L.F.; Purkait, T.K.; Veinot, J.G.C. Low temperature synthesis of silicon carbide nanomaterials using a solid-state method. *Chem. Commun.* **2013**, *49*, 7004–7006. [CrossRef]
42. Rejhon, M.; Zhou, X.; Lavini, F.; Zanut, A.; Popovich, F.; Schellack, L.; Witek, L.; Coelho, P.; Kunc, J.; Riedo, E. Giant Increase of Hardness in Silicon Carbide by Metastable Single Layer Diamond-Like Coating. *Adv. Sci.* **2023**, *10*, e2204562. [CrossRef]
43. Chaudhury, P.; Samantaray, S. Modelling and Optimization of Machining of SiC-CNT Conductive Ceramic Composite used for Micro and Nano Sensor by Electrical Discharge Machining. *J. Inst. Eng. Ser. D* **2021**, *102*, 437–452. [CrossRef]
44. Kaunisto, K.; Lagerbom, J.; Honkanen, M.; Varis, T.; Lambai, A. Evolution of alumina phase structure in thermal plasma processing. *Ceram. Int.* **2023**, *49*, 21346–21354. [CrossRef]
45. Busca, G. Structural, Surface, and Catalytic Properties of Aluminas. In *Advances in Catalysis*; Elsevier Inc.: Amsterdam, The Netherlands, 2014; Volume 57, pp. 319–404. ISBN 9780128001271.
46. Spiridigliozzi, H. Bottom-up Strategy for the Development of Optically Transparent and Nanostructured $MgAl_2O_4$ Spinel Ceramics. Master's Thesis, Université Sorbonne Paris Nord, Villetaneuse, France, 2024.
47. Aruna, S.T.; Balaji, N.; Shedthi, J.; Grips, V.K.W. Surface & Coatings Technology Effect of critical plasma spray parameters on the microstructure, microhardness and wear and corrosion resistance of plasma sprayed alumina coatings. *Surf. Coat. Technol.* **2012**, *208*, 92–100. [CrossRef]
48. Warlimont, H. Classes of Materials (Ceramics). In *Springer Handbook of Condensed Matter and Materials Data*; Martienssen, W., Warlimont, H., Eds.; Springer: Berlin/Heidelberg, Germany, 2005; pp. 431–476.

49. Abyzov, A.M. Aluminum Oxide and Alumina Ceramics (review). Part 1. Properties of $Al_2O_3$ and Commercial Production of Dispersed $Al_2O_3$. *Refract. Ind. Ceram.* **2019**, *60*, 24–32. [CrossRef]
50. Xu, C.; Liu, H.; Yang, H.; Yang, L. A Green Biocompatible Fabrication of Highly Porous Functional Ceramics with High Strength and Controllable Pore Structures. *J. Mater. Sci. Technol.* **2016**, *32*, 729–732. [CrossRef]
51. Ghanizadeh, S.; Bao, X.; Vaidhyanathan, B.; Binner, J. Synthesis of nano α-alumina powders using hydrothermal and precipitation routes: A comparative study. *Ceram. Int.* **2014**, *40*, 1311–1319. [CrossRef]
52. Ghanizadeh, S. Membrane Emulsification for Particle Production. Ph.D. Thesis, Loughborough University, Reading, UK, 2013.
53. Li, J.G.; Sun, X. Synthesis and sintering behavior of a nanocrystalline α-alumina powder. *Acta Mater.* **2000**, *48*, 3103–3112. [CrossRef]
54. Suchanek, W.L. Hydrothermal synthesis of alpha alumina (α-$Al_2O_3$) powders: Study of the processing variables and growth mechanisms. *J. Am. Ceram. Soc.* **2010**, *93*, 399–412. [CrossRef]
55. Bapat, R.A.; Yang, H.J.; Chaubal, T.V.; Dharmadhikari, S.; Abdulla, A.M.; Arora, S.; Rawal, S.; Kesharwani, P. Review on synthesis, properties and multifarious therapeutic applications of nanostructured zirconia in dentistry. *RSC Adv.* **2022**, *12*, 12773–12793. [CrossRef]
56. Hu, C.; Sun, J.; Long, C.; Wu, L.; Zhou, C.; Zhang, X. Synthesis of nano zirconium oxide and its application in dentistry. *Nanotechnol. Rev.* **2019**, *8*, 396–404. [CrossRef]
57. Cannizzaro, G.; Torchio, C.; Felice, P.; Leone, M.; Esposito, M. Immediate occlusal versus non-occlusal loading of single zirconia implants. A multicentre pragmatic randomised clinical trial. *Eur. J. Oral Implantol.* **2010**, *3*, 111–120.
58. Hisbergues, M.; Vendeville, S.; Vendeville, P. Review zirconia: Established facts and perspectives for a biomaterial in dental implantology. *J. Biomed. Mater. Res. Part B Appl. Biomater.* **2009**, *88*, 519–529. [CrossRef]
59. Sailer, I.; Zembic, A.; Jung, R.E.; Siegenthaler, D.; Holderegger, C.; Hämmerle, C.H.F. Randomized controlled clinical trial of customized zirconia and titanium implant abutments for canine and posterior single-tooth implant reconstructions: Preliminary results at 1 year of function. *Clin. Oral Implants Res.* **2009**, *20*, 219–225. [CrossRef] [PubMed]
60. Balagangadharan, K.; Viji Chandran, S.; Arumugam, B.; Saravanan, S.; Devanand Venkatasubbu, G.; Selvamurugan, N. Chitosan/nano-hydroxyapatite/nano-zirconium dioxide scaffolds with miR-590-5p for bone regeneration. *Int. J. Biol. Macromol.* **2018**, *111*, 953–958. [CrossRef]
61. Wang, J.; Huang, C.; Wan, Q.; Chen, Y.; Chao, Y. Characterization of fluoridated hydroxyapatite/zirconia nano-composite coating deposited by a modified electrocodeposition technique. *Surf. Coatings Technol.* **2010**, *204*, 2576–2582. [CrossRef]
62. Hakim, L.F.; George, S.M.; Weimer, A.W. Conformal nanocoating of zirconia nanoparticles by atomic layer deposition in a fluidized bed reactor. *Nanotechnology* **2005**, *16*, S375–S381. [CrossRef] [PubMed]
63. Zhu, Y.; Zhu, R.; Ma, J.; Weng, Z.; Wang, Y.; Shi, X.; Li, Y.; Yan, X.; Dong, Z.; Xu, J.; et al. In vitro cell proliferation evaluation of porous nano-zirconia scaffolds with different porosity for bone tissue engineering. *Biomed. Mater.* **2015**, *10*, 55009. [CrossRef]
64. Andreiotelli, M.; Wenz, H.J.; Kohal, R.J. Are ceramic implants a viable alternative to titanium implants? A systematic literature review. *Clin. Oral Implants Res.* **2009**, *20*, 32–47. [CrossRef]
65. Chopra, D.; Guo, T.; Gulati, K.; Ivanovski, S. Load, unload and repeat: Understanding the mechanical characteristics of zirconia in dentistry. *Dent. Mater.* **2024**, *40*, e1–e17. [CrossRef]
66. Rashad, M.M.; Baioumy, H.M. Effect of thermal treatment on the crystal structure and morphology of zirconia nanopowders produced by three different routes. *J. Mater. Process. Technol.* **2008**, *195*, 178–185. [CrossRef]
67. De Aza, A.H.; Chevalier, J.; Fantozzi, G.; Schehl, M.; Torrecillas, R. Crack growth resistance of alumina, zirconia and zirconia toughened alumina ceramics for joint prostheses. *Biomaterials* **2002**, *23*, 937–945. [CrossRef]
68. Ji, H.; Huang, Z.; Chen, K.; Li, W.; Gao, Y.; Fang, M.; Liu, Y.-G.; Wu, X. Synthesis of $Si_3N_4$ powder with tunable α/β-$Si_3N_4$ content from waste silica fume using carbothermal reduction nitridation. *Powder Technol.* **2014**, *252*, 51–55. [CrossRef]
69. Li, B.; Feng, Y.; Li, G.; Chen, H.; Chen, J.; Hou, X. Preparation of high-purity α-$Si_3N_4$ nano-powder by precursor-carbothermal reduction and nitridation. *Ceram. Int.* **2019**, *45*, 6335–6339. [CrossRef]
70. Heimann, R.B. Silicon Nitride Ceramics: Structure, Synthesis, Properties, and Biomedical Applications. *Materials* **2023**, *16*, 5142. [CrossRef] [PubMed]
71. Legut, D.; Wdowik, U.D.; Kurtyka, P. Vibrational and dielectric properties of α-$Si_3N_4$ from density functional theory. *Mater. Chem. Phys.* **2014**, *147*, 42–49. [CrossRef]
72. Van Cuong, T.; Duc Ha, N.; Quoc Khanh, D.; Manh Tuong, N.; Hung Vu, N.; Dao, V.-D. Preparation of $Si_3N_4$ powder by carbothermal reduction and nitridation of glucose and water glass. *Inorg. Chem. Commun.* **2024**, *162*, 112241. [CrossRef]
73. Leitch, S.; Moewes, A.; Ouyang, L.; Ching, W.Y.; Sekine, T. Properties of non-equivalent sites and bandgap of spinel-phase silicon nitride. *J. Phys. Condens. Matter* **2004**, *16*, 6469–6476. [CrossRef]
74. Katz, R.N. High-Temperature Structural Ceramics. *Science* **1980**, *208*, 841–847. [CrossRef]
75. Riley, F.L. Silicon nitride and related materials. *J. Am. Ceram. Soc.* **2000**, *83*, 245–265. [CrossRef]
76. Yang, J.H.; Chen, Y.X.; Liu, G.H.; Lin, Z.M.; Li, J.T. Effects of diazenedicarboxamide additive on the content of α-Si 3N 4 synthesized by combustion method. *Ceram. Int.* **2012**, *38*, 961–965. [CrossRef]
77. Du, X.; Lee, S.S.; Blugan, G.; Ferguson, S.J. Silicon Nitride as a Biomedical Material: An Overview. *Int. J. Mol. Sci.* **2022**, *23*, 6551. [CrossRef]

78. Matsumae, T.; Kurashima, Y.; Higurashi, E.; Nishizono, K.; Amano, T.; Takagi, H. Room temperature bonding of aluminum nitride ceramic and semiconductor substrate. *Ceram. Int.* **2020**, *46*, 25956–25963. [CrossRef]
79. Li, G.; Li, B.; Ren, B.; Chen, H.; Zhu, B.; Chen, J. Synthesis of Aluminum Nitride Using Sodium Aluminate as Aluminum Source. *Processes* **2023**, *11*, 1034. [CrossRef]
80. Wang, C.P.; Huang, Y.C.; Liu, H.Q. Efficiency improvement of power LED modules using a hybrid aluminum nitride substrate. *Microelectron. Eng.* **2020**, *223*, 111227. [CrossRef]
81. Huang, D.; Liu, Z.; Harris, J.; Diao, X.; Liu, G. High thermal conductive AlN substrate for heat dissipation in high-power LEDs. *Ceram. Int.* **2019**, *45*, 1412–1415. [CrossRef]
82. Varshni, Y.P. Temperature dependence of the energy gap in semiconductors. *Physica* **1967**, *34*, 149–154. [CrossRef]
83. Kato, T.; Sugawara, K. Low-Temperature Synthesis of Aluminum Nitride by Addition of Ammonium Chloride. *ACS Omega* **2019**, *4*, 14714–14720. [CrossRef]
84. Lee, K.B.; Kim, J.; Shim, C.H.; Kim, Y.; Choi, H.; Ahn, J.P. Low-Temperature synthesis of high-purity AlN from Al powder. *J. Mater. Res. Technol.* **2022**, *21*, 4526–4536. [CrossRef]
85. Molisani, A.L.; Yoshimura, H.N. Low-temperature synthesis of AlN powder with multicomponent additive systems by carbothermal reduction-nitridation method. *Mater. Res. Bull.* **2010**, *45*, 733–738. [CrossRef]
86. CeramTec Industrial. *Ceramic Material with Very High Thermal Conductivity*; CeramTec Industrial: Plochingen, Germany. Available online: https://www.ceramtec-industrial.com/en/materials (accessed on 21 November 2024).
87. Oliveros, A.; Guiseppi-Elie, A.; Saddow, S.E. Silicon carbide: A versatile material for biosensor applications. *Biomed. Microdevices* **2013**, *15*, 353–368. [CrossRef]
88. Hassanpour, P.; Panahi, Y.; Ebrahimi-Kalan, A.; Akbarzadeh, A.; Davaran, S.; Nasibova, A.N.; Khalilov, R.; Kavetskyy, T. Biomedical applications of aluminium oxide nanoparticles. *Micro Nano Lett.* **2018**, *13*, 1227–1231. [CrossRef]
89. Clinton, D.J.; Morrell, R. Hardness testing of ceramic materials. *Mater. Chem. Phys.* **1987**, *17*, 461–473. [CrossRef]
90. Bai, H.; Zhong, L.; Kang, L.; Liu, J.; Zhuang, W.; Lv, Z.; Xu, Y. A review on wear-resistant coating with high hardness and high toughness on the surface of titanium alloy. *J. Alloys Compd.* **2021**, *882*, 160645. [CrossRef]
91. Qadir, A.; Ali, S.; Dusza, J.; Rafaja, D. Predicting hardness of graphene-added $Si_3N_4$ using machine learning: A data-driven approach. *Open Ceram.* **2024**, *19*, 100634. [CrossRef]
92. Shin, J.H.; Choi, J.; Kim, M.; Hong, S.H. Comparative study on carbon nanotube- and reduced graphene oxide-reinforced alumina ceramic composites. *Ceram. Int.* **2018**, *44*, 8350–8357. [CrossRef]
93. Krstic, Z.; Krstic, V.D. Silicon nitride: The engineering material of the future. *J. Mater. Sci.* **2012**, *47*, 535–552. [CrossRef]
94. Ming-Chang, J.; Li-Yung, Y. Environmental effects on wear behaviour of alumina. *Wear* **1993**, *161*, 111–119. [CrossRef]
95. Correa Filho, L.; Fu, L.; Engqvist, H.; Xia, W.; Persson, C. Wear Performance of a Novel Silicon Nitride Ceramic for Biomedical Applications. *Biomed. Mater. Devices* **2023**, *1*, 990–999. [CrossRef]
96. Yazdani, B.; Xu, F.; Ahmad, I.; Hou, X.; Xia, Y.; Zhu, Y. Tribological performance of Graphene/Carbon nanotube hybrid reinforced $Al_2O_3$ composites. *Sci. Rep.* **2015**, *5*. [CrossRef]
97. Maros B., M.; Németh, A.K.; Károly, Z.; Bódis, E.; Maros, Z.; Tapasztó, O.; Balázsi, K. Tribological characterisation of silicon nitride/multilayer graphene nanocomposites produced by HIP and SPS technology. *Tribol. Int.* **2015**, *93*, 269–281. [CrossRef]
98. Sharma, N.; Alam, S.N.; Ray, B.C.; Yadav, S.; Biswas, K. Wear behavior of silica and alumina-based nanocomposites reinforced with multi walled carbon nanotubes and graphene nanoplatelets. *Wear* **2019**, *418–419*, 290–304. [CrossRef]
99. Shah, W.A.; Luo, X.; Rabiu, B.I.; Huang, B.; Yang, Y.Q. Toughness enhancement and thermal properties of graphene-CNTs reinforced $Al_2O_3$ ceramic hybrid nanocomposites. *Chem. Phys. Lett.* **2021**, *781*, 138978. [CrossRef]
100. Yazdani, B.; Porwal, H.; Xia, Y.; Yan, H.; Reece, M.J.; Zhu, Y. Role of synthesis method on microstructure and mechanical properties of graphene/carbon nanotube toughened $Al_2O_3$ nanocomposites. *Ceram. Int.* **2015**, *41*, 9813–9822. [CrossRef]
101. Liu, J.; Yan, H.; Reece, M.J.; Jiang, K. Toughening of zirconia/alumina composites by the addition of graphene platelets. *J. Eur. Ceram. Soc.* **2012**, *32*, 4185–4193. [CrossRef]
102. Anselmi-Tamburini, U.; Garay, J.E.; Munir, Z.A.; Tacca, A.; Maglia, F.; Spinolo, G. Spark plasma sintering and characterization of bulk nanostructured fully stabilized zirconia: Part I. Densification studies. *J. Mater. Res.* **2004**, *19*, 3255–3262. [CrossRef]
103. Shojaie-bahaabad, M.; Bozorg, M.; Najafizadeh, M.; Cavaliere, P. Ultra high temperature ceramic coatings in thermal protection systems (TPS). *Ceram. Int.* **2024**, *50*, 9937–9951. [CrossRef]
104. Koo, K.H.; Ha, Y.C.; Kim, S.Y.; Yoon, K.S.; Min, B.W.; Kim, S.R. Revision of ceramic head fracture after third generation ceramic-on-ceramic total hip arthroplasty. *J. Arthroplasty* **2014**, *29*, 214–218. [CrossRef]
105. Almansoori, A.; Majewski, C.; Rodenburg, C. Nanoclay/Polymer Composite Powders for Use in Laser Sintering Applications: Effects of Nanoclay Plasma Treatment. *JOM* **2017**, *69*, 2278–2285. [CrossRef]
106. Almansoori, A.; Abrams, K.J.; Ghali Al-Rubaye, A.D.; Majewski, C.; Rodenburg, C. Novel plasma treatment for preparation of laser sintered nanocomposite parts. *Addit. Manuf.* **2019**, *25*, 297–306. [CrossRef]
107. Bódis, E.; Cora, I.; Németh, P.; Tapasztó, O.; Mohai, M.; Tóth, S.; Károly, Z.; Szépvölgyi, J. Toughening of silicon nitride ceramics by addition of multilayer graphene. *Ceram. Int.* **2019**, *45*, 4810–4816. [CrossRef]
108. Maitra, S. *Nanoceramic Matrix Composites: Types, Processing and Applications*; Woodhead Publishing Limited: Cambridge, UK, 2014; ISBN 9780857091208.

109. Nevarez-Rascon, A.; Aguilar-Elguezabal, A.; Orrantia, E.; Bocanegra-Bernal, M.H. On the wide range of mechanical properties of ZTA and ATZ based dental ceramic composites by varying the $Al_2O_3$ and $ZrO_2$ content. *Int. J. Refract. Met. Hard Mater.* **2009**, *27*, 962–970. [CrossRef]
110. Parchovianský, M.; Galusek, D.; Švančárek, P.; Sedláček, J.; Šajgalík, P. Thermal behavior, electrical conductivity and microstructure of hot pressed $Al_2O_3$/SiC nanocomposites. *Ceram. Int.* **2014**, *40*, 14421–14429. [CrossRef]
111. Selvarajan, L.; Venkataramanan, K. Surface morphology and drilled hole accuracy of conductive ceramic composites $Si_3N_4$–TiN and $MoSi_2$–SiC on EDMed surfaces. *Wear* **2023**, *530–531*, 204973. [CrossRef]
112. Bayal, N.; Jeevanandam, P. Synthesis of $TiO_2$-MgO mixed metal oxide nanoparticles via a sol-gel method and studies on their optical properties. *Ceram. Int.* **2014**, *40*, 15463–15477. [CrossRef]
113. Yang, Y.; Li, B.; Zhang, C.; Wang, S.; Liu, K.; Yang, B. Fabrication and properties of graphene reinforced silicon nitride composite materials. *Mater. Sci. Eng. A* **2015**, *644*, 90–95. [CrossRef]
114. Kwon, S.M.; Lee, S.J.; Shon, I.J. Enhanced properties of nanostructured $ZrO_2$-graphene composites rapidly sintered via high-frequency induction heating. *Ceram. Int.* **2015**, *41*, 835–842. [CrossRef]
115. Liu, J.; Li, Z.; Yan, H.; Jiang, K. Spark plasma sintering of alumina composites with graphene platelets and silicon carbide nanoparticles. *Adv. Eng. Mater.* **2014**, *16*, 1111–1118. [CrossRef]
116. Román-Manso, B.; Figueiredo, F.M.; Achiaga, B.; Barea, R.; Pérez-Coll, D.; Morelos-Gómez, A.; Terrones, M.; Osendi, M.I.; Belmonte, M.; Miranzo, P. Electrically functional 3D-architectured graphene/SiC composites. *Carbon N. Y.* **2016**, *100*, 318–328. [CrossRef]
117. Cheng, Y.; Lyu, Y.; Han, W.; Hu, P.; Zhou, S.; Zhang, X. Multiscale toughening of $ZrB_2$-SiC-Graphene@$ZrB_2$-SiC dual composite ceramics. *J. Am. Ceram. Soc.* **2019**, *102*, 2041–2052. [CrossRef]
118. Yin, Z.; Yuan, J.; Chen, M.; Si, D.; Xu, C. Mechanical property and ballistic resistance of graphene platelets/B4C ceramic armor prepared by spark plasma sintering. *Ceram. Int.* **2019**, *45*, 23781–23787. [CrossRef]
119. Chen, J.; Li, N.; Wei, Y.; Han, B.; Zhang, Y. A low-cost approach to fabricate SiC nanosheets by reactive sintering from Si powders and graphite. *J. Alloys Compd.* **2019**, *788*, 345–351. [CrossRef]
120. Tapasztó, O.; Tapasztó, L.; Markó, M.; Kern, F.; Gadow, R.; Balázsi, C. Dispersion patterns of graphene and carbon nanotubes in ceramic matrix composites. *Chem. Phys. Lett.* **2011**, *511*, 340–343. [CrossRef]
121. Ramírez, C.; Belmonte, M.; Miranzo, P.; Osendi, M.I. Applications of ceramic/graphene composites and hybrids. *Materials* **2021**, *14*, 2071. [CrossRef] [PubMed]
122. Nakhanivej, P.; Park, S.K.; Shin, K.H.; Yun, S.; Park, H.S. Hierarchically structured vanadium pentoxide/reduced graphene oxide composite microballs for lithium ion battery cathodes. *J. Power Sources* **2019**, *436*, 226854. [CrossRef]
123. Peng, S.; Chen, T.; Lee, C.; Lu, H.; Lue, S.J. Optimal cobalt oxide ($Co_3O_4$): Graphene (GR) ratio in $Co_3O_4$/GR as air cathode catalyst for air-breathing hybrid electrolyte lithium-air battery. *J. Power Sources* **2020**, *471*, 228373. [CrossRef]
124. Han, X.; Zhang, Z.; Chen, H.; You, R.; Zheng, G.; Zhang, Q.; Wang, J.; Li, C.; Chen, S.; Yang, Y. Double-shelled microscale porous Si anodes for stable lithium-ion batteries. *J. Power Sources* **2019**, *436*, 226794. [CrossRef]
125. Liang, J.; Zhang, Z.; Yang, W.; Li, X.; Li, P.; Guo, X.; Jung, Y.; Dong, X. Nitrification protection of Si monocrystal nanoparticles into the graphene matrix as the high-performance anode material for lithium-ion batteries. *Mater. Chem. Phys.* **2020**, *249*, 123156. [CrossRef]
126. Tang, J.; Liu, J.; Huang, H. Dielectric, Piezoelectric and Ferroelectric Properties of Flexible 0–3 Type PZT/PVDF Composites Doped with Graphene. *J. Electron. Mater.* **2019**, *48*, 4033–4039. [CrossRef]
127. Mallada, C.; Menéndez, J.L.; Dura, O.J.; López de la Torre, M.A.; Menéndez, R.; Santamaría, R. Spark plasma sintered $BaTiO_3$/graphene composites for thermoelectric applications. *J. Eur. Ceram. Soc.* **2017**, *37*, 3741–3746. [CrossRef]
128. Kim, H.W.; Na, H.G.; Kwon, Y.J.; Kang, S.Y.; Choi, M.S.; Bang, J.H.; Wu, P.; Kim, S.S. Microwave-Assisted Synthesis of Graphene-$SnO_2$ Nanocomposites and Their Applications in Gas Sensors. *ACS Appl. Mater. Interfaces* **2017**, *9*, 31667–31682. [CrossRef]
129. Tyagi, P.; Sharma, A.; Tomar, M.; Gupta, V. A comparative study of RGO-$SnO_2$ and MWCNT-$SnO_2$ nanocomposites based $SO_2$ gas sensors. *Sens. Actuators B Chem.* **2017**, *248*, 980–986. [CrossRef]
130. Taleb, M.; Ivanov, R.; Bereznev, S.; Kazemi, S.H.; Hussainova, I. Graphene-ceramic hybrid nanofibers for ultrasensitive electrochemical determination of ascorbic acid. *Microchim. Acta* **2017**, *184*, 897–905. [CrossRef]
131. Liu, J.; Li, S.; Zhang, B.; Xiao, Y.; Gao, Y.; Yang, Q.; Wang, Y.; Lu, G. *Ultrasensitive and Low Detection Limit of Nitrogen Dioxide Gas Sensor Based on Flower-Like ZnO Hierarchical Nanostructure Modified by Reduced Graphene Oxide*; Elsevier B.V.: Amsterdam, The Netherlands, 2017; Volume 249, ISBN 8643185167.
132. Fu, H.; Jiang, Y.; Ding, J.; Zhang, J.; Zhang, M.; Zhu, Y.; Li, H. Zinc oxide nanoparticle incorporated graphene oxide as sensing coating for interferometric optical microfiber for ammonia gas detection. *Sens. Actuators B Chem.* **2018**, *254*, 239–247. [CrossRef]
133. Tan, Y.Q.; Luo, H.; Zhou, X.S.; Peng, S.M.; Zhang, H.B. Boron carbide composites with highly aligned graphene nanoplatelets: Light-weight and efficient electromagnetic interference shielding materials at high temperatures. *RSC Adv.* **2018**, *8*, 39314–39320. [CrossRef] [PubMed]
134. Wajahat, M.; Kim, J.H.; Ahn, J.; Lee, S.; Bae, J.; Pyo, J.; Seol, S.K. *3D Printing of $Fe_3O_4$ Functionalized Graphene-Polymer (FGP) Composite Microarchitectures*; Elsevier Ltd.: Amsterdam, The Netherlands, 2020; Volume 167, ISBN 8255280146.

135. Tang, X.; Zhu, C.; Cheng, D.; Zhou, H.; Liu, X.; Xie, P.; Zhao, Q.; Zhang, D.; Fan, T. Architectured Leaf-Inspired $Ni_{0.33}Co_{0.66}S_2$/Graphene Aerogels via 3D Printing for High-Performance Energy Storage. *Adv. Funct. Mater.* **2018**, *28*, 1805057. [CrossRef]
136. Li, S.; Chen, H.; Wang, X.; Dong, X.; Huang, Y.; Guo, D. Catalytic degradation of clothianidin with graphene/$TiO_2$ using a dielectric barrier discharge (DBD) plasma system. *Environ. Sci. Pollut. Res.* **2020**, *27*, 29599–29611. [CrossRef] [PubMed]
137. Trinh, D.T.T.; Channei, D.; Chansaenpak, K.; Khanitchaidecha, W.; Nakaruk, A. Photocatalytic degradation of organic dye over bismuth vanadate–silicon dioxide–graphene oxide nanocomposite under visible light irradiation. *J. Aust. Ceram. Soc.* **2020**, *56*, 1237–1241. [CrossRef]
138. Zhao, G.; Mo, Z.; Zhang, P.; Wang, B.; Zhu, X. Synthesis of graphene/$Fe_3O_4$/NiO magnetic nanocomposites and its application in photocatalytic degradation the organic pollutants in wastewater. *J. Porous Mater.* **2015**, *22*, 1245–1253. [CrossRef]
139. Zhou, M.; Lin, T.; Huang, F.; Zhong, Y.; Wang, Z.; Tang, Y. Highly Conductive Porous Graphene / Ceramic Composites for Heat Transfer and Thermal Energy Storage. *Adv. Funct. Mater.* **2012**, *23*, 2263–2269. [CrossRef]
140. Zhou, M.; Bi, H.; Lin, T.; Lü, X.; Wan, D.; Huang, F.; Lin, J. Heat transport enhancement of thermal energy storage material using graphene/ceramic composites. *Carbon N. Y.* **2014**, *75*, 314–321. [CrossRef]
141. Garman, P.D.; Johnson, J.M.; Talesara, V.; Yang, H.; Du, X.; Pan, J.; Zhang, D.; Yu, J.; Cabrera, E.; Yen, Y.C.; et al. Dual Silicon Oxycarbide Accelerated Growth of Well-Ordered Graphitic Networks for Electronic and Thermal Applications. *Adv. Mater. Technol.* **2019**, *4*, 1800324. [CrossRef]
142. Garman, P.D.; Johnson, J.M.; Talesara, V.; Yang, H.; Zhang, D.; Castro, J.; Lu, W.; Hwang, J.; Lee, L.J. Silicon Oxycarbide Accelerated Chemical Vapor Deposition of Graphitic Networks on Ceramic Substrates for Thermal Management Enhancement. *ACS Appl. Nano Mater.* **2019**, *2*, 452–458. [CrossRef]
143. Chan, K.F.; Zaid, M.H.M.; Mamat, M.S.; Liza, S.; Tanemura, M.; Yaakob, Y. Recent Developments in Carbon Nanotubes-Reinforced Ceramic Matrix Composites: A Review on Dispersion and Densification Techniques. *Crystals* **2021**, *11*, 457. [CrossRef]
144. Guo, C.; Luo, X.; Shah, W.A.; Huang, B.; Li, J.K.; Umer, M.A.; Yang, Y.Q. Mechanical and electrical properties of carbon nanotube-reinforced $Al_2O_3$ nanocomposites. *J. Mater. Sci.* **2020**, *55*, 8728–8740. [CrossRef]
145. Karagedov, G.R.; Shutilov, R.A.; Kolesov, B.A.; Kuznetsov, V.L. The effect of carbon nanotubes introduction on the mechanical properties of reaction bonded boron carbide ceramics. *J. Eur. Ceram. Soc.* **2021**, *41*, 5782–5790. [CrossRef]
146. Yadhukulakrishnan, G.B.; Rahman, A.; Karumuri, S.; Stackpoole, M.M.; Kalkan, A.K.; Singh, R.P.; Harimkar, S.P. Spark plasma sintering of silicon carbide and multi-walled carbon nanotube reinforced zirconium diboride ceramic composite. *Mater. Sci. Eng. A* **2012**, *552*, 125–133. [CrossRef]
147. Shahedi Asl, M.; Farahbakhsh, I.; Nayebi, B. Characteristics of multi-walled carbon nanotube toughened $ZrB_2$-SiC ceramic composite prepared by hot pressing. *Ceram. Int.* **2016**, *42*, 1950–1958. [CrossRef]
148. Wiśniewska, M.; Laptev, A.M.; Marczewski, M.; Leshchynsky, V.; Lota, G.; Acznik, I.; Celotti, L.; Sullivan, A.; Szybowicz, M.; Garbiec, D. Influence of carbon nanotubes on thermal and electrical conductivity of zirconia-based composite. *Ceram. Int.* **2023**, *49*, 15442–15450. [CrossRef]
149. Michálek, M.; Kašiarová, M.; Michálková, M.; Galusek, D. Mechanical and functional properties of $Al_2O_3$-$ZrO_2$-MWCNTs nanocomposites. *J. Eur. Ceram. Soc.* **2014**, *34*, 3329–3337. [CrossRef]
150. Ahmad, I.; Unwin, M.; Cao, H.; Chen, H.; Zhao, H.; Kennedy, A.; Zhu, Y.Q. Multi-walled carbon nanotubes reinforced $Al_2O_3$ nanocomposites: Mechanical properties and interfacial investigations. *Compos. Sci. Technol.* **2010**, *70*, 1199–1206. [CrossRef]
151. Anstis, G.R. Indentation and fracture toughness, I. *Transformation* **1981**, *46*, 533–538.
152. Cuadrado, N.; Casellas, D.; Anglada, M.; Jiménez-Piqué, E. Evaluation of fracture toughness of small volumes by means of cube-corner nanoindentation. *Scr. Mater.* **2012**, *66*, 670–673. [CrossRef]
153. *ASTM E399*; Annual Book of ASTM Standards: Standard Test Method for Plain Strain Fracture Toughness of Metallic Materials. ASTM: Philadelphia, PA, USA, 1991.
154. Balázsi, K.; Furkó, M.; Balázsi, C. Ceramic Matrix Graphene and Carbon Nanotube Composites. *Encycl. Mater. Tech. Ceram. Glas.* **2021**, *2*, 243–259. [CrossRef]
155. Zhao, W.; Sun, J.; Huang, Z. Three-dimensional graphene-carbon nanotube reinforced ceramics and computer simulation. *Ceram. Int.* **2021**, *47*, 33941–33955. [CrossRef]
156. Ahmad, I.; Yazdani, B.; Zhu, Y. Recent advances on carbon nanotubes and graphene reinforced ceramics nanocomposites. *Nanomaterials* **2014**, *5*, 90–114. [CrossRef] [PubMed]
157. Cheng, Y.; Zhang, Y.; Wan, T.; Yin, Z.; Wang, J. Mechanical properties and toughening mechanisms of graphene platelets reinforced $Al_2O_3$/TiC composite ceramic tool materials by microwave sintering. *Mater. Sci. Eng. A* **2017**, *680*, 190–196. [CrossRef]
158. Sun, D.; Jiang, X.; Su, L.; Sun, H.; Hu, C.; Song, T.; Luo, Z. Fabrication and mechanical properties of Al2O3–TiC ceramic composites synergistically reinforced with multi-walled carbon nanotubes and graphene nanoplates. *Ceram. Int.* **2020**, *46*, 20068–20080. [CrossRef]
159. Gurnani, L.; Mukhopadhyay, A. Development of Carbon Nanotube-Reinforced Ceramic Matrix Nanocomposites for Advanced Structural Applications. In *Handbook of Advanced Ceramics and Composites*; Springer International Publishing: Cham, Switzerland, 2020; pp. 929–974.
160. Shah, W.A.; Luo, X.; Yang, Y.Q. Microstructure, mechanical, and thermal properties of graphene and carbon nanotube-reinforced Al2O3 nanocomposites. *J. Mater. Sci. Mater. Electron.* **2021**, *32*, 13656–13672. [CrossRef]

161. Yazdani, B.; Xia, Y.; Ahmad, I.; Zhu, Y. Graphene and carbon nanotube (GNT)-reinforced alumina nanocomposites. *J. Eur. Ceram. Soc.* **2015**, *35*, 179–186. [CrossRef]
162. Silvestre, J.; Silvestre, N.; De Brito, J. An overview on the improvement of mechanical properties of ceramics nanocomposites. *J. Nanomater.* **2015**, *2015*, 106494. [CrossRef]
163. Dassios, K.G.; Alafogianni, P.; Antiohos, S.K.; Leptokaridis, C.; Barkoula, N.M.; Matikas, T.E. Optimization of sonication parameters for homogeneous surfactant assisted dispersion of multiwalled carbon nanotubes in aqueous solutions. *J. Phys. Chem. C* **2015**, *119*, 7506–7516. [CrossRef]
164. Estili, M.; Kawasaki, A.; Sakamoto, H.; Mekuchi, Y.; Kuno, M.; Tsukada, T. The homogeneous dispersion of surfactantless, slightly disordered, crystalline, multiwalled carbon nanotubes in α-alumina ceramics for structural reinforcement. *Acta Mater.* **2008**, *56*, 4070–4079. [CrossRef]
165. Rubel, R.I.; Ali, M.H.; Jafor, M.A.; Alam, M.M. Carbon nanotubes agglomeration in reinforced composites: A review. *AIMS Mater. Sci.* **2019**, *6*, 756–780. [CrossRef]
166. Rivero-Antúnez, P.; Cano-Crespo, R.; Esquivias, L.; de la Rosa-Fox, N.; Zamora-Ledezma, C.; Domínguez-Rodríguez, A.; Morales-Flórez, V. Mechanical characterization of sol-gel alumina-based ceramics with intragranular reinforcement of multiwalled carbon nanotubes. *Ceram. Int.* **2020**, *46*, 19723–19730. [CrossRef]
167. Tu, S.; Wang, Q.; Ramachandran, C.S. Parametric investigation of in-situ synthesis of carbon nanotubes on Al2O3 powder by the rotary chemical vapor deposition method. *Ceram. Int.* **2022**, *48*, 28258–28267. [CrossRef]
168. Inam, F.; Yan, H.; Reece, M.J.; Peijs, T. Structural and chemical stability of multiwall carbon nanotubes in sintered ceramic nanocomposite. *Adv. Appl. Ceram.* **2010**, *109*, 240–245. [CrossRef]
169. Suslova, E.; Savilov, S.; Egorov, A.; Shumyantsev, A.; Lunin, V. Carbon nanotube frameworks by spark plasma sintering. *Microporous Mesoporous Mater.* **2020**, *293*, 109807. [CrossRef]
170. Cho, J.; Boccaccini, A.R.; Shaffer, M.S.P. Ceramic matrix composites containing carbon nanotubes. *J. Mater. Sci.* **2009**, *44*, 1934–1951. [CrossRef]

*nanomaterials*

MDPI

*Review*

# Antimicrobial Nanotubes: From Synthesis and Promising Antimicrobial Upshots to Unanticipated Toxicities, Strategies to Limit Them, and Regulatory Issues

**Silvana Alfei [1,*] and Gian Carlo Schito [2]**

1 Department of Pharmacy (DIFAR), University of Genoa, Viale Cembrano, 4, 16148 Genoa, Italy
2 Department of Surgical Sciences and Integrated Diagnostics (DISC), University of Genoa, Viale Benedetto XV, 6, 16132 Genova, Italy; giancarlo.schito@unige.it
* Correspondence: alfei@difar.unige.it; Tel.: +39-010-355-2296

**Abstract:** Nanotubes (NTs) are nanosized tube-like structured materials made from various substances such as carbon, boron, or silicon. Carbon nanomaterials (CNMs), including carbon nanotubes (CNTs), graphene/graphene oxide (G/GO), and fullerenes, have good interatomic interactions and possess special characteristics, exploitable in several applications because of the presence of sp2 and sp3 bonds. Among NTs, CNTs are the most studied compounds due to their nonpareil electrical, mechanical, optical, and biomedical properties. Moreover, single-walled carbon nanotubes (SWNTs) have, in particular, demonstrated high ability as drug delivery systems and in transporting a wide range of chemicals across membranes and into living cells. Therefore, SWNTs, more than other NT structures, have generated interest in medicinal applications, such as target delivery, improved imaging, tissue regeneration, medication, and gene delivery, which provide nanosized devices with higher efficacy and fewer side effects. SWNTs and multi-walled CNTs (MWCNTs) have recently gained a great deal of attention for their antibacterial effects. Unfortunately, numerous recent studies have revealed unanticipated toxicities caused by CNTs. However, contradictory opinions exist regarding these findings. Moreover, the problem of controlling CNT-based products has become particularly evident, especially in relation to their large-scale production and the nanosized forms of the carbon that constitute them. Important directive rules have been approved over the years, but further research and regulatory measures should be introduced for a safer production and utilization of CNTs. Against this background, and after an overview of CNMs and CNTs, the antimicrobial properties of pristine and modified SWNTs and MWCNTs as well as the most relevant in vitro and in vivo studies on their possible toxicity, have been reported. Strategies and preventive behaviour to limit CNT risks have been provided. Finally, a debate on regulatory issues has also been included.

**Keywords:** carbon nanotubes (CNTs); mechanical, electrical, and optical properties; CNTs biomedical applications; antibacterial nanotubes; nanotubes toxicity; CNT regulation

## 1. Introduction

Nanotechnology is one of the most exciting 21st century technologies [1,2], with the capability to observe, measure, control, assemble, and manufacture materials at the nanoscale, and it translates the theory of nanoscience into practical applications [3–5]. Nanotechnology is a discipline of global interest, which is constantly growing [6–8]. It deals with a variety of materials produced at <100 nm through different chemical and physical methods [9]. It is conducted at the nanoscale (1–100 nm) and has unique phenomena that enable novel applications in a wide range of fields, including chemistry, physics, biology, medicine, engineering, and electronics [10,11]. To keep up with the rate of advancement in science and technology, new types of nanomaterials with unconventional edge and innovative properties are incessantly required [12]. Nanotubes (NTs) belong to a promising group of nanomaterials that allow us to approach several new electronic, magnetic, optical, and mechanical properties [13].

Many of the nanotube structures that have been extensively studied contain boron, silicon, and molybdenum atoms; carbon nanotubes (CNTs) are the most researched group among existing NT structures [14]. In fact, carbon nanomaterials (CNMs), including CNTs, graphene/graphene oxide (G/GO), and fullerenes, are allotropes of sp2 and sp3 hybridized carbon atoms and their sp2 and sp3 bonds and are thought to be responsible for their unique characteristics [12]. They have good interatomic interactions [14] and are very interesting nanostructures in terms of their special properties and possible applications [15]. CNTs are composed of graphite sheets rolled up in an unbreakable and non-stop hexagonal-like lattice structure, in which carbon atoms appear at the tops of the hexagon-type forms. Based on the number of carbon sheets, CNTs are categorized as single-wall carbon nanotubes (SWCNTs), double-wall carbon nanotubes (DWCNTs), and multi-wall carbon nanotubes (MWCNTs) [13]. Although MWCNTs have also been extensively studied for their demonstrated antimicrobial potential, SWCNTs have displayed excellent antibacterial action, according to some research, due to the smaller size of these materials, which play a significant part in the inhibition of microbes [12]. Through a search made on Scopus on 1 March 2025, using "nanotubes" as a keyword, for the last fifty years (2010–2025), we found 87,798 documents concerning "nanotubes" and 80,535 documents concerning "carbon nanotubes" (Figure 1). Through an additional survey and using the keywords "single-walled nanotubes", we found 17,225 publications (Figure 1).

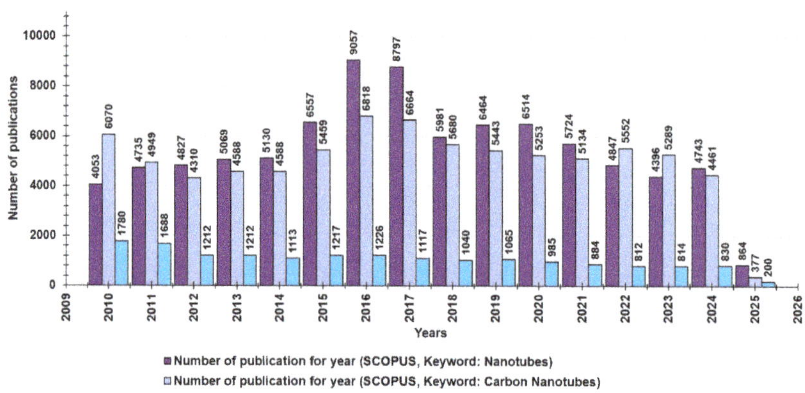


**Figure 1.** The number of publications per year during the last 15 years, according to Scopus, concerning nanotubes (purple bars), carbon nanotubes (light purple bars), and SWNTs (blue bars).

Generally, CNMs offer great surface area, mechanical resistance, thermal conductivity, photoluminescence, transparency, and constructional durability, in addition to antibacterial activities against pathogens and remarkable electrical conductivity [16]. These properties increasingly encourage the application of CNMs in several nanocomposites, such as in thin-film transistors, transparent conducting electrodes, photovoltaics, supercapacitors, biosensors, drug delivery systems, tissue engineering, photothermal therapy, and antimicrobial food packaging. As a part of the larger family of CNMs, CNTs are also materials with extraordinary properties that are useful across a variety of new, state-of-the-art applications in sensors, printed electronics, e-readers, flexible displays, energy storage medical treatments, and more. Specifically, since their discovery in 1991 by Ijima [17], SWNTs have influenced a significant amount of activity in both research and industry across the world. Furthermore, SWNTs have stimulated considerable investment in manufacturing methods, characterization techniques, and application development. However, the main drawback of CNTs is immediately recognized; it consists of their scarce solubility in most solvents, which limits their application [16]. To address this issue and amplify CNT use, researchers have focused on surface modifications [14–16]. Surface modifications, together with several other factors, including their chemical composition, target bacteria, and reaction environment, can also affect the CNMs' antimicrobial activity [18]. The antimicrobial effects of CNMs are derived mainly from their capacity to physically isolate pathogen cells from their supportive environment. Anyway, their capacity to penetrate microbial cell walls/membranes, thus causing irreversible structural damage, supports their antimicrobial activity to a large extent [12]. Additionally, the interaction of CNMs with bacterial cells stimulates the generation of noxious substances, including reactive oxygen species (ROS), triggering oxidative stress (OS) in cells and thus promoting their death. Additionally, the interactions between CNMs and microorganisms result in an electron transfer, which induces ROS-independent OS and causes the biological death of pathogens [19]. To confirm that the OS plays an additional role in CNT and SWCNT antimicrobial mechanisms [20], Haung et al. investigated the mechanical effects that influenced the antimicrobial properties of CNTs, including low wear rates, low friction coefficients, favourable tribological characteristics, and high corrosion resistance [21]. Among different types of CNMs and CNTs, the antimicrobial activity of SWCNTs is found to be higher due to their advantageous physicochemical properties [22–25]. In this regard, Kang and co-workers, in their first report on the antimicrobial activity of purified SWCNTs and MWCNTs, demonstrated that both materials showed significant antimicrobial effects. Moreover, those of SWCNTs seemed to be stronger than those of MWCNTs [26]. They found that CNTs' antibacterial effect mainly depended on their ruinous impact on the integrity of bacterial membranes upon direct contact; the higher activity of SWCNTs could be due to their small size, which provides a larger surface area to facilitate the membrane perturbation [26,27]. The morphology and metabolic activities of pathogens were also compromised [27]. Also, based on studies conducted by Chen et al., the SWCNTs played a significant role as "nano-darts", which penetrated bacterial cell walls, reduced membrane potential, released intracellular constituents (DNA and RNA), and ultimately disrupted bacterial membranes [28]. Unfortunately, despite their possible toxicity being initially disregarded due to their harmless carbon-based composition, early and recent investigations have unexpectedly unveiled that CNTs could be toxic to humans, animals, and the environment, depending on their concentrations and exposure times [29]. In fact, even if CNTs are made of normally innocuous and intrinsically safe carbon and do not have nano-dimensioned length, they should be considered as nanomaterials, which, thanks to their nanosized diameter, can penetrate various cells, including those in bacteria, yeast, and mammals [30]. It has been found that, mainly due to their poor biodegradability, accumulation, and persistence in the body, surface structure, and residual impurities, CNTs can be remarkably toxic, genotoxic, and cancerogenic to several cells, organs, and animal models. In this regard, amyloid deposits have been found in the brain, liver, lungs, and kidneys of exposed animals [29,31]. Additionally, investigations prompted by the similarity of the needle-like CNT fibres with those of asbestos evidenced that they can be responsible for pulmonary toxicity, mainly in CNT manufacturers, which results from contact with and/or inhalation of CNT fibres [32,33]. Also, toxicity to the cardiovascular system, immune and reproductive systems, embryos, and neurotoxicity have been reported [33]. Based on these findings, several strategies using advanced synthetic methods, highly efficient purification procedures, post-synthesis chemical modifications, as well as preventive behavioural conducts have already been developed to limit CNTs' toxic effects [34]. Anyway, more and incessant studies should be implemented to make their production and employment safer. On the other hand, despite significant advancements in terms of regulations concerning CNT production and possible applications for health and safety at the production sites being implemented, several unsolved regulatory issues survive, which need urgent solutions [34]. The increasing production and commercialization of differently structured CNTs and related nanocomposites and their daily application have stimulated the development of more rigid regulatory standards concerning their manufacturing and utilization [34]. Anyway, substantial concern regarding toxicity and environmental safety persists, mainly due to contradictory outcomes of toxicity investigations [35]. However, it must be recognized that CNTs, despite their still not fully clear risk to living organisms and the environment and limited regulation, are paradoxically ubiquitously exploitable for improving the quality of their life. In this regard, with this review, we aimed at stimulating further research on these nonpareil materials to better understand their potential as antimicrobial agents and the mechanisms supporting their cytotoxic effects on pathogens, which could also help to clarify the basis of their toxicity to humans. Such improved knowledge could allow their safer, large-scale production and utilization to enhance environment equilibrium and our quality of life. Against this background, and after an overview of CNMs and CNTs, the antimicrobial properties of pristine and modified SWCNTs and MWCNTs, as well as the most relevant in vitro and in vivo studies on their possible toxicity, have been discussed. The strategies already developed and the suggested preventive behaviour to limit CNT toxicity have also been provided and discussed. Finally, an extensive debate on the regulatory issues and the currently available standard guidelines by relevant International Organizations to produce, characterize, and manage CNTs and related nanocomposites have also been included.

This work is a thorough and contemporary review, which thoroughly explores and discusses different aspects concerning CNTs in a single document, including the synthesis, structure, applications, antimicrobial properties, functionalization strategies, toxicological concerns, and regulatory issues. This review affords systematic and well-organized information on CNTs and their potential applications. This study is supported by a substantial body of literature research and several case-specific tables, which enhance its readability and clarity, showing the double-sided impact that CNT use could have on our lives and providing future perspectives for the most challenging aspects of CNTs.

## 2. Approaching Carbon Nanotubes

Carbon is a chemical element that has atomic number 6. It is one of the most abundant elements in the Universe by mass, capable of providing approximately 10 million different pure organic compounds. Such compounds possess the uncommon ability to form polymers at earth temperatures, thus being the chemical basis of all known life [35].

CNTs are thin and long cylinders made of carbon, which were found for the first time in 1991 by Sumio Iijima [17]. Many physical properties of CNTs are still obscure and need to be disputed, as well as technical and not technical hurdles, which still limit the CNTs' success and their extensive application. Anyway, it is generally recognized that CNTs possess a plethora of thermal, electronic, mechanical, and structural properties that can vary based on their different existing forms [35]. In this regard, CNTs differ mainly in their diameter, length, chirality, or twist. Table A1, in Appendix A and reproduced from our previous work [35], collects the key properties and potential uses of CNTs, as well as the current remaining technical and non-technical hurdles. References in Table A1 do not belong to this work but are those contained in Alfei and Schito, 2022 [35]. We have decided on this solution so as not to overly burden the already significantly long list of references in this work. In addition to those reported in Table A1, a plethora of other possible applications for CNTs exist, including solar collectors, conductive and/or waterproof paper, catalyst supports, nano-porous filters, and coatings of all types. Since CNTs possess the capability to absorb infrared light, they may be applied to the I/R Optics Industry. We are confident that many unexpected applications for these nonpareil materials will be found in the next years, which will reinforce the belief that CNTs could be the most important and valuable nanomaterial ever used until now.

### 2.1. Synthesis of CNTs

Table A2 in Appendix A, reproduced by our previous work [35] collects the descriptions of the three most used methods to synthetize CNTs. Briefly, they include arc discharge (AD) method [36,37], finalized to the production of fullerenes, before 1991 [17], laser ablation (LA) [38,39], developed by Dr. Richard Smalley and co-workers at Rice University, and chemical vapor deposition (CVD) [40–46], which is the most widely used method to produce CNTs [35]. An advanced CVD technique is named plasma-enhanced chemical vapor deposition (PECVD) and consists of generating plasma by the application of a strong electric field during CNTs growth [47]. By adapting the reactor's geometry, it is possible to synthesize vertically aligned CNTs (VACNTs) [35], whose morphology is of interest to researchers attracted by electron emission from CNTs. Other methods have been described in detail in our previous work [35]. They include high-pressure carbon monoxide (HiPco) process, optimized at Rice University [35], super-growth CVD (SGCVD) [48,49], introduced by Kenji Hata, Sumio Iijima and col-leagues at AIST (Japan), plasma torch (PT) [50,51], an invention by Olivier Smiljanic in the year 2000, working at Institut National de la Recherche Scientifique (INRS) (Varennes, Canada), subsequently modified in IPT procedure by researchers from Sherbrooke University and the National Research Council of Canada and liquid electrolysis method (LEM) [52,53], which allowed to obtain MWCNTs by electrolysis of molten carbonates [35]. In addition to all these artificial inventions by human re-searcher, CNTs can form naturally in commonly originated flames emitted by burning methane [54], ethylene [55], and benzene [56]. Highly irregular in dimensions and low-quality CNT-based structures can form in smoke from both indoor and outdoor air [57]. Anyway, lacking the high degree of uniformity, due to uncontrolled conditions, necessary to satisfy the many needs of both research and industry, their practical ap-plication is hampered. Despite this, efforts focus on the strategies to control environ-mental flames by theoretical models to produce less irregular CNTs [58–62], thus going towards large-scale, low-cost CNTs synthesis competitive with large scale CVD pro-duction.

### 2.2. Architecture and Other Features of CNTs

CNTs deserved the prefix "nano" due to their very small size of 1/50,000th of the diameter of a human hair.

Pure carbon can develop in several carbon-based architectures, and CNTs are one of them. For example, diamond is the tetrahedron crystalline structure of carbon, while graphite or graphene owns a planar structure, with linked carbon atoms forming hexagons [35]. Buckyballs were named after F. Buckminister Fuller, who first designed geodesic spheres; they are also called fullerenes [63]. They are made of hexagons of pure carbon atoms linked to form a sphere. The typical structure of CNTs instead derives from the rolling up of a sheet of carbon atoms associated to form hexagons (graphene), thus forming cylinders (Figure 2a).

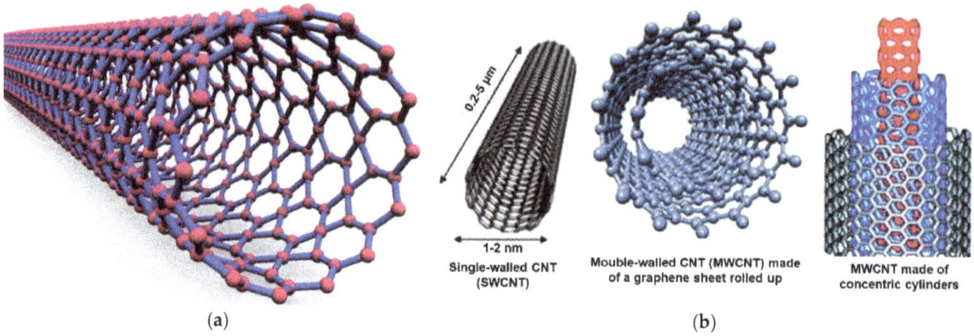


**Figure 2.** Structure of CNTs (**a**); structures of a SWCNT (left), of two different shapes of a MWCNT (center and right) (**b**). Specifically, an MWCNT made of a single graphene sheet rolled up (center) and an MWCNT made of more than two graphene cylinders with one inside the other (right). These images, by an unknown author, are licensed under CC BY and have been reproduced from our previous work [35].

Different structures that CNTs can assume are represented in Figure 2b. While SWCNTs (left side) are not naturally formed CNTs, MWCNTs are more complex CNTs formed naturally [63].

Additionally, when CNTs consist of two or three graphene complete cylinders with one inside the other, they are termed double- and triple-walled carbon nanotubes (DWCNTs and TWCNTs). Figure 3 shows the representative structure of a DWCNT.

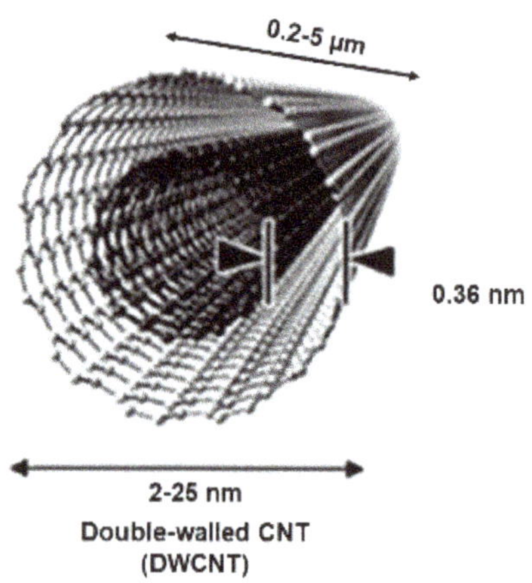


**Figure 3.** Structure of a DWCNT. This image by an unknown author is licensed under CC BY and was reproduced from our previous work [35].

CNTs are, atomically speaking, very stable, while SWCNTs are more stable than MWCNTs. Researchers in the field consider tubes with diameters less than 100 nm as CNTs [64]. However, no intrinsic limit exists on how long CNTs can develop [35]. However, their length is usually much larger than their diameter [35]. Table A3 in Appendix A summarizes the physical limits of CNTs. The references in the table do not belong to this work but are those contained in Alfei and Schito, 2022 [35]. We have decided on this solution for the reasons previously described. The production of SWCNTs is more difficult than that of MWCNTs, which can exhibit behaviours different from SWCNTs. In this regard, despite the dispersion of SWCNTs to obtain nanocomposites being more difficult than that of MWCNTs, generally, SWCNTs demonstrated better properties than MWCNTs. For this reason, scientists are focused mainly in finding more practical ways to mass-produce SWCNTs. Concerning the production of CNT-based nanocomposites, the possible junctions between two or more CNTs have been widely discussed theoretically [65,66], while those between CNTs and graphene have been considered both theoretically [67] and experimentally [68] (Table A4, Appendix A, references in the Table do not belong to this work but are those contained in Alfei and Schito, 2022 [35], according to reasons previously described). Junctions between CNTs were observed in CNTs prepared by AD or CVD methods. Lambin et al. were the first experts who studied theoretically the electronic properties of such junctions, starting from the assumption that a connection between a metallic tube and a semiconducting one could form a component of a CNT-based electronic circuit [69]. CNTs-graphene junctions are instead the basis of pillared graphene, characterized by three-dimensional (3D)-carbon nanotube (3D-CNTs) architectures, studied as building blocks to fabricate three-dimensional macroscopic structures [70]. Using these materials, it is possible to obtain free-standing, porous scaffolds made only of carbon, possessing macro-, micro-, and nano-structured pores and tailorable porosity with several applications. They could be used for the fabrication of the next-generation energy storages, supercapacitors, field emission transistors, high-performance catalysts, photovoltaics, and biomedical devices, such as implants and biosensors [71–73].

Relationships Structure/Properties

The study of CNT formations remains a cutting-edge field where new discoveries are being made regularly. Among the possible intimate structures that a SWCNT can assume, the armchair, zigzag, and chiral configurations are extensively recognized, based on the graphene mode to package into the CNTs cylinder (Figure 4).

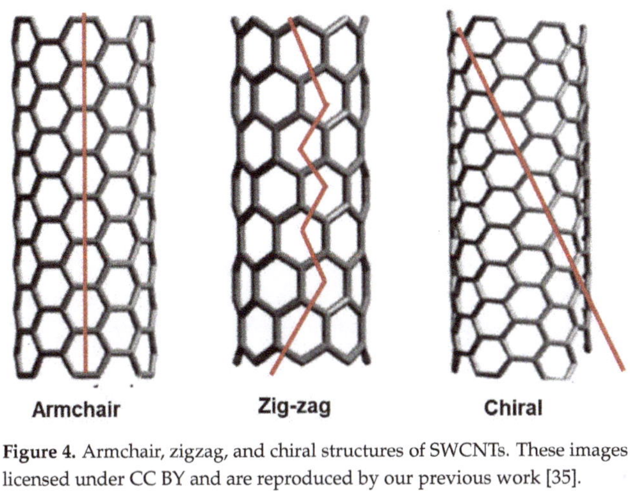


**Figure 4.** Armchair, zigzag, and chiral structures of SWCNTs. These images by unknown author are licensed under CC BY and are reproduced by our previous work [35].

The structural conformation of CNTs has a direct effect on their properties, mainly on the mechanical and electrical ones. For example, in MWCNTs, the outer walls of MWCNTs can protect the inner carbon tubes from chemical interactions with outside materials. Over the years, possible structure-related mechanical properties and structure-related electrical and thermal properties of CNTs have been reported in several studies. Structure-related mechanical investigations established that CNTs are the greatest nanomaterials discovered so far. Particularly, CNTs were better than other nanomaterials in terms of elastic modulus and tensile strength [35,74–76]. Anyway, it has also been reported that defects in the structure of CNTs due to atomic vacancies or rearrangements of the carbon bonds can cause weak points in small segments of the CNT, which reduce their elasticity and weaken their tensile strength [35]. On the other hand, structure-related electrical and thermal investigations showed that the structure of CNTs affect their conductive potency, both in terms of electrical and thermal conductivity [35,77–82].

## 3. Biomedical Applications of CNTs

CNTs belonging to all categories, including SWCNTs, DWCNTs, and MWCNTs, can differ in purity, length, and functionality. Anyway, all own a plethora of properties, such as high electrical conductivity, high tensile strength, light weight [83–85], and high biocompatibility [86]. All possess the capability to load molecules for their transport and delivery, large surface area, and chemical inertness. All can be enriched with functional groups and demonstrated good elasticity, thermal conductivity, capability to expand, electron emission capacity, and high aspect ratio [83–85].

These properties, added to peculiar composition and geometry, make CNTs suitable for numerous potential applications, including energy storage, biomedical uses, and air and water filtration. Also, CNTs could become molecular electronics, thermal materials, structural materials, electrical conductors, fabrics and fibres, catalyst supports, conductive plastics, conductive adhesives, as well as ceramics [83].

For this reason, scientists working in this sector are involved in efforts aimed at increasingly lowering the costs of CNT production to commercially viable levels by scaling up their synthesis.

Interestingly, CNTs have great potential to be applied in nanomedicine for disease diagnosis and drug targeting, as well as to transport various biomolecules such as proteins, DNA, RNA, immune-active compounds, and lectins [87–89].

CNTs are also extensively applied to develop electrochemical sensors, DNA-based sensors, as well as piezoelectric and gas sensors [83]. Additionally, opportunely structured CNTs, such as cationic CNTs, have been revealed to possess interesting antibacterial and antifungal activity [83]. Table A5 in Appendix A, reproduced from our previous article [35] summarizes the several applications that CNTs could have in the biomedical area. References in the Table do not belong to this work but are those contained in Alfei and Schito, 2022 [35]. We have decided on this solution for the reasons previously described.

### *3.1. Antimicrobial Properties of Carbon Nanotubes*

#### 3.1.1. Mechanisms of Action and Influencing Factors

It has been reported that CNTs, especially SWCNTs and MWCNTs, possess excellent antibacterial and antifungal activity. Several mechanisms have been suggested to justify the CNTs' toxicity against bacteria. Figure 5A shows the most recognized antimicrobial mechanisms [90].

Kang et al. reported, for the first time in 2007, that SWCNTs demonstrated strong antibacterial activity against *Escherichia coli*, by cell membrane impairment via direct contact, thus inhibiting 80% of bacterial cells [26]. Another study by the same authors, using *E. coli*, on the gene expression analysis, disclosed that the potency of antibacterial activity of CNTs mainly depended on their size and that the a-specific disruption of the bacterial membrane was the main mechanism of their antibacterial effects [27]. In the same year, the same authors demonstrated that when exposed to MWCNTs, *E. coli* revealed significant oxidative stress (OS), complemented by cell membrane disruption, cell lysis, and release of intracellular contents [20]. Nagai and Toyokuni, who studied the differences and similarities between carbon nanotubes and asbestos fibres' effects on cells during mesothelial carcinogenesis, also reported that the mechanism by which CNTs enter non-phagocytic cells was mainly based on the impairment of cell membrane by direct pore formation on their surface [91]. On the other hand, Kang et al. reported that the length of CNTs is pivotal for their interactions with bacterial membranes, where shorter tubes showed higher toxicity against bacteria [27]. Additionally, Aslan et al. remarked that shorter SWCNTs were more toxic to target cells due to the higher density of open tube ends [92]. A similar observation was also reported by Johnson, showing that nanotubes with smaller diameters manage to impair membranes of target cells via easier interactions with their surface. More specifically, concerning the inhibition of some bacteria and fungi, including *E. coli, Bacillus subtilis, Staphylococcus aureus, Pseudomonas aeruginosa,* and *Candida albicans* caused by exposure to SWNT, DWNT, and MWNT, it was demonstrated that their death was induced by the accumulation of CNTs following their entrapment onto the microbial cell surface [93,94]. Differently, Arias et al. observed that SWCNTs functionalized with -OH and -COOH groups used in buffered suspension caused *Salmonella* cell aggregation and the formation of SWCNTs-pathogens aggregates with an increasing size depending on SWCNTs concentrations [95]. Although the authors admitted that additional mechanisms for the toxicity of SWCNTs to bacterial cells could exist, this was assumed to be the main one. The toxicity of SWCNTs to bacterial cells was also dependent on a buffer of suspensions [95]. Many studies have confirmed that SWCNTs are more toxic to different pathogens than MWCNTs and cause their membrane disruption [20,28]. Kang et al. found that while SWCNTs kill most *E. coli* bacterial cells after one hour of treatment, incubation with MWCNTs leaves bacteria alive [20,27]. The authors also evidenced that *E. coli* exposed to SWCNTs displayed higher levels of stress-related genes compared to MWCNT treatments. On the contrary, Young et al. demonstrated that the MWCNTs are more toxic than SWCNTs to *E. coli* [96]. Also, Saleemi et al., who have assessed the antimicrobial effects of DWCNTs and MWCNTs against *S. aureus, P. aeruginosa, Klebsiella pneumoniae,* and *C. albicans,* demonstrated that non-covalently dispersed MWCNTs exhibited higher antimicrobial activity than DWCNTs [97]. Despite the contradictory reports, which need further investigations, CNTs are very competitive compared to other nanomaterials in inhibiting a broad range of microorganisms, including both Gram-positive and Gram-negative species and fungi, such as *S. aureus, E. coli, Enterococcus faecalis, Lactobacillus acidophilus,* and *Bifidobacterium adolescentis* [20,26,28]. As reported in Figure 5B, several key factors can affect the antimicrobial efficacy of CNTs, including their composition, geometry, surface modification, intrinsic properties, and electronic structure [90]. In this regard, metallic SWCNTs possess higher antimicrobial effects than semiconducting CNTs due to their capability to cause oxidation of pathogens intracellular components [98]. Anyway, the antibacterial activity of CNTs may also depend on the species and morphology of microorganisms [28,99], the mechanical properties of cell surfaces [99], and their growth state (planktonic or sessile) (Figure 5B) [100]. Gram-positive bacteria, such as *B. subtilis* and *S. aureus,* are more susceptible to CNTs piercing due to their softer surface, thus leading to higher bacteria death rates [28,99]. Chen et al. demonstrated that CNTs work better against spherical-shaped pathogens than rod-shaped ones [28]. Moreover, as observed for traditional or innovative antibiotics, when microorganisms protect themselves within the structure of biofilm, they are difficult to reach, thus becoming less susceptible also to the effects of CNTs [100]. Anyway, it has been reported that longer immobilized CNTs prevent bacterial settlement and biofilm growth [101], while vertically aligned arrays of carbon nanotubes (VACNTs), consisting of tubes much smaller than the usual size of a bacterial cell, can reduce biofilm formation [102]. However, since the overall mechanisms by which CNTs prevent biofilm establishment and maturation have not yet been fully clarified, more extensive research is required to design robust and effective CNT-based coating materials to impede biofilm formation [90].

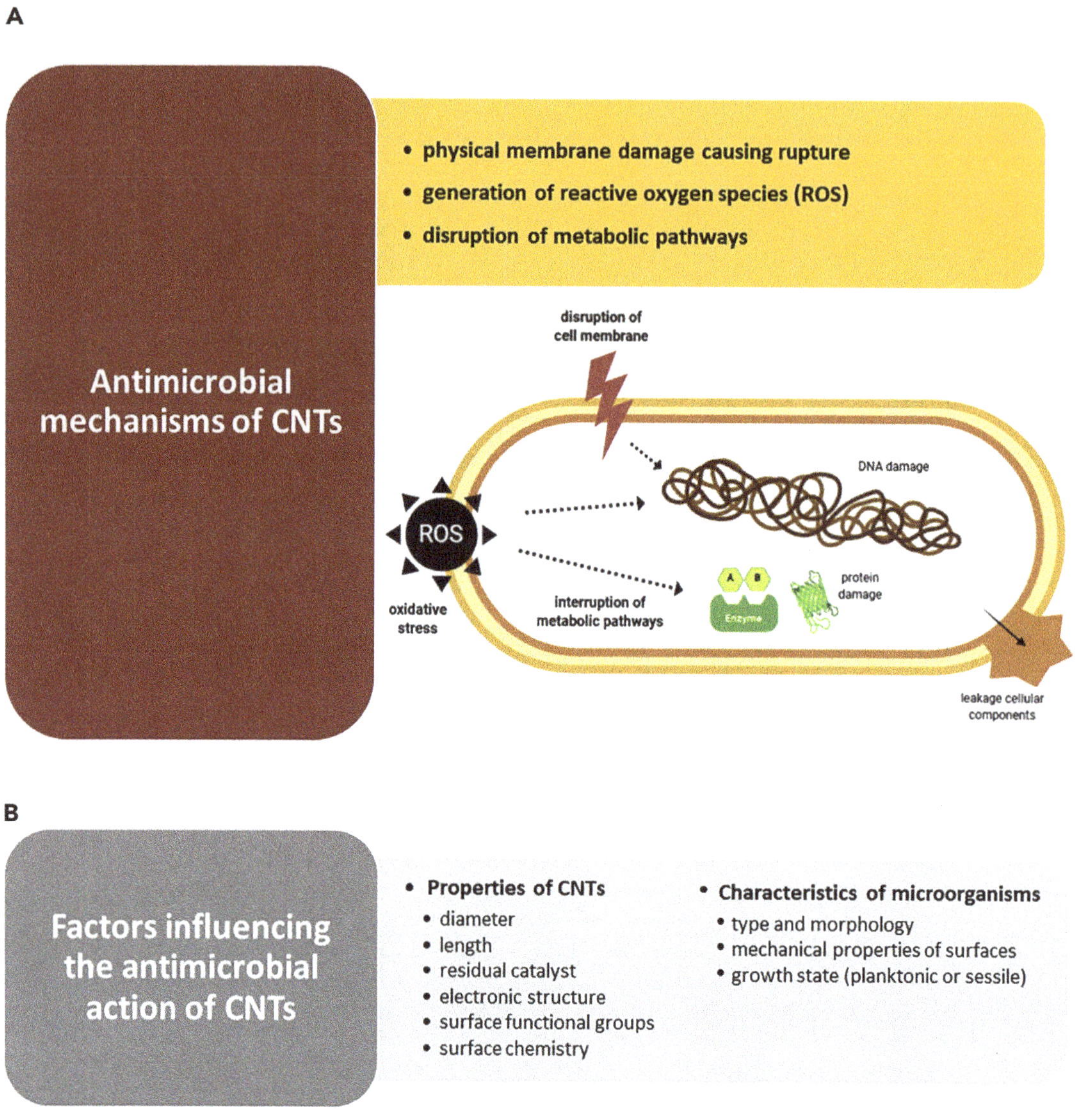


**Figure 5.** (**A**) Possible mechanisms by which CNTs exert antimicrobial effects: physically piercing of outer membrane, with irreversible membrane damage, cell lysis and release of cytoplasmic content; reactive oxygen species (ROS) overproduction causing oxidative stress (OS), which leads to DNA damage, protein and lipid peroxidation, and impairment of metabolic pathways, which results in cell death. (**B**) Factors influencing the antimicrobial performance of CNTs. This image has been reproduced by an article published in open access [90] and available under the Creative Commons CC-BY-NC-ND (CC BY-NC-ND 4.0, https://creativecommons.org/licenses/by-nc-nd/4.0/, accessed on 18 March 2025) license and permits non-commercial use of the work as published, without adaptation or alteration provided the work is fully attributed.

### 3.1.2. Most Relevant Case Studies on the Antimicrobial Effects of Non-Modified CNTs

Table 1 below collects some relevant case studies reported over the years about the antimicrobial properties of non-modified CNTs.

**Table 1.** The antimicrobial performance of pristine carbon nanotubes in different studies.

| Types of CNTs | Synthesis Method | Concentration | Species | Main Findings | Ref. |
|---|---|---|---|---|---|
| SWCNTs | CO disproportionation | 5 μg/mL | *E. coli* | Releasing intracellular content due to irrecoverable outer membrane damage | [26] |
| SWCNTs | CO disproportionation | 5 μg/mL | *E. coli* | Microbial cells lost their cellular integrity | [27] |
| MWCNTs | CVD method | 5 μg/mL | *E. coli* | Many of the bacterial cells remain intact and preserve their outer membrane | [20] |
| SWCNTs/MWCNTs | CVD method | 20, 50, 100 μg/mL | *L. acidophilus*, *E. coli* *B. adolescentis*, *E. faecalis* *S. aureus* | Antimicrobial mechanism associated with length-dependent wrapping and diameter-dependent piercing upon microbial cell membrane damage and the release of intracellular contents | [28] |
| MWCNTs | Nanocycle productions | 1.5–100 mg/L | *E. coli* | Low microbial toxicity | [103] |
| MWCNTs | - | - | *E. coli*, *B. subtilis* *P. aeruginosa* | 2-log cell density reduction in viability of pathogens | [104] |
| DWCNTs/MWCNTs | NE scientific productions | 20/100 μg/mL | *S. aureus*, *P. aeruginosa* *K. pneumoniae*, *C. albicans* | MWNTs' antimicrobial activity is higher than DWNTs | [97] |
| MWCNTs | Nanotech Labs productions | 20 mg/20 mL | *P. fluorescens* | 44% of inactivated bacteria MWNTs showed a significant effect on the inhibition of microbial adhesion due to the electrochemical potential | [105] |
| SWCNTs | - | 5 μg/mL | *E. coli*, *B. subtilis* | No physical destruction was observed below 10 nN of applied force | [94] |
| SWCNT/DWCNT MWCNT | Electric arc discharge and CCVD | 100 μg/mL | *S. aureus*, *P. aeruginosa*, *C. albicans* | Microbial death induced by the aggregation of CNTs trapped on the microbial cell surface | [93] |
| SWCNTs/MWCNTs | - | 0.2 mg/mL | *E. coli* | Control bacteria grow by laser-activated CNTs | [106] |
| SWCNT-OHs | - | 50 to 250 μg/mL | *Salmonella* | −7log reduction in cell viability at 200–250 μg/mL | [95] |

CO = Carbon oxide; CVD = chemical vapor deposition; CCVD = catalytic chemical vapor deposition.

Collectively, the results from studies reported in Table 1 established that both pristine SWCNTs and MWCNTs prepared by different methods possess moderate to remarkable antimicrobial and microbicide effects (1.5–250 μg/mL) against several species of Gram-positive and Gram-negative bacteria and *C. albicans*. As shown above in Figure 5A, most studies have confirmed that CNTs inhibit pathogens mainly by causing irreversible damage to their outer membrane. Upon aggregation on their surface via electrostatic interactions, severe impairments of microbial cell integrity occur by length-dependent wrapping and diameter-dependent piercing, determining cell lysis and lethal release of intracellular contents. Additionally, CNTs have shown the capability to impede the adhesion of bacteria on surfaces, thus paving the way for their promising application to prevent biofilm formation.

Generally, the antimicrobial activity of MWCNTs was higher than that of DWCNTs but lower than that of SWCNTs.

### 3.1.3. Most Relevant Case Studies on the Antimicrobial Effects of Modified CNTs

Several other studies investigating the antimicrobial properties of SWCNTs and MWCNTs, have been focused specifically on assessing the effect of their functionalization with different chemical groups and/or their modification by combination with metal nanoparticles (MNPs), polymers, antimicrobial peptides (AMPs), dendrimers, antibodies, etc., on their antimicrobial ability toward various bacterial and fungi strains. In this regard, Figure 6 shows the main synergistic effects observed by associating AMPs, MNPs, and polymers with CNTs [90].

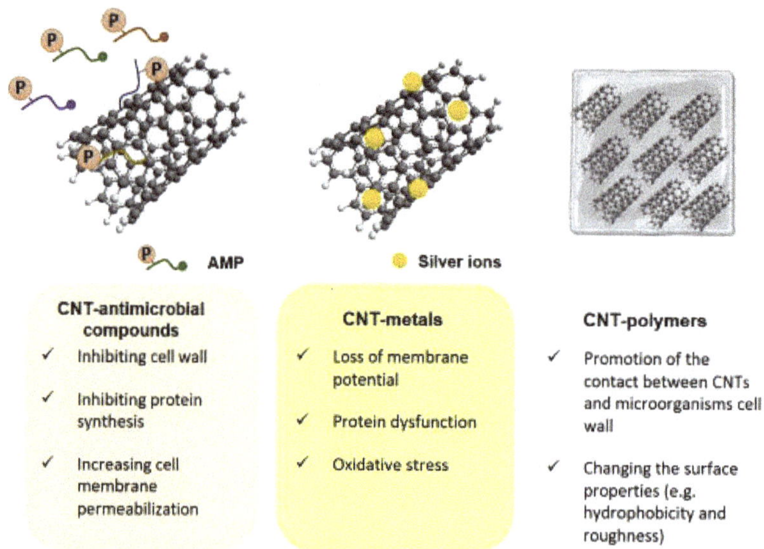


**Figure 6.** Synergistic effects observed by associating CNTs to AMPs, MNPs, and polymers. This image has been reproduced by an article published in open access [90] and available under the Creative Commons CC-BY-NC-ND (CC BY-NC-ND 4.0, https://creativecommons.org/licenses/by-nc-nd/4.0/, accessed on 18 March 2025) license and permits non-commercial use of the work as published, without adaptation or alteration provided the work is fully attributed.

In the following, a brief description of the most relevant results observed using modified CNTs is reported.

#### Modified SWCNTs

CNTs can be functionalized by both covalent and non-covalent methods. The functionalization of the surface of CNTs could have different goals, including the avoidance of desorption processes and of undesired absorption of molecules from the biological media, as well as enhancing the effectiveness of their antimicrobial activity [107–111]. Correlations between the toxicity to bacteria and the physicochemical properties or the agglomeration status of functionalized SWCNTs (f-SWCNTs) were studied by Pasquini et al., discovering that no direct correlation was identified between the bacterial cytotoxicity and thermal, physicochemical, and structural properties of f-SWCNTs [112]. On the contrary, they found that the aggregation of nanoparticles had more incidence than the single chemical and physical properties of functional groups on the f-SWCNTs cytotoxicity [112]. Arias et al.

functionalized SWCNTs with different groups and tested them in suspension to evaluate the possible increase in their interaction with pathogens and the effects on their antimicrobial activity [95]. Specifically, the authors studied the effects of various surface functional groups, including -$NH_2$, -COOH, and -OH ones, on their antimicrobial effects against *S. aureus*, *B. subtilis*, and *Salmonella typhimurium*. SWCNTs bearing the cationic -$NH_2$ group inhibited bacterial growth only at high concentrations. On the contrary, SWCNTs bearing anionic -COOH and neutral -OH groups reduced the viability of pathogens by 7 log CFU/g. To explain this fact, the authors assumed that the long carbon chain used to attach the $NH_2$ groups to SWCNTs surface impedes the cylindrical shape of SWCNTs from being in close direct contact with microbial cell walls, thus probably reducing their inhibitory effects. On the contrary, -COOH and -OH groups were derived directly from the surface of SWCNTs, thus allowing the direct contact of bacteria with SWCNTs surface, thus sustaining and enhancing their antimicrobial potency. Several authors have reported on the antimicrobial activity of silver nanoparticles (AgNPs) and other metal oxides together with their inhibitory effects on the infections [113]. Chaudhari et al. observed that the antimicrobial properties of silver-coated SWCNTs on *S. aureus*, when applied in a skin model, can be modified with antimicrobial peptides (AMPs) [114]. In this regard, they observed that the proliferation of bacteria was reduced by $10^5$ CFU/g by silver-coated functionalized CNT after skin treatment [114]. Generally, AgNPs bind and penetrate the bacterial cell membrane, causing cell death by altering the permeability of the membrane and increasing ROS induction [115], while AMPs have shown antimicrobial effects on various fungi, bacteria, and viruses by the same mechanisms [116]. Therefore, the synergistic effects of AgNPs with AMPs increase the toxic effects of SWCNTs, thus allowing for the possible development of novel antimicrobial therapies [114]. The same authors tested their functionalized SWCNTs against *S. aureus*, *Streptococcus pyogenes*, *Salmonella enterica* serovar Typhimurium, and *E. coli* [117], finding that the conjugation led to a strong synergistic antibacterial effect of TP359 with SWCNTs-adsorbed AgNPs. Kumar et al. reported the excellent antibacterial potency of SWCNTs enriched with AgNPs inserted in cotton fabrics against *S. aureus* and *E. coli* [118]. Moreover, AgNPs embedded in the silica-coated SWCNT substrate inactivated *E. coli* growth better, with respect to AgNPs plasma-treated SWCNT substrates [119]. Eco-designed biohybrids based on liposomes containing cholesterol, mint–nano-silver, and carbon nanotubes demonstrated antioxidant and antimicrobial properties against *S. aureus*, *E. coli*, and *E. faecalis* [120]. Chang et al. used a simple one-step procedure to synthesize nanocomposites encompassing CNTs, graphene oxide, and AgNPs effective against *E. coli* and *S. aureus*, exhibiting high disinfection properties [121]. Further investigation proved that the nanocomposite of Chang and colleagues was able to induce $O_2$-based OS on bacteria, which damaged cell membrane integrity, thus triggering cell death. SWCNTs coated with Ag-doped $TiO_2$ nanoparticles demonstrated strong antibacterial activity against both *E. coli* and *S. aureus*, although *S. aureus* was more tolerant than *E. coli* under illumination by UV light [107]. Park et al. manufactured pegylated SWCNTs (pSWCNTs) covered AgNPs, and their antibacterial effects were investigated on foodborne pathogenic bacteria [122]. A significant reduction in proteins associated with bacterial biofilm formation and quorum sensing, as well as of proteins necessary for the conservation of cellular structure, were observed. This was associated with a decrease in cell motility in the foodborne pathogens that remained alive. Also, by a simple and low-cost one-pot synthetic procedure, Singh et al. produced a SWCNTs/Ag/PPy-based nanocomposite which succeed in fully inhibiting the growth of *S. aureus*, *P. aeruginosa*, *E. coli*, and *B. cereus* completely within 24 h treatment [123]. A new antimicrobial nano system made of mesoporous silica and AgNP-coated SWCNTs (SWCNTs@m$SiO_2$-TSD@Ag), which demonstrated strong antimicrobial activity against multidrug-resistant (MDR) bacterial strains by impairing cell membrane and releasing of Ag ions, was projected by Zhu et al. [124]. Yun et al. manufactured CNTs-Ag and GO-Ag, which demonstrated antibacterial effects against both Gram-positive and Gram-negative species, even if the effects of CNTs-Ag were higher than those of GO-Ag nanocomposites [125]. Additionally, a carbon-Ag nanosized complex inhibited the microbial growth of methicillin-resistant *S. aureus*, *K. pneumoniae*, *Acinetobacter baumannii*, *Yersinia pestis*, and *Burkholderia cepacian* [126]. SWCNTs were also modified with natural enzymes, such as lysozyme (LSZ) succeeding in enhancing their antibacterial impact against different bacterial species, including *S. aureus* and *Micrococcus lysodeikticus*, by causing cell wall lysis via hydrolytic breck of the β-1, 4 bonds in peptidoglycan [127–129]. Chemical modifications of SWCNTs were mainly finalized to enhance their dissolution properties and chemical compatibility. On the contrary, the functionalization of SWCNTs with polymers, achieving deposited aggregates and membrane coatings, improved their dispersibility and solubility and increased the interfacial interaction to polymeric matrices in their composites, thus demonstrating high chemical stability and high toxicity toward the microbes. In contrast, pristine SWCNTs, polymer-modified SWCNTs are less expensive and provide an enlarged range of mechanical, degradation, and structural properties, thus being suitable for developing ideal antimicrobial nanomaterials. Several case studies have been available in the literature reporting on the antibacterial activity of these carbon-based nanocomposites. Aslan et al. ideated SWCNTs with poly-(lactic-co-glycolic acid) (PLGA) and explored their antibacterial effects against *S. epidermidis* and *E. coli*, observing a 98% viability reduction and a significant decrease in the metabolic activity of the bacteria [92]. On the other hand, the SWCNTs modified with polyvinyl-N-carbazole demonstrated a 90 and 94% inhibition of planktonic cells for *B. subtilis* and *E. coli*, respectively, while reducing their biofilm formation [130]. Nanocomposite deriving by refashioning SWCNTs with poly-(L-glutamic acid) and poly-(L-lysine) resulted in a 90% inhibition rate of *E. coli* and *S. epidermidis* [131]. Goodwin et al. synthesized a SWCNTs-poly-(vinyl alcohol) nanosized composite, which gradually inactivated *P. aeruginosa* cells depending on increasing concentrations of SWCNTs [132]. SWCNTs reformulated with porphyrin, which possessed appreciable antibacterial effects against *S. aureus* in the presence of visible light and used a tungsten-halogen lamp, were reported by Sah et al. [133]. The activity of these materials was based on a photochemical reaction, which ultimately transferred an electron to the atmospheric molecular oxygen to form ROS, which destructed the bacterial cell wall, thus driving bacterial death [133]. Also, SWCNTs covalently bound with polyamide membranes showed a 66% inactivation of bacteria, translating into a delay in membrane biofouling [134]. The well-known poly-(ethylene glycol) (PEG) coating agents, recognized for their high hydrophilicity and biocompatibility, were attached in their linear and branched form by Cajero-Zul et al. to the surface of CNTs, achieving a nanoplatform suitable to manufacture medical devices [135]. Even if SWCNTs-copolymer of star-shaped PEG and poly-(ε caprolactone) (PCL) did not possess antimicrobial activity, they demonstrated thermal and mechanical characteristics superior to those of polymeric matrix. On the contrary, the star-shaped PCL-PEG copolymer structure prevented bacterial growth [135].

Modified MWCNTs

MWCNTs with surface enriched with -COOH group caused a 30, 27, 20–40, and 15~50% reduction in the viability of bacteria, including *B. subtilis*, *P. aeruginosa*, *E. coli*, and *S. aureus* [136–138]. Chen et al. showed that MWCNTs possessing -OH, or -COOH functional groups showed a significant dose-dependent antimicrobial effect on microbes such as *E. coli*, *S. aureus*, *Enterococcus faecalis*, *Lactobacillus acidophilus*, and *B. adolescentis* [28]. Ding et al. observed similar properties against *Vibrio parahaemolyticus* [139], while

Arias et al. discovered that MWCNTs functionalized with -COOH, -$NH_2$, and -OH did not exert appreciable antimicrobial effects with respect to analogous SWCNTs [95]. Various studies evaluated the antimicrobial action of non-covalently dispersed DWCNTs and MWCNTs against *S. aureus*, *K. pneumoniae*, *P. aeruginosa*, and *C. albicans*, finding time- and concentration-dependent mechanisms [97]. MWCNTs decorated with ethanolamine suppressed the microorganism's growth to a large extent, compared with non-modified MWCNTs [140]. Another study showed that MWCNTs modified with oxygen groups could increase their antimicrobial properties [141]. MWCNTs added with Ag NPs revealed significant antimicrobial performance, which is like the SWCMT counterpart. Specifically, silver/MWCNTs complex caused 94–99, 57, 100 and 70% inactivation of *S. epidermidis* and *E. coli*, *S. aureus*, *Sphingomonas* spp., and *Methylobacterium* spp., as well as *P. aeruginosa*, respectively [138,142,143]. Silver-coated-MWCNTs complexed with amphiphilic poly-(propylene imine) (PPIs) dendrimers inactivated >90% of *S. aureus*, *B. subtilis*, and *E. coli* strains [136]. When the silver/MWCNTs complex was functionalized with polymer colloids, it displayed strong antimicrobial effect against *S. aureus* and *E. coli* [144]. Also, silver sulphide ($Ag_2S$) quantum dots associated to poly-(amidoamine) (PAMAMs)-grafted MWCNTs showed better antimicrobial activity than cadmium sulphide quantum dot-coated-MWCNTs resulting in a microbial growth inhibition by 56, 98 and 79% when *S. aureus*, *E. coli*, and *P. aeruginosa* were exposed [138]. Aiming at more favourable results, researchers also amalgamated MWCNTs with copper nanoparticles, thus decreasing the bacteria viability by 75% [145]. Zinc oxide evidenced strong antimicrobial activity against *E. coli* [146], titanium, and gold is observed to have remarkable microbial growth inhibition against *B. subtilis*, *K. pneumoniae*, *S. aureus*, *C. albicans*, *Streptococcus pneumoniae*, *Proteus vulgaris*, and *Shigella dysenteriae* [147]. Very recently, Kim et al. modified MWCNTs with ZnO NPs, obtaining an antibacterial nanocomposite that demonstrated remarkably antibacterial effects against *E. coli*, *P. aeruginosa*, *E. faecalis*, *S. aureus* (Figure 7B,D). When incorporated in a urinary catheter, it inhibited *E. coli* and *P. aeruginosa* biofilm formation by 53.42 and 56.44% after 120 h of incubation, respectively (Figure 7A,C) [148].

Moreover, titanium alloy-coated-MWCNTs combined with rifampicin inhibited the formation of biofilm for up to 5 days [149]. As in the case of SWCNTs, enzymes like chloroperoxidase (CPO) and laccase were impaled on the MWCNTs' surface, resulting in nanocomplexes capable of reducing the viability of *S. aureus* and *E. coli* by 99%. It was observed that the laccase-MWCNTs were even capable of reducing the microbial growth and spore formation of *B. cereus* and *B. anthracis* by >99% [150]. Additionally, it was demonstrated that CPO in CPO-MWCNTs oxidated chloride into HOCl by $H_2O_2$ in acidic conditions, driving the formation of singlet oxygen. It and HOCl functioned as strong oxidants capable of inhibiting, controlling, or reducing microbial growth of *E. coli* and *S. aureus* [150]. On the contrary, it was understood that the main antimicrobial agent in the laccase-methyl syringate (MS) system was the hydroxyl radical produced mainly by the Haber–Weiss reaction starting from $H_2O_2$ and superoxide radical, which instigated detrimental and irreversible impairments in the bacterial cells leading to their destruction [150]. When MWCNTs merged with polymers, the antimicrobial activity was maintained. The bacterial growth of *E. coli*, *B. subtilis*, and *S. aureus* was inhibited by 87% and 97% using MWCNTs modified with amphiphilic PPIs dendrimer synthesized by Murugan et al. [136]. Similarly, Neelgund et al. formulated MWCNTs with an aromatic polyamide dendrimer, which exerted good antimicrobial effects on *P. aeruginosa* (65.2%) and *E. coli* by inhibiting their growth by 65 and 73%, respectively [137]. Also, MWCNTs functionalized with PAMAMs inhibited the growth of several selected bacteria in an NT concentration-dependent way [138], as also reported by Goodwin et al. for MWCNT-poly (vinyl alcohol) nanocomposites [132]. A current trend is increasingly driving the interest of researchers toward the evaluation of the antimicrobial activity of MWCNT-based chitosan hydrogels due to the biocompatibility of the hydrogel-based materials. In this regard, a robust antimicrobial activity of MWCNT-based chitosan hydrogels was confirmed against *S. aureus*, *E. coli*, and *C. tropicalis* [151], while Mohamed et al. informed that MWCNT-based chitosan possessed a broad-spectrum antimicrobial activity [152]. The following tables, Tables 2 and 3, schematically collect the most relevant case studies concerning the antimicrobial effects of modified SWCNTs and MWCNTs, respectively.

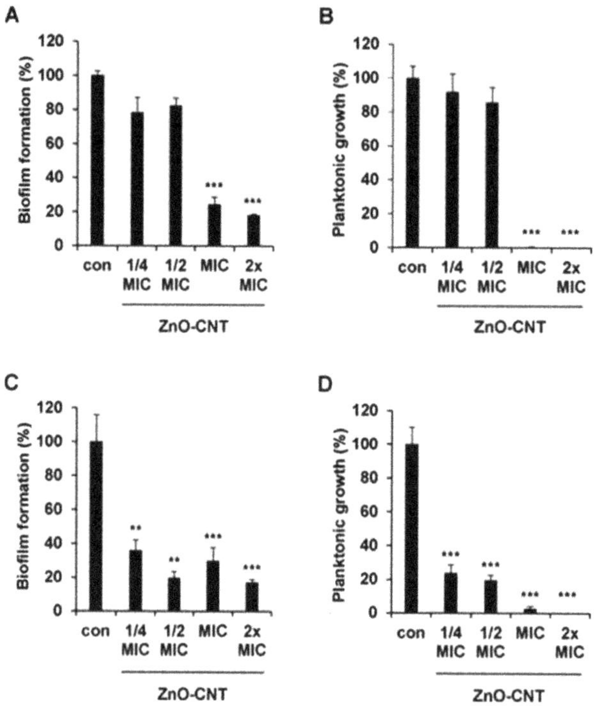


**Figure 7.** ZnO-CNT nanocomposite with different concentrations, 1/4, 1/2, 1, and 2× MIC, showed biofilm inhibition against (**A**) *E. coli* and (**C**) *P. aeruginosa*. Planktonic growth inhibitions against (**B**) *E. coli* and (**D**) *P. aeruginosa* were determined. Error bars represent the mean ± SD from the average of triplicate experiments. *** $p < 0.001$, ** $p < 0.01$. Reprinted (adapted) with permission from "Copyright 2025 American Chemical Society" [148].

**Table 2.** Overview of the antimicrobial activity of functionalized SWCNT-based nanocomposites in different studies.

| Material Blend | Concentration | Species | Main Findings | Ref. |
| --- | --- | --- | --- | --- |
| SWCNTs-OH, -COOH, -$NH_2$ | 50–200 µg/mL | *S. aureus*, *B. Subtilis*, *S. typhimurium* | SWCNTs-OH and -COOH showed higher microbial inhibition rates (7-log reduction) than SWCNTs-$NH_2$ | [95] |
| Ag-SWCNTs containing TP226, TP359, TP557 peptides | 5 µg/mL | *S. aureus* | In skin models treated with silver-SWCNTs antimicrobial activity of only 1-log reduction was observed | [114] |
| SWCNTs functionalized with DNA and LSZ | ~25 mg/L | *S. aureus*, *M. lysodeikticus* | DNA- and LSZ-SWCNTs caused 84% microbial reduction | [127] |

**Table 2.** *Cont.*

| Material Blend | Concentration | Species | Main Findings | Ref. |
|---|---|---|---|---|
| SWCNTs-PLGA complexes | <2% by weight | *E. coli, S. epidermidis* | SWCNTs-PLGA caused a 98% reduction in metabolic activity | [92] |
| SWCNTs- PVK nanocomposite | 3 wt.% | *E. coli, B. subtilis* | SWCNTs-PVK induced 90% and 94% of *B. subtilis* and *E. coli* inhibition in the planktonic cells and showed a significant reduction of biofilm formation | [130] |
| SWCNTs-PGA/PLL (layer-by-layer) | <2% by weight | *E. coli, S. epidermidis* | SWCNTs/PGA/PLL showed a 90% reduction in pathogens | [92] |
| Oxidized SWCNTs-PVOH) nanocomposite | 0–10% (*w*/*w*) | *P. aeruginosa* | O-SWCNTs-PVOH gradually decreased viability of cells with an increase in nanotube loading | [132] |
| SWCNTs/porphyrin composite | 0.04 mg/mL | *S. aureus* | SWCNTs/porphyrin caused visible light-induced damage to the cell membrane | [133] |
| SWCNTs-PEG)/poly-(ε caprolactone) composites | 0.5–1.0 wt.% | *P. aeruginosa, S. aureus* | SWCNT/copolymer complex caused lower bacteria inhibition than pure polymer complex | [135] |
| SWCNTs-polyamide membranes | 0.1–0.2 mg/mL | *E. coli* | Nanocomposite inactivated the microbial cells by 60% after 1 h of contact time | [133] |

LSZ = lysozyme; PLGA = poly-(lactic-co-glycolic acid); PVK = polyvinyl-N-carbazole; PGA = poly-(L-glutamic acid); PLL = poly-(L-lysin); PVOH = poly-(vinyl alcohol); PEG = poly-(ethylene glycol).

**Table 3.** Overview of the antimicrobial activity of functionalized MWCNT-based nanocomposites in different studies.

| Material Blend | Concentration | Species | Main Findings | Ref. |
|---|---|---|---|---|
| MWCNTs-OH, -COOH, -$NH_2$ | 50–200 μg/mL | *S. aureus, B. subtilis S. typhimurium* | MWCNTs-OH and -COOH did not significantly induce antimicrobial activity | [95] |
| | 25 μg/mL | *E. coli, B. subtilis, S. aureus* | MWCNTs-COOH caused 30, 40, 50% inactivation for *B. subtilis, E. coli, S. aureus*, respectively | [136] |
| | 20 μg/mL | *S. aureus, E. coli P. aeruginosa* | MWCNTs-COOH caused 27, 34, 23% inactivation for *P. aeruginosa, E. coli, S. aureus* | [137] |
| | 20 mg/20 mL | | MWCNTs-COOH inactivated the bacterial cells by 27, 20, 15% for *P. aeruginosa, E. coli, S. aureus* | [138] |
| | 20, 50, 100 μg/mL | *E. coli, S. aureus, E. faecalis L. acidophilus, B. adolescentis* | MWCNTs-COOH and MWNTs-OH induced dose-dependent microbial inhibition | [97] |
| | 1000 μg/mL | *V. parahaemolyticus* | Time-dependent antimicrobial activity CNTs did not pierce the cell membrane CNTs wrapped around the surface of pathogens | [153] |
| Surfactant-functionalized MWCNTs with SDBS, SC, SDS TX-100, DTAB, CTAB, PVP | 0–100 mg/mL | Group A *Streptococcus* | Carboxylate-MWCNTs with antibodies mitigate soft tissue infections | [154] |
| | 20–100 μg/mL | *S. aureus, P. aeruginosa K. pneumoniae, C. albicans* | Non-covalently dispersed CNTs inhibited bacteria by a time and concentration-dependent mechanism | [97] |

**Table 3.** *Cont.*

| Material Blend | Concentration | Species | Main Findings | Ref. |
|---|---|---|---|---|
| Surfactant-functionalized MWCNTs with SDBS, SC, SDS TX-100, DTAB, CTAB, PVP | 1.0, 0.5, 0.25, 0.125 mg/mL | *S. mutans* | Functionalized MWCNTs caused cell membrane rupture via direct contact | [155] |
| | 0.1, 0.5, 1 mg/mL | *E. coli* | Functionalized MWCNTs penetrated the bacterial cell membrane due to electrostatic forces | [156] |
| AgNP-coated MWCNTs | 2–30 wt.% | | Cell membrane of bacteria damaged via direct contact | [157] |
| ZnO-MWCNTs | 0.0078, 1 wt.% 0.00625, 0.0039 wt.% | *E. coli, P. aeruginosa E. faecalis, S. aureus* | ZnO-CNT urinary catheter inhibited biofilm formation by 53.42% and 56.44% after 120 h against *E. coli* and *P. aeruginosa* | [148] |
| MWCNTs-lysine | 0.01875–0.6 mg/mL | *S. aureus, E. coli, S. agalactiae S. typhimurium, S. dysgalactiae K. pneumoniae* | Electrostatic adsorption | [158] |
| MWCNTs-PPI | 25 μg/mL | *E. coli, B. subtilis, S. aureus* | MWCNT nanocomposite caused 97, 87% inactivation for *S. aureus/B. subtilis*, and *E. coli* | [136] |
| MWCNTs-aromatic dendrimer polyamide | 20 μg/mL | *S. aureus, E. coli, P. aeruginosa* | MWCNT nanocomposite caused 36, 65, 73% inactivation for *S. aureus, P. aeruginosa, E. coli* | [137] |
| PAMAM-grafted MWCNTs | 20 mg/20 mL | *S. aureus, E. coli, P. aeruginosa* | MWCNT nanocomposite caused 60, 34, 23% for *P. aeruginosa, E. coli, S. aureus* | [138] |
| Oxidized MWCNTs/PVOH nanocomposite | 0–10% (*w/w*) | *P. aeruginosa* | MWCNTs–PVOH reduced bacteria viability by increasing concentrations | [132] |
| MWCNTs–chitosan hydrogels | 25, 50, 100 mg/40 mL | *S. aureus, E. coli, C. tropicalis* | MWCNTs–chitosan hydrogels exhibited higher antimicrobial activity against *S. aureus* and *C. tropicalis* than *E. coli* | [151] |
| | 0.01%, 0.1% and 0.2% (*w/w*) | *E. coli, S. pneumoniae S. racemosum, C. albicans P. aeruginosa G. candidum, A. fumigatus* | MWCNT nanocomposite showed a higher microbial inhibition rate against Gram-positive bacteria | [152] |

PPI = poly-(propyleneimine); PAMAM = poly(amidoamine); PVOH = poly-(vinyl alcohol); SDBS = sodium dodecylbenzene sulfonate; SC = sodium cholate; SDS = sodium dodecyl sulphate; TX-100 = triton X-100 (TX-100); DTAB = dodecyltrimethylammonium bromide; CTAB = cetyltrimethylammonium bromide; PVP = polyvinylpyrrolidone.

From data reported in Table 2, we noted that the functionalization of SWCNTs with oxygenated groups (-OH, -COOH) translated to more enhanced bactericidal activity, while the improvement derived by amine group insertion was inferior. Combination of SWCNTs in nanocomposites led to remarkable antibacterial effects (>90% bacteria inactivation) when PLGA, PVK, and PLG/PLL polymers were used, while the antibacterial activity was moderate when PEG/poly-ε-caprolactone or polyamide polymers were utilized. Interestingly, SWCNTs-PVK induced 90% and 94% of *B. subtilis* and *E. coli* inhibition in the planktonic cells and a significant reduction of biofilm formation, thus being promising to counteract this difficult-to-treat form of pathogens resistance, causing orthopaedic implants-associated severe infections. Concerning modified MWCNTs, they generally demonstrated a time- and concentration-dependent antimicrobial activity toward both bacteria and fungi. In some cases, modified MWCNTs did not pierce the pathogen's membrane or the non-penetrated cells, but they wrapped around their surface, causing severe injury without lysis. Anyway, in most cases, including MWCNTs modified with gold (Au) NPs, upon contact via electrostatic interactions, MWCNTs inhibited microbes by damaging their external membrane up to its disruption and cell penetration. Functionalization to provide MWCNTs-OH and -COOH materials did not significantly induce antimicrobial activity or caused a moderate to scarce inactivation of 15–50% for *B. subtilis*, *E. coli*, *P. aeruginosa*, and *S. aureus*. Better antibacterial activity was observed against *S. aureus*, *B. subtilis*, and *E. coli* when MWCNTs were merged with PPI (87–97% inactivation). Differently, limited antibacterial effects were displayed by MWCNTs against *E. coli*, *P. aeruginosa*, and *S. aureus* when modified with an aromatic polyamide dendrimer (36–73% bacteria inactivation) or a PAMAM one (23–60% bacteria inactivation). Generally, modified CNTs exhibited higher antimicrobial activity against Gram-positive bacteria due to their softer surface and lower resistance to piercing and damage.

## 4. Impediments to the Extensive Application of CNTs: Toxicity Issues

Since the invention of CNTs in 1991 [17], researchers have not thought about the possible negative effects of their use and contact with humans and animals or about their environmental impact, and so they were not investigated [159]. This attitude was mostly due to the limited production processes available at that time, which forced researchers to generate CNTs only on a laboratory level and at a high cost [160]. Only when novel methods, such as CVD, made their production possible on a massive scale since 2000 did the impact of CNTs on health become progressively studied through toxicological studies [161]. It was mainly their shape, like that of asbestos fibres, that prompted the very first health warnings. In fact, asbestos fibres were already identified as material causing inflammation and leading to lung cancer, which starts when macrophage cells try to absorb the needle-like asbestos fibres but fail because they are too long, thus causing the activation of the so-called giant cells. The constant activation of these cells can drive to the establishment of a nodular tissue (granuloma), which may translate into mesothelioma over a long latency period of thirty to forty years. On the other hand, since a huge range of CNTs exists, their possible negative biological effects depend mainly on their shape. To check this assumption, most studies included and include experiments on cells and animal models using rats or mice. Animals were exposed to CNTs and inhaled them through the nose or were subjected to fibres application directly to their lungs or thoraxes. The surface characteristics of CNTs also affect the possible biological outcomes deriving from cells and animals' exposure to these materials. An injection of 400 μg/mouse of water-soluble SWCNTs has been shown to be quickly expelled through the kidneys with a half-life of 3 h. Similarly, other researchers reported that hydroxylated SWCNTs, regardless of the form of administration, were quickly removed from the body through the urine [162], thus establishing that the contact of the body with these non-bio persistent CNTs could be very short and not worrying [159].

### 4.1. Introduction to Environmental and Human Safety

Generally, the results from some reported case studies concerning the impact of exposure of humans and the environment to CNTS are contradictory. Indeed, fruit fly larvae nourished with a CNTs-rich diet developed normally [35], while, despite the fish foetus appearing normal, CNTs delayed embryo development in zebrafish [163]. Inflammation of the lung like that caused by asbestos fibres was perceived for a few months in mice exposed to CNTs [164]. Certain human tumour cells proliferated more rapidly when exposed to CNTs [165]. CNT-based solar cells need a cadmium-telluride mixture coating, which is extremely toxic, thus impeding the widespread use of such solar devices [165]. This establishes that frequently, coatings applied to the CNTs, rather than CNTs themselves, are environmentally dangerous. A highly worrying concern about the massive use of CNTs regards their slow biodegradation, which leads to the release of tubes in the environment, their passage into our food supply, and from food into our bodies, with consequences not fully clear until now. It must, however, be considered that CNT applications in electronics are unlikely to be very risky because of the small volumes involved. Collectively, more inspections and regulation measures are necessary in this field, and it would be suggestable to treat CNTs as a new chemical material rather than as an allotrope of inert carbon [165].

Other Hurdles Are Close to Solution

Although the potential of CNTs is huge, obstacles to their production and application on a large scale endure. First, CNT-based devices were manufactured at prohibitive costs for typical consumers [83]. For years, both researchers in industry and academic laboratories have been making efforts to develop automated systems for growing CNTs endowed with uniform and predictable properties, thus making it possible to reduce their unclear toxicity. Indeed, until they realize a method of ideally structurally perfect CNT production on a huge scale, most of their possible applications will remain in the research laboratory, and silicon will manage most technologies, including the computing one [83]. Fortunately, nowadays, this era is closer than we might think. Very recently, a carbon copilot (CARCO), an artificial intelligence (AI)-driven platform that integrates transformer-based language models, robotic chemical vapor deposition (CVD), and data-driven machine learning models, has been developed by Li et al. [166]. Employing CARCO, the authors found a new titanium-platinum bimetallic catalyst for high-density horizontally aligned carbon nanotube (HACNT) array synthesis. This catalyst outperformed traditional ones and was treasured in millions of virtual experiments; an unprecedented 56% precision in synthesizing predetermined densities of HACNT arrays was achieved [166].

### *4.2. In Vitro and In Vivo Studies*

#### 4.2.1. In Vitro Studies

Over the years, increasing in vitro investigations about the cytotoxicity and genotoxicity of CNTs have been reported. Following is a collection of early studies developed in the years 2003–2011. The cytotoxic effect of titanium oxide ($TiO_2$) nanoparticles (NPs) and MWCNTs was studied in A549 human pneumocytes by Simon-Deckers et al., who evaluated cell viability and intracellular accumulation of both nanomaterials and compared results [167]. Both CNTs and $TiO_2$ NPs were capable of entering cells and passing into the cytoplasm. Anyway, $TiO_2$ NPs showed lower toxicity. Moreover, it was found that neither the presence of metallic impurities nor the length of CNTs influenced their cytotoxicity [167]. The capability of CNTs to penetrate the cellular membrane of both NR8383 rat macrophages and human pneumocytes A549, thus causing alteration of physiology and cellular functions, was reported by Pulskamp et al. [168]. Disfunctions are derived from a dose-dependent increase in intracellular reactive oxygen species (ROS) and from a decrease in the potential of the mitochondrial membrane in both rat NR8383 cells and human A549 lung ones. Interestingly, CNTs with deprived or with reduced metal content due to purification work-up demonstrated few or none of these undesired effects [168]. Collectively, several researchers proposed that the main response to the negative biological effects of commercial CNTs could be metal impurities. Moreover, it was reported that the cytotoxic effects of CNTs could heavily depend on their different degrees of agglomeration. Studies on human MSTO-211H cells revealed that the dispersion of CNTs in surfactant was less toxic compared to agglomerated CNTs, especially when CNTs agglomerated in string forms, which are more voluminous, more rigid, and more solid [169]. In this regard, Montero et al. studied the possible per se toxicity of different surfactants, finding that Pluronic F127 was less toxic [170]. MWCNTs, as well as MWCNTs coated with Pluronic F127, were tested on human keratinocytes, evidencing that the uncoated MWCNTs were found to be more cytotoxic [170]. Based on results from several studies, which confirmed the capability of CNTs to cross cell membranes [167,171], the intracellular distribution of modified SWCNTs was analysed in human fibroblasts 3T6 and murine 3T3 cells by Pantarotto et al. [171]. The results of this study demonstrated that functionalized CNTs could cross the cytoplasmic membrane and accumulate in cytosol or reach the nucleus without toxicity for the cells up to 10 μM concentrations [171]. Pantarotto et al. and Manna et al. investigated the toxicity of SWCNTs in HaCaT, HeLa, H1299, and A549 cells at different concentrations and exposure timings, finding a major loss of cell viability in keratinocytes (HaCaT), probably due to the activation of keratinocyte-specific transcription factor NF-κB by increased oxidative stress (OS) [171,172]. Anyway, Shvedova et al. have already explained that the unpleasant consequences demonstrated by SWCNTs in HaCaT and bronchial epithelial (BEAS-2B) cells, including alterations in cell ultrastructure and morphology, loss of cellular integrity and cellular apoptosis, OS, and depletion of antioxidants in the cells is specifically based on the alteration of certain genes following exposure to SWCNTs. Additionally, the authors declared that exposure to SWCNTs can trigger cutaneous and pulmonary toxicity [173]. Later, the same authors administered mouse macrophages (RAW 264.7) with different concentrations of SWCNTs, finding that even well-dispersed SWCNTs penetrated through the alveolar epithelium, caused interstitial fibrosis and alveolar wall thickening [174]. The production of TGF-β1 was triggered similarly to the effect of zymosan. Anyway, neither oxidative response nor nitric oxide production or cellular apoptosis was induced by CNTs [174]. The uptake of SWCNTs enriched with fluorescein-isothiocyanate by lymphocytes and macrophages reported by Dumortier et al. [175], upon in vitro experiments, did not lead to any changes in cell viability. These findings confirmed previous reports on in vitro cytotoxicity studies using human dermal fibroblasts exposed to modified SWCNTs, which demonstrated toxic effects lower than those observed using not modified SWCNTs [176]. Unexpectedly, while streptavidin-conjugated SWCNTs exhibited low toxicity in HL60 cells, the SWCNT–biotin–streptavidin complex caused cell death [177]. However, despite the functionalization of SWCNTs generally enhancing their potential by reducing their toxic effects, a general increase in toxicity was observed for the MWCNTs bearing carbonyl (C=O), carboxyl (COOH), and/or hydroxyl (OH) groups [178]. Moreover, indirect cytotoxicity of CNTs was reported deriving from their capability to stimulate immune-mediated cytotoxicity in various human cells, even at low concentrations (0.001–0.1 mg/mL). In this regard, it was reported that CNTs at lower concentrations may stimulate the secretion of cytokines with consequent activation of lymphocytes and upregulation of the NF-κB expression in immune cells [179]. The effect of SWCNTs on human epidermal keratinocytes (HEK293T) was investigated by Cui et al. using several techniques [180]. A dose- and time-dependent decrease in cell proliferation and adhesive ability was observed, with the formation of nodular structures, induction of G1 arrest, and apoptosis. SWCNTs (10–20 nm in diameter) showed higher toxicity than MWCNTs of the same dimensions, while fullerenes did not show any cytotoxicity when they were administered to alveolar macrophages in guinea pigs. These results suggested that the geometric structures of carbon nanomaterials are discriminant for cytotoxicity [181]. Pacurari et al. [182] demonstrated that SWCNTs induced ROS generation, increased cell death, and enhanced DNA damage and H2AX phosphorylation due to their fibrous characteristics involved in ROS generation. The DNA damage occurring upon exposure of lung V79 fibroblasts to acid-purified SWCNTs was confirmed by Kisin et al. [183]. According to

Francis et al., Patapovich found that, while iron-rich SWCNTs activated the macrophages and catalysed the transformation of extracellular $O^{2-}$ into detrimental hydroxyl radicals, thus encouraging IL-6 production, and causing OS and inflammation, the iron-deprived SWCNTs stimulated TGF-β production, thus promoting apoptosis [184]. MWCNTs were found capable of causing the release of proinflammatory cytokines (TNF-α) in rat peritoneal macrophages [185]. The cytotoxic effects of MWCNTs in rat and mouse peritoneal macrophages (J774.1) were assessed by Muller et al. and Hirano et al. They found higher cytotoxicity of CNTs with respect to that of crocidolite in mouse macrophages due to the action of MWCNTs with MARCO (macrophage receptor with collagenous structure), resulting in the disruption of the plasma membrane [185,186]. Human embryonic kidney 293 cells (HEK-293) exposed for prolonged times to high concentrations of MWCNTs showed reduced viability, a significant increase in IL-8, and altered expression of several proteins, supported by an increasing penetration of MWCNTs via cell membrane over time [187,188]. Also, MWCNTs induced cytotoxicity through GSH depletion, ROS generation, OS, cell inflammation, membrane leakage, lipid peroxidation, and protein release [189]. When human aortic endothelial cells were treated with CNTs, an increase in the messenger RNA of MCP-1, VCAM-1, and IL-8 was observed [184], while when human skin fibroblasts (HSF42) were exposed to MWCNTs and multi-walled carbon nano-onions (MWCNO), inhibition of the cellular cycle and an increase in apoptosis and necrosis, were detected [190]. Several intracellular signalling pathways were altered after exposure to CNTs, depending on the material tested and the dose, establishing that MWCNTs are more toxic than MWCNO by an interferon and p38/ERK-MAPK mediated toxicity. Soto et al., using chrysotile as a positive control, explained that the cellular response to MWCNTs aggregates was like that observed for the chrysotile-induced response [191]. Similarly, Murr et al. showed that SWCNTs aggregates, two types of MWCNTs aggregates, and the chrysotile aggregate provoked cell death starting at 2.5 μg/mL with similar cytotoxic response in mouse alveolar macrophages [192]. Also, Bottini et al. compared the time- and dose-dependent toxicity to T-lymphocytes caused by not modified MWCNTs- and oxidized MWCNTs finding that the latter induced significant apoptosis, being more toxic than hydrophobic pristine MWCNTs, which can result toxic only at high concentrations [193]. The studies reported on the cytotoxicity of CNTs are summarized in Table 4.

**Table 4.** In vitro cytotoxicity of CNTs for early experiment of years 2003–2011.

| Nanomaterials | Cell Lines | Observation | Ref. |
|---|---|---|---|
| Metal oxide nanoparticles and SWCNTs | A549 | Penetrate the cell | [167] |
| CNTs | Rat macrophages (NR8383), A549 | Increase in intracellular reactive oxygen species | [168] |
| CNTs | MSTO-211H | Agglomerated CNTs show greater cytotoxicity | [169] |
| MWCNTs | HEK | Decrease in cell viability and increase in IL-8 | [170] |
| Functionalized SWCNTs | Human fibroblasts 3T6 and murine 3T3 | Pass through the cellular membrane and concentrate in the cytoplasm | [171] |
| SWCNTs | HaCaT, HeLa, H1299, and A549 | Increase in oxidative stress and inhibition of cell proliferation | [172] |
| SWCNTs | HaCaT | Cell death, oxidative stress, and increase in lipid peroxides | [173] |

**Table 4.** *Cont.*

| Nanomaterials | Cell Lines | Observation | Ref. |
|---|---|---|---|
| SWCNTs | HaCaT and BEAS-2B | Loss of cellular integrity and cellular apoptosis | [184] |
| SWCNTs | Lymphocytes and macrophages | Uptake of SWCNTs | [175] |
| MWCNTs | HEK | Cell-cycle inhibition | [188] |
| Functionalized SWCNTs | Human dermal fibroblasts | Less toxicity | [176] |
| SWCNTs-streptavidin complex | HL60 | Low toxicity | [177] |
| CNTs | Lymphocytes | Increase the secretion of cytokines | [179] |
| SWCNTs | HEK293T | Inhibition of cell proliferation and decrease in cell adhesive ability | [180] |
| SWCNTs and MWCNTs | Alveolar macrophages | SWCNTs showed higher toxicity than MWCNTs | [181] |
| Purified SWCNTs | Lung fibroblast V79 | DNA damage | [183] |
| SWCNTs | RAW 264.7 | Production of TGF-β1 | [174] |
| Iron-rich SWCNTs | Macrophages | Phagocytosis of the SWCNTs and conversion of extracellular $O^{2-}$ into hydroxyl radicals | [184] |
| Ground MWCNTs | Rat peritoneal macrophages | Cytotoxicity and overproduction of proinflammatory cytokines | [185] |
| MWCNTs | J774.1 | Cytotoxic effects by the rupturing of plasma membrane | [186] |
| CNTs | Human aortic endothelial cells | Increase in the MCP-1, VCAM-1, and IL-8 | [184] |
| MWCNTs | Alveolar macrophages | Cell death | [192] |
| Pristine MWCNTs and oxidized MWCNTs | T lymphocytes | Apoptosis | [193] |

CNT = carbon nanotube; HEK = human epidermal keratinocyte; MWCNT = multi-walled carbon nanotube; SWCNT = single-walled carbon nanotube.

#### 4.2.2. In Vivo Studies: Pulmonary Toxicity

Over the years, experts in the field have developed increasing specialization in devising synthetic and characterization methods for obtaining improved CNTs in large scale. At the same time, the application possibilities of CNTs have grown exponentially to the point of making more studies on their possible cytotoxicity and genotoxicity imperative, also in vivo. The almost daily use of CNTs in novel nanotube-based products greatly increased the possibility of contact with humans, animals, and the environment. CNTs can enter the body through various routes of exposure, including derma, mouth, and nose. Among the possible contact opportunities, CNTs can enter the respiratory airways of the workers and accumulate in the lungs first during manufacture. Also, since CNTs are used as fillers in food packaging products, they can reach the gastrointestinal tract (GIT) of the consumers. Several reports from in vivo studies on rodents have demonstrated that the ingestion of SWCNTs and MWCNTs is toxic. Based on these results, the possibility that CNTs could also be risky to humans has become an increasingly worrying reality, and confirming or refuting it is a necessity. Pulmonary toxicity of raw CNTs, containing iron impurities, acid-purified CNTs, and CarboLex CNTs-rich in nickel and yttrium impurities, was investigated by Lam et al. They dispersed all samples in the serum and in carbon black, or quartz particles were used as negative and positive controls [194]. The samples were administered intratracheally to mice at different concentrations. After 7 days, in the mice group treated with CNTs, epithelioid granulomas and interstitial inflammation were observed in a dose-dependent way, which augmented after 90 days. Moreover, peri-bronchial inflammation and necrosis were also discovered in the lungs of animals that were treated with CNTs via alveolar septa, thus establishing that CNTs are more toxic than carbon black and quartz, which led to negative health effects only in chronic inhalation exposures [194]. Pulmonary toxicity of pristine SWCNTs was assessed by Warheit et al., which used quartz particles as positive control and carbonyl iron particles as negative control, using rodents as animal models [195]. Rats were administered intratracheally with SWCNTs at 1 and 5 mg/kg. Upon administration of the latter dose, 15% of rats died within 24 h, probably due to the mechanical obstruction of the airways by the SWCNTs. Anyway, transient inflammation and cell damage were found in surviving animals, while multifocal granulomas, attributable to tissue reactions against the foreign body, were observed. These early findings confirmed the cytotoxic power of SWCNTs while demonstrating the huge necessity for more chronic study. Shvedova et al. showed that pharyngeal aspiration of SWCNTs induced anomalous pulmonary effects in C57BL/6 mice, which brought acute inflammation, rapid progressive fibrosis, granuloma, and alveolar wall thickening in the lungs [174]. These pathological lesions were associated with functional respiratory deficiencies and decreased bacterial clearance. To complete this worrying scenario, increased clinical marker values were detected in bronchoalveolar lavage (BAL) fluid, which endorsed that the SWCNTs were more toxic than crystalline silica [174]. Even higher inflammatory response, OS, collagen deposition, and fibrosis were observed in C57BL/6 mice when SWCNTs were inhaled rather than pharyngeal aspirated. Anyway, pharyngeal aspiration of SWCNTs provoked faster atherosclerotic plaque formation in ApoE mice [196]. Mutlu et al. administered intratracheally raw aggregated and highly dispersed SWCNTs in 1% Pluronic F 108NF to mice at a 40-µg dose, which was higher than the dose used by Shvedova et al. [174] to induce pulmonary fibrosis in mice [197]. Lung inflammation was induced by aggregated SWCNTs in PBS, while highly dispersed SWCNTs do not cause any adverse phenomena. Muller et al. reported the respiratory toxicity of both MWCNTs and ground MWCNTs suspended in sterile saline (0.9% NaCl), which was observed after 2 months of exposure in the form of pulmonary lesions characterized by collagen-rich granulomas, dose-related pulmonary fibrosis, and increased production of TNF-$\alpha$ [185]. The dangerousness of the inflammatory effect produced by MWCNTs was anyway halved with respect to asbestos and carbon black [185]. Mitchel et al. reported that the exposure of mice to MWCNTs at high doses caused immunosuppression after 14 days of whole-body inhalation. On the contrary, inflammation and granuloma formation were not observed, which was different from what was previously reported [198]. Later, Muller et al. administered intratracheally to rats (2 mg/rat) MWCNTs in the forms of both tubes heated to 600 °C and 2400 °C and ground MWCNTs heated at 2400 °C [199]. Observations revealed that the pulmonary toxicity of MWCNTs was less compared to those of ground MWCNTs, thus indicating that the toxicity of the tubes mostly depends on the imperfect sites in their carbon architecture. However, the group of Donaldson showed that the toxicity of ground MWCNTs may depend either on their larger dispersion in the lungs or on the effects of the released metals upon grinding [200]. Poland and his group reported that mice showed an asbestos-like pathogenic behaviour after exposition to long MWCNTs, which translated into inflammation and granuloma formation. Such a pathogenic scenario was also observed in the mice administered directly into the abdominal cavity with CNTs [32]. The additional exposure of lungs, with allergen-based inflammation, to SWCNTs caused increased pulmonary toxicity characterized by increased lung protein levels of T helper cytokines and chemokines, augmented the level of OS-related biomarkers and accessory allergen-specific IgG1 and IgE activity [201]. By light microscopic examination technique, Kobayashi et al. showed that MWCNT aggregates, accumulated in the lungs, were phagocytized by alveolar macrophages and remained up to 6 months post-exposure [202]. Granulomatous lesions or collagen accumulations were observed only in animals exposed to highly dispersed MWCNTs through intratracheal administration but not in those administered with MWCNTs by inhalation, thus demonstrating that MWCNTs produce pulmonary lesions, based on the route of administration and dose [203]. Also, Grubek-Jaworska et al. showed that four different commercial MWCNTs and SWCNTs, characterized by very low content of iron intratracheally administered, caused pneumonia with an interstitial non-specific focal reaction in guinea pigs, treated for 3 months [204]. Significantly increased concentrations of IL-8 in the bronchoalveolar lavage (BAL) fluids and augmented macrophages and eosinophils were instead caused by other types of CNTs. Additionally, carbon sediments were found mainly in the bronchioles, while the alveolar ducts and the alveoli were free. No granulomas were detected in lungs, while CNT aggregates and mechanical obstructions were found in certain airways of some animals. More recently, Francis et al. showed that the single exposure of rats to MWCNTs per se induced inflammation, epithelial cell membrane damage, and cell lysis, confirmed by augmented inflammation markers concentrations, including TNF-$\alpha$, IL-4, LDH, WBC, and increased ALP activity [205]. Histopathological experiments revealed that dispersion of MWCNT in the lungs caused fibrosis and granuloma. Since occupational and intentional exposure to MWCNTs has exponentially increased over the years with large-scale production of CNTs, it is mandatory to investigate their toxicity upon prolonged exposure timings. For an easier understanding by the readers of the scenario of experiments carried out to accredit the possible pulmonary toxicity of CTNs, the early pulmonary toxicity studies discussed in this section on CNTs have been summarized in Table 5.

**Table 5.** In vivo case studies on pulmonary toxicity of CNTs.

| Nanomaterials | Animals | Observation | Reference |
|---|---|---|---|
| SWCNTs | Mice | Peri-bronchial inflammation and necrosis | [194] |
| Pristine SWCNTs | Rat | Inflammation and multifocal granulomas | [195] |
| SWCNTs/$SiO_2$ | Mice | Granulomas and lung fibrosis | [174] |
| MWCNTs | Rat | Pulmonary fibrosis | [185] |
| SWCNTs | Mice | Inflammatory response, OS, collagen deposition | [196] |
| MWCNTs | Mice | Inflammation and granulomas | [32] |
| MWCNTs | Rat | Granuloma and collagen depositions | [203] |
| MWCNTs/SWCNTs | Guinea pigs | Pneumonitis | [204] |
| MWCNTs | Rat | Inflammation, granuloma, and lung fibrosis | [205] |

CNT = carbon nanotube; MWCNT = multi-walled carbon nanotube; SWCNT = single-walled carbon nanotube; OS = oxidative stress.

CNTs Cytotoxic Effects to Other Tissues

Initial animal studies reported previously in this section evidenced that exposure of rodents to different forms of CNTs can cause acute and chronic pulmonary inflammation, based on increasing OS and on the assumption of inflammatory effects of atherosclerosis and air pollution [174,185,194,195]. Based on these results, it was presumed that the toxic effects of CNTs could also affect other tissues. In this regard, some authors, such as Simenova and Erdely [206], as well as Li et al. [196], investigated the cardiovascular toxicity of CNTs in rodents by different ways of administration. The studies revealed that CNTs are capable of activating the blood cells via inflammatory markers release, which leads to adverse cardiovascular effects. Damage to the mitochondrial DNA of the aorta was observed in a stress-dependent and dose-dependent mode. The effects caused by CNTs may predispose to atherogenesis. An activated oxidative marker was found in the aorta and cardiac tissues of mice with high blood cholesterol levels exposed to SWCNTs, which also developed aortic DNA damage. Atherosclerotic plaques on the surface of the aorta and increased atherosclerotic lesions in the brachiocephalic arteries were observed in mice that had been exposed to SWCNTs [184,196,206]. Pietroiusti et al. [207], Bai et al. [208] and Philbrook et al. [209] studied the toxic effects of CNTs on the reproductive and developmental system. It was reported that functionalized CNTs administered at low doses were capable of triggering high resorption processes, progressive malformations in the survived infant rodents, and induced ROS generation in the placentas of exposed animals. A higher percentage of resorptions, as well as huge morphological defects and skeletal abnormalities in foetuses, were discovered in *Drosophila melanogaster* and CD-1 mice exposed to functionalized CNTs. CNTs accumulated in the testicles, causing OS, tissue damage, and alterations, which raised concerns about possible adverse effects on male fertility [207–209]. A collection of more recent case studies on in vitro and in vivo CNT toxicity has been reported in Table 6, while Table 7 collects case studies on CMT toxicity in various organs. Tables 6 and 7 already contain detailed explanations of observed outcomes and do not need additional discussion.

**Table 6.** In vitro and in vivo studies for assessing the possible cytotoxicity and genotoxicity by exposure to CNTs.

| Theme of the Study | Tested Cells | Toxic Effects | Findings | Refs. |
|---|---|---|---|---|
| Genotoxicity of MWCNTs in human cells | HeLa, MCF7 Human respiratory epithelial cell | Genotoxicity | DNA damage, chromosomal aberrations, OS by exposure to CNTs | [210,211] |
| Inhalation exposure to CNTs induces pulmonary toxicity | Mice | Pulmonary toxicity | Inflammation, granuloma formation, lung fibrosis upon inhalation ↑ Levels of cytokines | [194,212] |
| Skin exposure to CNTs to assess dermal toxicity | Murine epidermal cells (JB6 P+), and immune-competent hairless SKH-1 mice | Dermal toxicity | Direct skin exposure to CNTs led to irritation and inflammation Penetration into deeper skin layers is still debated | [213] |
| Impact of CNTs on immune system | Lymphocytes, T cells, monocytes, and dendritic cells | Immunotoxicity | CNTs affected immune responses Altered cytokine production and immune cell functions | [214,215] |
| Cardiovascular effects of CNTs in animal models | Mice | Cardiovascular toxicity | Inflammation- and OS-induced increased blood pressure and cardiovascular diseases | [216] |
| Liver toxicity/biodistribution of CNTs in mice | Mice | Hepatoxicity | Liver damage and alterations in liver enzyme levels | [217] |
| Renal toxicity of MWCNTs in rats | Human embryonic kidney (HEK293) cells | Renal toxicity | CNT accumulation in the kidneys, renal inflammation, OS Kidney functional impairment | [218] |
| OS induced by CNTs | Alveolar macrophage (AM) | OS | CNTs generate ROS, OS, cell and tissue damage | [181] |
| In vitro neurotoxicity of CNTs | Mammalian cell lines | Neurotoxicity | Neuroinflammation and neurons damage | [219] |

OS = Oxidative stress; ROS = reactive oxygen species; ↑ indicated increased.

**Table 7.** A quick overview of CNTs' toxicity on various organs by in vivo and in vitro studies.

| CNTs | Tested Cells | Affected Organ | Results | Ref. |
|---|---|---|---|---|
| MWCNTs | Male Sprague Dawley rats | Nervous system | Dramatic alterations of sympathetic and parasympathetic nervous system's equilibrium | [220] |
| MWCNTs | Mice | Nervous System and BBB | Acute lung exposure to MWCNTs damaged BBB integrity, induced nerve inflammatory responses | [221] |
| CNTs | Male NMRI mice | Nervous system | Behavioral toxicity with manifestation of sadness or anxiety | [222] |
| SWCNTs | PC-12 cells | Nervous system | Toxicity to PC-12 cells ↑ Harmful to differentiated PC-12 cells | [223] |
| SWCNTs | Male C57BL/6 mice | Pulmonary immune system | CNTs make people more vulnerable to respiratory virus infections | [224] |
| SWCNTs | Six-week-old specific-pathogen-free ICR mice | Immune system Reproductive system | CNTs affected development and reproduction and produced immunological toxicity | [225] |
| SWCNTs | BALB/c macrophage cells J774A and BALB/c mice | Immune system | Immune toxicity * | [226] |
| MWCNTs | T lymphocytes | Immune system | Concentration-dependent harmful of CNTs to human T cells | [193] |
| SWCNTs | 6/8 weeks old females of the CD1 outbred strain Mouse ES cell line D3 NIH3T3 cells | Embryos | SWCNTs can cause harm to mammalian embryos | [207] |
| CNTs | Kunming mice | Embryos | CNTs harmed fetuses, cause miscarriage, and were harmful to embryos | [227] |
| MWCNTs | Zebrafish embryo | Embryos | Significant length-dependent development risk | [228] |
| CNTs | Mouse embryonic fibroblasts (MEFs) and C57BL/6J mice | Embryos | Induction of hereditary embryotoxicity | [229] |
| O-SWCNTs | Artemia salina | Embryos | Large amount generation of ROS and deformed Artemia salina | [230] |
| CNTs | Male BALB/c mice | Genitals | CNTs poison mice's reproductive organs | [208] |
| SWCNTs/MWCNTs | MeT-5A and BEAS 2B cells | Genome | DNA damage in MeT-5A cells by both MWCNTs and SWCNTs | [231] |

ROS = reactive oxygen species; ↑ indicated increased, high, higher; * there was a negative correlation between immune toxicity and SWCNT dispersion; O-SWCNTs = Oxidized SWCNTs.

## 5. Possible Strategies to Moderate CNTs' Toxic Effects: Future Perspective and Preventive Actions

CNTs cause a cascade of events that are noxious to human and animal body tissues. In this regard, Lettiero et al. and Moghimi et al. have reported that inhibitors of such events, achievable by CNTs surface modifications and functionalization, could prevent the activation of the toxic system in the body tissues [232,233]. Generally, PEG-modified SWCNTs were not toxic and biocompatible to mice, achieving the longest blood circulation in 1 day and being fully eliminated from the vital tissues of mice in about 2 months [234]. Longer circulation time, minor uptake in the reticule-endothelial system, and reduced accumulation in the spleen and liver were observed for PEG-CNTs by Yang et al. [235]. Coating CNTs with C1q recombinant globular proteins and adding functional groups of different types to their side walls are other methods to minimize CNT toxicity [176,236]. Silva et al. reported that the onset of CNT toxicity is strictly correlated with the route of administration [237]. CNTs administered by inhalation provided no sign of inflammation after 1 day, but inflammation appeared after 21 days. On the contrary, instilled CNTs caused inflammation after 1 day of exposure, which disappeared after 21 days [238].

Adding catalytic horseradish peroxidase that promoted -COOH CNTs degradation in an acidic medium led to no toxic or inflammatory effects on the lungs of mice used for toxicity experiments.

Collectively, limiting the toxicity of CNTs remains a serious aspect of their safer use in various applications. Table 8 collects some relevant strategies proposed over the years up until now to mitigate potential harmful effects, which could arise from an extensive exposure to CNTs.

**Table 8.** Strategies to reduce CNT toxicity.

| Strategy | Goal | Modifying Agents/Methods | Results | Refs. |
| --- | --- | --- | --- | --- |
| Modify CNTs surface with biocompatible materials or other molecules | ↑ Dispersion in biological fluids<br>Influenced cellular uptake<br>↑ Solubility<br>↓ Toxicity | Proteins, surfactants | ↑ Tumor targeting<br>↑ Therapeutic benefits<br>↓ Toxicity | [238–240] |
| | | Folic acid | ↑ In vivo tumor targeting<br>↑ Therapeutic benefits<br>↓ Toxicity | [241] |
| | | Polyacrylamide hydrogels *<br>Biomaterial, $TiO_2$ | 100% survival of L929 mouse fibroblast | [242] |
| Application of coatings to CNTs | ↑ CNTs biocompatibility<br>↓ Potential toxicity<br>Prevent direct contact with biological systems<br>↑ CNTs solubility | Curcumin lysine ** | ↓ IL-6, IL-8, IL-1β, TNFα, N-FκB<br>↑ Antioxidant enzyme catalase<br>↓ ROS generation<br>Recovery of mitochondrial membrane potential<br>↓ Cell death | [243] |
| Encapsulation of CNTs<br>Using CNTs to entrap bioactive molecules | ↓ Direct cells exposure to CNTs Control of CNTs release<br>↓ CNTs impact on tissues | PEG (entrapping agent)<br>Oxaliplatin (entrapped agent) | PEGylation delayed oxaliplatin release rate<br>↑ Drug's anticancer effects on HT-29 cells | [244] |
| Tailor the diameter size and length of CNTs | ↓ Toxicological impact | N.A. | ↑ Specific surface area<br>↑ Transmembrane mobility<br>↓ Toxicity | [245] |
| | | | ↓ Harm to lysosomes for large-diameters MWCNTs | [246] |

**Table 8.** *Cont.*

| Strategy | Goal | Modifying Agents/Methods | Results | Refs. |
|---|---|---|---|---|
| Optimization of purification processes | Remove metal impurities<br>Remove residual catalysts | Chemical/electrochemical oxidation [247]<br>High chlorine partial pressure [248]<br>Microwave-assisted digestion [249]<br>Incandescent annealing | ⬇ Lower harmful effects | [250] |
| Engineering controls<br>Suitable PPE | ⬇ Inhalation exposure | Proper ventilation/ respiratory protection | ⬇ Risk of respiratory toxicity | N.R. |
| Co-administration of CNTs with antioxidants | ⬇ Potential OS<br>⬇ Cellular damage | Quercetin | Prevention of the oxidative damage<br>⬇ Inflammatory effects<br>⬇ Immuno-toxic effects | [251] |

* Encapsulation agent for CNTs-COOH; ** used to coat MWCNTs; N.A. = not applicable; N.R. = not reported; ⬇ indicates minor reduction, lower, decreased, decrease; ⬆ indicates improved, increase, increased, major; PPE = personal protective equipment; OS = oxidative stress.

Proper modification of the surface of CNTs with biocompatible materials or selected molecules can enhance their dispersion in biological fluids and reduce toxicity. Functionalization can also influence cellular uptake and interactions [252,253]. One of the main causes of the hazardous effects of CNTs on the human and animal body is their poor water solubility. This is a problem that may be solved by CNT surface modification [254,255]. A current trend followed by scientists in the sector consists of searching for biocompatible substances that may be smeared with CNTs to improve their characteristics [256]. Much research has been done on the detrimental effects of CNTs on pulmonary systems, with confirming results on A549 cells [257,258]. By covering MWCNTs with curcumin, their propensity to induce OS, inflammation, and cell death was significantly reduced [243]. Wu et al. enriched MWCNTs-COOH and pegylated MWCNTs with oxaliplatin, a cisplatin-derived chemotherapeutic, by both their surface functionalization with the drug and its encapsulation in CNTs cavities [244]. The authors demonstrated that, in an aqueous environment, the entrapped oxaliplatin could easily exit MWCNT-COOH and MWCNT-polyethylene glycol (PEG) and that Pegylated MWCNTs were capable of an oxaliplatin-controlled release, thus augmenting its therapeutic effects.

Shorter and well-dispersed CNTs with a high specific surface area and robust transmembrane mobility showed lower toxicity compared to longer aggregates [245]. These substances can harm proteins and cellular components, leading to macrophage malfunction or even death [222]. Sohaebuddin et al. demonstrated that MWCNTs with large diameters provided minimal harm to lysosomes, while those with tiny diameters might create membrane instability [246]. Highly pure CNTs that do not contain residual metals or catalysts induce lower harmful effects [250]. Considerable progress has been achieved in the purification of CNTs [259]. Also, CNTs with structures that favour biodegradation over time could reduce their persistence in circulation, thus reducing tissue toxicity [260]. Concerning CNT-based delivery systems, the choice of alternative administration routes, including topical or transdermal ones, can minimize the direct exposure of sensitive organs or tissues to CNTs, thus reducing their systemic harmful impact.

Quercetin has good bioavailability and possesses formidable analgesic, antioxidant, anti-inflammatory, and anti-atherosclerotic properties [261]. Additionally, quercetin reduces reactive oxygen species (ROS) generation and increases the antioxidant enzyme activity [262]. It could be exploited as an adjuvant to prevent oxidative damage, inflammation, and immune-toxic effects of MWCNTs [251].

Despite the several strategies developed so far to limit the toxic outcomes of extensive use of CNTs to humans and the environment, it is mandatory to conduct preventive behaviour to limit the large diffusion of their possible cytotoxic impacts. Biocompatibility assays using appropriate in vitro and in vivo models are always necessary before a CNT's widespread use. Knowing the individual biological responses to different types of CNTs is critical to assessing their specific risk [263]. Although a more specific regulation is urgently needed for these nonpareil nanomaterials, adhering to existing regulations and guidelines on nanomaterial safety is obligatory. The responsible development and use of CNTs could be assured only by ensuring compliance with regulatory standards [264]. Education training for workers who produce, handle, and apply CNTs, thus making them aware of potential risks which could derived by CNTs identified until now and implementing proper safety protocols can help to minimize harmless exposure. Concerning the environment, monitoring programs to evaluate the dispersion and potential impact of CNTs in workplaces and surrounding environments can help update risk management strategies. The invention and development of environmentally friendly or "green" synthesis procedures for creating CNTs as alternative options to the complicated machineries in use could reduce the usage of dangerous chemicals and toxic precursors and can lead to CNTs with fewer impurities and lower toxicity. Duraia et al. recently proposed a straightforward, simple method to generate CNTs on graphitic layers using corn seeds [265]. Alternative nanomaterials to CNTs with similar functionalities but lower toxicity risks should be considered. It is the case of graphene, which could be less hazardous than CNTs [266]. Indeed, as graphene is more robust than CNTs to immune cell destruction, thus limiting the possible release of hazardous compounds deriving from such degradation, it could be a promising option for CNTs. In particular, graphene oxide (GO) was less vulnerable to macrophage destruction than SWCNTs. Unfortunately, efficient standardized techniques to assess the toxicity of these nanomaterials are still challenged by difficulties and restrictions [267]. Permanent research and collaboration between scientists, engineers, and regulators are pivotal to developing safe routines for CNT production and use in various industries.

## 6. Regulatory Considerations

The National Institute for Occupational Safety and Health (NIOSH) is the most important federal agency in the United States, which conducts research and offers supervision for occupational safety and the consequences on human health from the extensive applications of and exposure to nanomaterials. Studies have revealed that nanoparticles (NPs) may pose a greater health risk than bulk materials due to their greater surface area/unit mass ratio [35]. In 2013, NIOSH published a Current Intelligence Bulletin detailing the potential hazards of CNTs and recommended the exposure limit for CNTs and nanosized fibres [268]. Subsequently, in October 2016, SWCNTs were registered within the European Union's Registration, Evaluation, Authorization, and Restriction of Chemicals (REACH) regulations as compounds to be studied concerning their possible dangerous effects. As a direct consequence, the commercialization of SWCNT in UE was permitted up to 10 metric tons. REACH has a centralized database that records all chemicals produced or imported into the EU in quantities of one ton per year or more per supplier or importer. A chemical safety report must be submitted to REACH when the quantity of its production is ten tonnes or more. There have been frequent complaints about REACH not having taken into account the nano-specific characteristics of materials, but only the amount of their production. Fundamentally, substances of a specific interest may be subjected to REACH restrictions. If such restraints can be fulfilled by the European Chemicals Agency (ECHA), a thorough comment and consultation procedure must be carried out, and the approval of the Member States' committee must be pursued for that substance. A Candidate List of Very High Concern Substances (SVHC) containing almost 40 substances was compiled in 2010 by ECHA, but no compound was a nanomaterial. This was mainly due to the high thresholds of production quantity for nanomaterials—1 ton per year or more to impose their registration and 10 tonnes per year or more to make it necessary to submit a chemical safety report. Chemical Regulation in the EU and the US (TSCA; Toxic Substances Control Act) used the chemical composition of a material to define regulation measures without considering particle size or nano-specific features. The problem of controlling carbon-based products has become particularly apparent, especially with the advent of their large-scale production. In REACH, carbon was initially considered a non-issue ("minimum risk owing to its intrinsic features"). This status of immunity was eliminated at the end of the year 2008 when a clear differentiation was made between carbon and carbon in its nano-scale forms. In this regard, C60-fullerene received a distinct CAS (Chemical Abstracts Service) registration number according to the worldwide norms of the chemical denomination in relation to carbon, carbon black, and graphite since these materials, despite their same molecular character, necessitate to be dissociated from each other. (Nano)tubes that would have a diameter smaller than 1 nm and a length exceeding 100 nm were not considered as nanomaterials until 2011 in EU. To address this potential omission, Recommendation 2011/696/EU included in the definition both fullerenes, graphene flakes, and single-wall carbon nanotubes, with one or more external dimensions below 1 nm, as nanomaterials [269]. Unlike the EU, the US introduced CNTs under the Toxic Substances Control Act to strict notification requirements in October 2010. Their production, import, or usage (also as a continuation of ongoing activity) shall be notified to the authority, which will decide whether to import or process the substance within 90 days. The only exclusion was agreed to CNTs stalwartly integrated into a matrix. Concerning health and safety on the job, in 2011, the European Agency of Safety at Work (EU-OSHA) compiled a literature review in the field of CNTs, dealing with occupational exposure, toxicology, and protective measures in the work environment [270]. It was also a compilation of advice and guidelines for technical occupational safety and personal protective equipment. The toxicological data available for SWCNTs and MWCNTs have also been summarized [270]. The purpose of the overview was to provide the Swedish Work Environment Authority with information and a basis for various types of measures [270]. In the United Kingdom—the Health and Safety Executive (HSE), referred to the "10 Recommendations" to be considered for health and safety purposes at the job, published by the Royal Society and the Royal Academy of Engineering, 25 years earlier on the application of nanotechnologies. Based on these rules, the HSE advised on precautionary measures and risk avoidance practices. In 2009, HSE considered CNTs as "extremely concerned substances". The British Standards Institution (BSI) suggested an exposure limit of 0.01 fibres/mL, even if investigations to monitor this limit test were time-consuming [159]. In the US, the National Institute for Occupational Safety and Health (NIOSH) recommended an occupational exposure limit of less than 7 μg CNTs per $m^3$ of air [159]. This exposure limit is the smallest concentration that can be correctly measured [271]. On the other hand, the exposure limit for 5 $mg/m^3$ of airborne graphite dust set by the EU Occupational Safety and Health Administration (EU-OSHA) is about 1000 times higher [159]. It should also be stressed that dust particles are bigger and, therefore, heavier. Anyway, despite only a few types of CNTs being utilized, the precautionary principles should provide indications for health, safety, and risk limitations to be applied to all types of CNT-based existing products. Indeed, since CNT handling cannot be fully prevented, it should aim at a high standard of safety and control as well [272].

Although the determination of a correct occupational limit for CNT exposure is mainly contingent on exposure data and measurement technology, the US limits concentrations dictated for CNTs are considered by NIOSH high-risk to develop adverse effects on respiratory health. In this context, the promotion of research in this area is essential and mandatory. Also, measures to limit hazards by CNTs need improvement, and measures to maintain airborne CNTs to a minimum level and below the limits of exposure should be developed. Studies have demonstrated that high concentrations of needle-shaped, long, thin, and bio-persistent CNTs have a serious adverse effect on the respiratory tract. In this context, the mechanisms of action and the dose–effect relationship for as high a number of CNT types as possible need to be clarified to be capable of compiling a dangerous chart. Also, the life cycle of CNTs in the environment still has not been thoroughly assessed, and argumentative information on CNT ecotoxicity and credible exposure data are limited and need improvement to better assess their environmental risk. Although legislation on hazardous chemicals and regulations on occupational health and safety exists, it still does not offer specific requirements for handling CNTs. Even though specifications of REACH on the volume thresholds for registration (1 JT) and for the obligatory chemical safety report (10 JT) exist, they have been set at levels that do not include all CNT manufacturers and do not consider CNT-specific features. Although efforts are proceeding at international levels in occupational health and safety to set limits for occupational airborne exposure to CNTs, specific regulations are still not adopted. Analytical and detection methods need improvements through further research and development. Lawmakers, regulators, and environmental activists' interest in nanomaterials has incessantly increased over the years, and the interrogation of the correct nanotechnology regulations is a daily challenge. The incessant development of nanomaterials and the daily application of nanotechnology have led to rigid regulatory standards in both animal and human manufacturing and utilization. The US Environmental Protection Agency (EPA) reported that CNTs encompass substantial regulatory concerns regarding toxicity and environmental safety [159]. The US National Environmental Policy Act (NEPA) has published a series of rules regarding the main factors for controlling nanomaterials, including quality control and manufacturing and product safety evaluation, like toxicity, clearance, biodistribution, and metabolism. In the following sections, we have briefly introduced the most relevant international organizations that work to develop and publish technical standards, which cover procedures for testing, classification, and characterization of several sorts of materials. Then, we have focused on the main standards concerning nanomaterials and CNTs published by the International Organization for Standardization (ISO) and the American Society for Testing and Materials (ASTM) International Organization.

### 6.1. *International Standards Concerning Nanomaterials and CNTs*

Different international organizations, including ISO (1946), ASTM, BSI (1901), IEC (1906), DIN (1917), ANSI (1918), and AFNOR (1926), were created to develop and publish key standards to chemically and physically characterize materials of every sort. In the following, we consider ISO and ASTM.

#### 6.1.1. ISO Standards

The International Organization for Standardization (ISO) brings global experts together to agree on the best way of operating in several sectors, including the preparation methods to achieve a product, its characterization, and how to manage a process. As one of the oldest non-governmental international organizations, ISO has enabled trade and cooperation between people and companies all over the world since 1946. The International Standards published by ISO must be approved by all ISO members and serve to make life easier, safer, and better (available online at https://www.iso.org/about, accessed 15 April 2025). ISO has provided several standards, guidelines, methods, and requirements concerning differently structured nanomaterials, CNTs, and CNT-containing nanocomposites. The most recent ones concerning CNTs and CNT-containing nanocomposites are available online at https://www.iso.org/search.html?PROD_isoorg_en[query]=carbon%20nanotubes (accessed on 15 April 2025) and have been included in the following Table 9.

**Table 9.** Published and under development * ISO standards concerning CNTs and CNT-containing nanocomposites.

| Standard | Description |
|---|---|
| ISO/TS 19808:2020 | Specifies the characteristics of suspensions containing MWCNTs to measure and the corresponding measure methods |
| ISO/TS 23690:2023 | Specifies a mild oxidation method by TGA to determine the content of carbon impurities (amorphous carbon, other types of structured carbon) |
| ISO/TS 11308:2020 | Guidelines for the characterization of CNT-containing samples and impurities by TGA in inert or oxidizing environment |
| ISO/TR 23463:2022 | Reviews correct characterization of CNT and CNF aerosols for inhalation Establishes characterization required to assess their inhalation toxicity |
| ISO/TS 11888:2017 | Describes methods for characterizing mesoscopic shape factors of MWCNTs by SEM, TEM, viscometry, DLS |
| ISO/TR 10929:2012 | Identifies the basic properties of MWCNTs and the content of impurities<br>Highlights the available measurement methods to industry for their determination |
| ISO/TS 10797:2012 | Establishes methods for characterizing the morphology of SWCNTs, identifying the elemental composition of other materials in SWCNT samples by TEM and EDXRD |
| ISO/TS 10867:2019 | Gives guidelines for the determination of the chiral indices of the semi-conducting SWCNTs by NIR-PL |
| ISO/TS 10868:2017 | Provides methods to characterize the diameter, purity, and other characteristics of compounds containing SWCNTs by OAS |
| ISO/TS 10798:2011 | Establishes methods to characterize morphology and identify the elemental composition of catalysts and other inorganic impurities in RSWCNTs and PSWCNT powders and films by SEM, EDXRD |
| ISO/TS 13278:2017 | Provides methods for the determination of residual elements other than carbon in samples of SWCNTs and MWCNTs by ICPMS |
| ISO/TS 11251:2019 | Specifies method for the characterization of evolved gas components in SWCNT samples by EGA/GCMS |
| ISO 20523:2017 | Specifies classification, designations, and short names for CNT-based films deposited by PVD or CVD |
| ISO/AWI 11308 * | Gives guidelines for the characterization of CNT-containing samples by TGA in inert or oxidizing environment<br>Provides guidelines on the purity assessment of the CNT samples through quantitative measures of the carbon and non-carbon species present |
| ISO/AWI TS 21497 * | Describes a method to remove CNTs, NG, and CNHs from industrial and research wastewater by using hypochlorite compounds |

CNHs = Carbon nano-horns; NG = nano graphene; TGA = thermogravimetric analysis; PVD = physical vapor deposition; CVD = chemical vapor deposition; EGA/GCMS = evolved gas analysis/gas chromatograph mass spectrometry; ICPMS = inductively coupled plasma mass spectroscopy; RSWCNTs = raw SWCNTs; OAS = optical absorption spectroscopy; NIR-PL = near infrared photoluminescence spectroscopy; EDXRD = energy dispersive X-ray diffraction; SEM = scanning electron microscopy; TEM = transmission electron microscopy; DLS = dynamic light scattering analysis; * under development (all other standards are published).

#### 6.1.2. ASTM Standards

American Society for Testing and Materials (ASTM) international organization is a nonprofit organization that develops and publishes approximately 12,000 technical standards covering the procedures for testing and classification of materials of every sort. It encompasses 30,000 members from 135 countries. ASTM also serves as the administrator for the US TAGs (United States Technical Advisory Group) to an enormous amount of ISO/TCs (International Organization for Standardization/Technical Committee) and to their subcommittees. Specifically, ASTM Committee E56 on Nanotechnology was formed in 2005.

Nowadays, it comprises one hundred and eighty members from twenty-two countries and is formed by nine sub-committees with twenty-eight standards developed, which are available in Volume 14.02 of the Annual Book of ASTM standards. This Committee addresses issues related to standards and guidance materials for nanotechnology and nanomaterials, as well as the coordination of existing ASTM standardization related to nanotechnology needs. These international standards address topics including nano-enabled consumer products, physical and chemical characterization, and intellectual property issues (available online at https://www.astm.org/membership-participation/technical-committees/committee-e56, accessed on 15 April 2025). As an example, some key standards comprise the following:

- E2490 Standard Guide for Measurement of Particle Size Distribution of Nanomaterials in Suspension by Photon Correlation Spectroscopy (PCS);
- E2865 Standard Guide for Measurement of Electrophoretic Mobility and Zeta Potential of Nanosized Biological Materials;
- E2864 Standard Test Method for Measurement of Airborne Metal and Metal Oxide Nanoparticle Surface Area Concentration in Inhalation Exposure Chambers;
- E2524 Standard Test Method for Analysis of Haemolytic Properties of Nanoparticles;
- E2526 Standard Test Method for Evaluation of Cytotoxicity of Nanoparticulate Materials in Porcine Kidney Cells and Human Hepatocarcinoma Cells;
- E2535 Standard Guide for Handling Unbound Engineered Nanoscale Particles in Occupational Settings;
- E3025 Standard Guide for Tiered Approach to Detection and Characterization of Silver Nanomaterials in Textiles;
- E3275-21 Standard Guide for Visualization and Identification of Nanomaterials in Biological and Nonbiological Matrices Using Darkfield Microscopy/Hyperspectral Imaging (DFM/HSI) Analysis;
- E3172-18 Standard Guide for Reporting Production Information and Data for Nano-Objects;
- D8208-19 Standard Practice for Collection of Non-Fibrous Nanoparticles Using a Nanoparticle Respiratory Deposition (NRD) Sampler.

Anyway, we have only found a recent document specifically concerning CNTs: ASTM-D8526. This is the Standard Test Method for Analytical Procedure Using Transmission Electron Microscopy for the Determination of the Concentration of Carbon Nanotubes and Carbon Nanotube-containing Particles in Ambient Atmosphere (available online at https://www.document-center.com/standards/show/ASTM-D8526, accessed on 15 April 2025).

Despite ASTM also serving as the administrator to a huge amount of ISO/TCs and to their sub-committees, and its involvement in the development of standards about nanomaterials, our findings have evidenced that its specific interest in CNTs is extremely limited compared to that of ISO, thus confirming the great gap between the work and opinion of the different organizations worldwide about CNTs, which limit their regulation and hamper their safer production and utilization.

## 7. Conclusions and Perspectives

In this article, the main knowledge and findings gathered so far about CNTs have been reviewed, and we find that these tube-like nanomaterials, which are made only of carbon atoms or functionalized carbonaceous tubes, possess nonpareil properties, which enable them to be applied in several sectors with excellent results and fantastic improvements. Particular attention has been paid to their antimicrobial effects and toxic aspects on humans, animals, and the environment. These characteristics were confirmed by several case studies reporting in vitro and in vivo experiments developed over the years, which were inserted in reader-friendly tables. First, by a survey on the Scopus database, we have shown that research interest for CNTs in the years 2010–2025 has remained constant, with peaks of publications in 2016 and 2017. This is a feature probably due to the passage of CNTs' larger-scale production and the development of more efficient production procedures. Our specific research about their biomedical applications has revealed that both variants of CNTs, referred to as SWCNTs and MWCNTs, find application in numerous areas of nanomedicine. Also, concerning the more restricted field of their applicability as antimicrobial agents, our study has highlighted that these carbon-based nanomaterials (CNMs) are very promising as novel antimicrobial agents due to their extensively reported detrimental effects against both Gram-positive and Gram-negative bacteria, fungi, and biofilm. Due to their inhibiting effects on biofilm and excellent mechanical properties, CNTs could represent new optional ingredients for orthopaedical implant construction. Unfortunately, although the results from toxicological tests are often contrasting, several early and recent investigations have unveiled that CNTs could be toxic to humans, animals, and the environment. CNT toxicity is mainly due to their surface construct, nanosized diameters, needle-like structures, and residual metal impurities, and it depends on concentration and exposure times. Initially, the CNTs' possible toxicity was disregarded because they were made mainly of normally harmless carbon. Anyway, investigations prompted by the similarity of the needle-like CNT fibres with those of asbestos evidenced that they can be remarkably toxic and genotoxic to several cells, organs, and animal models. As with asbestos, the higher toxicity has resulted in pulmonary injuries, which is particularly worrying for CNT manufacturers who have long-time exposure to them and are at risk of inhaling their fibres. Toxicity to the cardiovascular system, as well as immune and reproductive systems, embryos, and neurotoxicity, have also been reported. Fortunately, several strategies, including preventive behaviours, which have been discussed here, have already been developed to limit the CNTs' toxic effects, but more and incessant studies should be implemented to make their production and employment safer. This study has also shown that despite the significant advancements in terms of regulations concerning CNT production and possible applications, as well as about health and safety at the production sites, several unsolved regulatory issues remain, which need urgent solutions. The incessant development of nanomaterials and CNTs and their daily application has led to rigid regulatory standards developed by different international organizations in both animal and human manufacturing and utilization, but standards and guidelines specific to CNTs are still very limited. Substantial regulatory concerns regarding CNT toxicity and environmental safety remain. Future perspectives should consider improved research finalized to the development of standardized protocols for toxicity assessment, which would allow for more dependable and comparable outcomes. More sophisticated in vitro and in vivo models that closely mimic human biological systems are needed. These will enable more precise forecasts of the behaviour and toxicity of CNTs in real-world situations, as well as the predictions of the long-term health effects resulting from exposure to CNTs. Surely, incessant CNT safety evaluations, including toxicity, clearance, biodistribution, and metabolism, can help in CNT inspection and in reducing conflicting results, thus allowing for a more reliable toxicological assessment. Additionally, the practical challenges of scaling up CNT-based technologies still remain unsolved. Low-cost, eco-friendly, and sustainable synthetic methods for preparing CNTs with precise and pure architecture on a macroscopic scale are necessary. Currently, the fabrication of CNTs requires expensive and energy-intensive professional methods like CVD or AD, generates a combination of different kinds of CNTs, and necessitates long secondary purifying procedures. High-purity, homogenous CNTs can be challenging to produce, thus resulting in harmful materials to human health.

Despite these drawbacks, current research also uses artificial intelligence (AI) and strives to resolve these problems and realize the full potential of CNTs in various applications.

In fact, it must be recognized that CNTs, despite their still unclear toxicity and insufficient regulation, are ubiquitously exploitable for improving the quality of human life. SWCNT catalyst support can enhance the selectivity of copper (Cu)-based electrocatalysts in the $CO_2$ conversion into valuable chemicals, which is a key strategy for achieving net-zero carbon emissions. Moreover, CNTs can be used as co-catalysts, as well as catalyst supports, for enhanced fuel cell applications and water splitting, which are excellent techniques for alternative clean energy systems due to zero carbon dioxide release, recyclability, and high gravimetric energy densities. Additionally, CNT-based single-atom catalysts (SACs) for electrochemical nitrogen reduction reactions (NRR) have received increasing attention due to their sustainable, efficient, and green advantages. In this regard, we hope that this review could stimulate further research on these materials, which are aimed at allowing their safer large-scale production and utilization.

**Author Contributions:** Conceptualization, investigation, writing—original draft preparation, supervision, project administration, S.A. Writing—review and editing, S.A. and G.C.S. All authors have read and agreed to the published version of the manuscript.

**Funding:** This research received no external funding.

**Data Availability Statement:** Not applicable.

**Conflicts of Interest:** The authors declare no conflicts of interest.

## Appendix A

**Table A1.** Key properties and potential uses of CNTs, as well as the residual current hurdles that limit CNTs' extensive applications.

<table>
<tr><th>Key Properties</th><th>Potential Uses</th><th>Current Hurdles</th></tr>
<tr><td rowspan="3">Size [61]</td><td rowspan="3">Nano-electronics [64]</td><td>Electronic heterogeneity [76]</td></tr>
<tr><td>Crystallographic defects</td></tr>
<tr><td>Stone–Wales defects</td></tr>
<tr><td>Ø < 100 nm<br>Thickness = 1–2 nm<br>Ideally infinite length</td><td>Conductors (SWCNTs)<br>Superconductors<br>Semi-conductors [1]<br>Supercapacitors</td><td>Not tunable conductivity</td></tr>
<tr><td>Electrical and electrochemical properties</td><td>Interconnects</td><td>⇩ Controllable orientation</td></tr>
<tr><td>⇧ Electrical conductivity<br>Constant resistively [63]<br>Electrons emission capacity</td><td rowspan="2">Chips manufacture [65]<br>Conductivity-enhancing components</td><td>⇩ Organized configuration</td></tr>
<tr><td>Thermal Properties</td><td rowspan="2">⇧ Variation in size and density [76]</td></tr>
<tr><td>⇧ Heat conductivity<br>Expansion</td><td>Optics and photonics</td></tr>
<tr><td>Mechanical properties</td><td>Light-emitting diodes (LEDs) [66,67] [2]</td><td rowspan="3">Export restrictions</td></tr>
<tr><td>⇧ Tensile strength<br>⇧ Elastic modulus</td><td>Photodetectors [68] [2]</td></tr>
<tr><td>Optical Properties</td><td>Bolometers [69] [2]</td></tr>
</table>

**Table A1.** *Cont.*

| Key Properties | Potential Uses | Current Hurdles |
| --- | --- | --- |
| Absorption properties | Optoelectronic memory devices [70] [2] | Export restrictions |
| Photoluminescence (fluorescence) | | |
| Raman spectroscopy properties | | |
| **Others** | **Batteries [71]** | Environmental concerns [76] |
| Easily modifiable structure | | |
| Presence of functional groups | Supercapacitors production<br>Ultra-thin flexible batteries<br>Implantable medical devices | |
| Chemical inertness | | |
| Easily optimizable solubility and dispersion | **Cleaning up polluted environments** | |
| | Water filtration<br>Air filters, such as smokestacks [72] | |
| | **Others** | |
| | Transistors production [73,74] | |
| | Energy production | |
| | Solar cells [75] | |
| Light weight | Energy Storage | |
| ⇧ Biocompatibility | CNT based fibers and fabrics | |
| Capability of molecules immobilization | CNT based ceramics | |
| Transport of protein, DNA, RNA | | |
| Large surface area | Sources of light | |
| High capability of absorbing chemicals from their surroundings | | Entrenched dominance of other material |

[1] Conductive capacity of CNTs depends on the chirality degree, i.e., the twist degree and dimension of the diameter of the actual nanotube; [2] SWCNTs; ⇧ = high, wide; ⇩ = scarcely.

**Table A2.** The three most used methods to synthetize CNTs.

| Method | Process Type | Products Purity | Conditions Yields (%) | Advantages | Disadvantages |
| --- | --- | --- | --- | --- | --- |
| Arc discharge (AC) | The carbon in the negative electrode sublimates due to the high discharge temperatures (T) | SWCNTs short tubes<br>Ø = 0.6–1.4 nm<br>MWCNTs short tubes<br>Inner Ø = 1–3 nm<br>Outer Ø = 10 nm<br>Medium purity | Argon/$N_2$<br>500 torr<br>T $\leq$ 4000 °C<br>20–100% | Few structural defects | Length $\leq$ 50 µm |
| Laser ablation (LA) | Graphite samples are vaporized in a reactor at ⇧ T by a pulsed laser in presence of an inert gas<br>CNTs grow up on the cooler reactor's surfaces as the vaporized carbon condenses | SWCNTs 5–20 µm<br>Ø = 1–2 nm<br>MWCNTs<br>Low purity | 500–1000 °C at ⇧ energy laser beam<br>25–1000 °C<br>$\leq$70% | Controllable Ø by the reaction temperature | ⇧ Expensive than AC, CVD |

**Table A2.** *Cont.*

| Method | Process Type | Products Purity | Conditions Yields (%) | Advantages | Disadvantages |
|---|---|---|---|---|---|
| Chemical Vapor Deposition (CVD) | Layered metal catalyst particles are heated in a reactor where a process gas and a carbon-containing gas are bled into | SWCNTs long tubes Ø = 0.6–4 nm MWCNTs long tubes Ø = 10–240 nm Medium-High purity | Low pressure inner gas (argon) 500–1200 °C 60–90% | ⇧ Economic and simple than AD and LA Synthesis at ⇩ temperature and AP ⇧ Yield and purity than AD and LA Versatile in the control of CNTs structure/architecture Suitable for scale up | ⇩ Crystallinity than that by AD and LA Removal of the catalyst support |

⇧ Denotes high, higher, increase, major; ⇩ denotes low, lower, decrease, minor; AP = ambient pressure.

**Table A3.** Physical limits of CNTs produced so far.

| Physical Property | Largest | Smallest | Refs. |
|---|---|---|---|
| Diameter | N.R. | 0.30 nm (MWCNT) [1]<br>0.43 nm (SWCNT) | [148] |
| Length | 0.5 m [2] | Cycloparaphenylenes (2.49 Å) [3] | [149,150] |
| Density | 1.6 g/cm * | N.R. | [151,152] |

[1] Armchair structure; [2] [149]; [3] [150]; * ohmic conductivity, lowest resistance of 22 kΩ.

**Table A4.** Possible associations between the different known allotropes of carbon and particular morphologies of CNTs.

| Associations Morphologies | Type of Interaction Arrangement | Possible Products | Applications/Properties | Refs. |
|---|---|---|---|---|
| CNTs/CNTs * | Junctions | Metallic CNTs/semiconducting CNTs | Component of NT-based electronic circuits | [153,154,157] |
| CNTs/GF * | Junctions | Pillared GF (3D-CNTs) | Energy storage<br>Supercapacitors<br>Field emission<br>Transistors<br>High-performance catalysis<br>Photovoltaics<br>Biomedical devices<br>Implants<br>Sensors | [158–161] |
| 2 CNTs/fullerenes * | Covalent Bond | Fullerene-like nanobuds | Field emitters | [162] |
| CNT/fullerenes * | Entrapment | Carbon peapods (CPPs) | Heating devices<br>Irradiating devices<br>Oscillators | [163–166] |
| Doughnut shape ** | CNTs twisted into a toroid (Annulus shape) | Nano tori (NTRs) | ⇧ magnetic moment<br>⇧ stability | [167,168] |
| GF/MWCNTs * | GFs integrated MWCNTs | GFCNTs | ⇧ Surface area<br>3D-framework<br>⇧ Total charge capacity per unit of nominal area | [169,170] |

**Table A4.** *Cont.*

| Associations Morphologies | Type of Interaction Arrangement | Possible Products | Applications/Properties | Refs. |
|---|---|---|---|---|
| 1D carbon structures ** | Stacking microstructure of GF layers | Cup-stacked CNTs (CSCNTs) | Semiconductors | [171] |

GF = graphene; * Associations; ** particular morphologies; ⇧ denotes high, higher, increase, major.

**Table A5.** Main biomedical application of CNTs.

| Applications | | Principle | Detection Target | Year Refs. |
|---|---|---|---|---|
| Sensors | Electrochemical sensors | Made of AuNPs + MWCNTs Mannan-Os Adducts | Dopamine | 2015 [9] |
| | | Made of glassy carbon electrodes modified with MWCNTs and CuMPs dispersed in PEIs | Amino acids and glucose | 2016 [10] |
| | | Based on MWCNTs | Clostridium tetani | 2016 [11] |
| | | Based on AgNPs/Bi NPs/MWCNTs nafion modified | Lead and Cadmium | 2020 [12] |
| | | Based on carbon sensor fabricated with coalesced Ru–$TiO_2$ NPs and MWCNTs | Cetrizine | 2019 [13] |
| | | Based on glassy carbon sensor modified with MWCNT in pH 9.0 PBS | Methdilazine | 2019 [14] |
| | | Based on (Ru–$TiO_2$) NPs and MWCNTs | Flufenamic acid (FFA) Mefenamic acid (MFA) | 2019 [15] |
| | | Based on Bi-MWCNT/MCPE at physiological pH | Gallic Acid (GA) | 2020 [16] |
| | | | **Function** | |
| | Piezoelectric sensors | Based on MWCNTs on PDMS as substrate | N.R. | 2017 [17] |
| | | Based on graphene, CNTs and graphene-CNTs composite | N.R. | 2018 [18] |
| | | Based on MWCNT on PDMS as substrate | For developing robotic hands (rehabilitation) Strain-detecting needle for tissue characterization | 2019 [19] |
| | | Based on MWCNTs on thermoplastic urethane as substrate | Sensors integrated into gloves and bandages for assessing specific human functions | 2019 [20] |
| | | Based on MWCNTs on PDMS substrate | To measure pressure directly without the use of deformation materials. | 2019 [21] |
| | | | **Detection Target** | |
| | Gas sensors | Based on a resonant electromagnetic transducer in microstrip technology | Volatile Organic Compounds (VOCs) | 2019 [22] |
| | | Based on dye functionalized matrix anchored onto MWCNTs ammonia | Sulfur dioxide and chlorine | 2018 [23] |
| | | Based on $W_xO_y$ nano-bricks and MWCNTs | Ammonia gas | 2019 [24] |

**Table A5.** *Cont.*

| Applications | Principle | Detection Target | Year Refs. |
|---|---|---|---|
| | | **Transported Drugs** | |
| Drug Targeting | Based on MWCNT and pH-responsive gel of chitosan-coated magnetic nanocomposites | Doxorubicin (DOX) to U-87 Glioblastoma cells | 2019 [25] |
| | Based on stimuli-responsive CNTs using Ag nanowires to stimulate the drug release from the core of NTs | Cisplatin | 2017 [26] |
| | Based on electro-responsive polymer-MWCNT hybrid hydrogel | Sucrose | 2013 [27] |
| | Based on MWCNTs biodegradable, biocompatible nanocomposite hydrogel | Diclofenac sodium | 2016 [28] |
| Cancer diagnosis and treatment | Based on MTX-loaded MWCNTs releasing drugs by enzymatic cleavage | Methotrexate to in vitro breast cells | 2010 [29] |
| | Based on DOX loaded dendrimer modified MWCNTs releasing drugs at low pH | DOX | 2013 [30] |
| | Based on cationic MWCNTs-NH3þ for direct intertumoral injections | Apoptotic siRNA against polo-like kinase (siPLK1) in calu6 tumor xenografts | 2015 [31] |
| | | **Target bacteria/Applications** | |
| Antibacterial agents | Based on chitosan/MWCNTs nanocomposites | *Enterococcus faecalis Staphylococcus epidermidis Escherichia coli* | [32–38] |
| | Based on $Fe_3O_4$/SWCNTs | *E. coli* | 2018 [39] |
| | Based on Ag–$Fe_3O_4$/SWCNTs | N.R. | 2018 [40] |
| | Based on CDX/Ag/MWCNTs | N.R. | 2014 [41] |
| | Based on MWCNTs containing carboxylic functions | N.R. | 2015 [42] |
| | Based on PA/graphene/CNTs | *Staphylococcus aureus, E. coli* | 2018, 2012 [43,44] |
| | Based on MWCNTs | Treatment of drinking water through removal and inactivation of viruses and bacteria | 2011 [45] |
| | Based on MWCNTs functionalized with mono-, di-, and tri-ethanolamine | *E. coli, Klebsiella pneumonia Pseudomonas aeruginosa Salmonella typhimurium, Bacillus subtilis S. aureus, Bacillus cereus Streptococcus pneumonia* | 2014 [46] |
| | Based on dispersion SWCNTs/TABM derivative with carboxy groups | *E. coli, S. aureus* | 2019 [47] |
| | | **Target fungi** | |
| Antifungal agents | Based on chitosan/MWCNTs | *Aspergillus niger, Candida trobicalis C. neoformans* | 2000 [48] |

**Table A5.** *Cont.*

| Applications | Principle | Detection Target | Year Refs. |
|---|---|---|---|
| Antifungal agents | Based on functionalized CNTs | *A. niger, C. albicans, A. fumigatus, Penicillium chrysogenum Saccharomyces cerevisiae, Fusarium culmorum Microsporum canis, Trichophyton mentagrophytes T. rubrum, P. lilacinum* | 2013 [49] |
| | Based on dispersion SWCNTs/TABM derivative with carboxy groups | *C. albicans* | 2019 [47] |

AuNPs = silver nanoparticles (AuNPs); CuMPs = copper microparticles; PEIs = polyethylenimine; Bi = bismuth; Ru–$TiO_2$ = ruthenium-doped titanium dioxide; PBS = phosphate buffer solution; Bi-MWCNT/MCPE = BiNPs decorated MWCNTs modified carbon paste electrode; PDMS = polydimethylsiloxane; $W_xO_y$ = tungsten oxide; MTX = methotrexate; DOX = doxorubicin; $Fe_3O_4$ = iron oxide; CDX = cyclodextrin; PA = polyaniline; TABM = tetra-arylbimesityl.

## References


1. Ifijen, I.H.; Maliki, M.; Omorogbe, S.O.; Ibrahim, S.D. Incorporation of Metallic Nanoparticles Into Alkyd Resin: A Review of Their Coating Performance. In *TMS 2022 151st Annual Meeting & Exhibition Supplemental Proceedings*; Springer: Cham, Switzerland, 2022; pp. 338–349.
2. Omorogbe, S.O.; Ikhuoria, E.U.; Ifijen, H.I.; Simo, A.; Aigbodion, A.; Maaza, M. Fabrication of Monodispersed Needle-Sized Hollow Core Polystyrene Microspheres. In *TMS 2019 148th Annual Meeting & Exhibition Supplemental Proceedings*; Springer: Cham, Switzerland, 2019; pp. 155–164.
3. Ifijen, H.I.; Ikhuoria, E.U.; Omorogbe, S.O. Correlative Studies on the Fabrication of Poly(Styrene-Methyl-Methacrylate-Acrylic Acid) Colloidal Crystal Films. *J. Dispers. Sci. Technol.* **2019**, *40*, 1023–1030. [CrossRef]
4. Ifijen, I.H.; Ikhuoria, E.U.; Omorogbe, S.O.; Aigbodion, A.I. Ordered Colloidal Crystals Fabrication and Studies on the Properties of Poly (Styrene–Butyl Acrylate–Acrylic Acid) and Polystyrene Latexes. In *Nanocomposites VI: Nanoscience and Nanotechnology in Advanced Composites*; Springer: Cham, Switzerland, 2019; pp. 125–135.
5. Ifijen, I.H.; Maliki, M.; Ovonramwen, O.B.; Aigbodion, A.I.; Ikhuoria, E.U. Brilliant Coloured Monochromatic Photonic Crystals Films Generation from Poly (Styrene-Butyl Acrylate-Acrylic Acid) Latex. *J. Appl. Sci. Environ. Manag.* **2019**, *23*, 1661. [CrossRef]
6. Omorogbe, S.O.; Aigbodion, A.I.; Ifijen, H.I.; Simo, A.; Ogbeide-Ihama, N.L.; Ikhuoria, E.U. Low-Temperature Synthesis of Superparamagnetic $Fe_3O_4$ Morphologies Tuned Using Oleic Acid as Crystal Growth Modifiers. In *TMS 2020 149th Annual Meeting & Exhibition Supplemental Proceedings*; Springer: Cham, Switzerland, 2020; pp. 619–631.
7. Ifijen, I.H.; Ikhuoria, E.U.; Maliki, M.; Otabor, G.O.; Aigbodion, A.I. Nanostructured Materials: A Review on Its Application in Water Treatment. In *TMS 2022 151st Annual Meeting & Exhibition Supplemental Proceedings*; Springer: Cham, Switzerland, 2022; pp. 1172–1180.
8. Ifijen, I.H.; Aghedo, O.N.; Odiachi, I.J.; Omorogbe, S.O.; Olu, E.L.; Onuguh, I.C. Nanostructured Graphene Thin Films: A Brief Review of Their Fabrication Techniques and Corrosion Protective Performance. In *TMS 2022 151st Annual Meeting & Exhibition Supplemental Proceedings*; Springer: Cham, Switzerland, 2022; pp. 366–377.
9. Ciambelli, P.; La Guardia, G.; Vitale, L. Nanotechnology for Green Materials and Processes. In *Studies in Surface Science and Catalysis*; Elsevier Science: Amsterdam, The Netherlands, 2020; pp. 97–116.
10. Omorogbe, S.O.; Aigbodion, A.I.; Ikhuoria, E.U.; Ifijen, I.H. Impact of Varying the Concentration of Tetraethyl-OrthoSilicate on the Average Particle Diameter of Monodisperse Colloidal Silica Spheres. *Chem. Sci. J.* **2018**, *9*, 1. [CrossRef]
11. Ifijen, I.H.; Maliki, M.; Odiachi, I.J.; Aghedo, O.N.; Ohiocheoya, E.B. Review on Solvents Based Alkyd Resins and Water Borne Alkyd Resins: Impacts of Modification on Their Coating Properties. *Chem. Afr.* **2022**, *5*, 211–225. [CrossRef]
12. Ifijen, I.H.; Omonmhenle, S.I. Antimicrobial Properties of Carbon Nanotube: A Succinct Assessment. *Biomed. Mater. Devices* **2024**, *2*, 113–120. [CrossRef]
13. Rathinavel, S.; Priyadharshini, K.; Panda, D. A review on carbon nanotube: An overview of synthesis, properties, functionalization, characterization, and the application. *Mater. Sci. Eng. B* **2021**, *268*, 115095. [CrossRef]
14. Su, Q.; Gan, L.; Liu, J.; Yang, X. Carbon Dots Derived from Pea for Specifically Binding with Cryptococcus Neoformans. *Anal. Biochem.* **2020**, *589*, 113476. [CrossRef]
15. Salari, S.; Jafari, S.M. Application of Nanofluids for Thermal Processing of Food Products. *Trends Food Sci. Technol.* **2020**, *97*, 100–113. [CrossRef]

16. Abd-Elsalam, K.A. Carbon Nanomaterials: 30 Years of Research in Agroecosystems. In *Carbon Nanomaterials for Agri-Food and Environmental Applications*; Elsevier: Amsterdam, The Netherlands, 2020; pp. 1–18.
17. Iijima, S. Helical Microtubules of Graphitic Carbon. *Nature* **1991**, *354*, 56–58. [CrossRef]
18. Rahman, M.A.; Abul Barkat, H.; Harwansh, R.K.; Deshmukh, R. Carbon-based Nanomaterials: Carbon Nanotubes, Graphene, and Fullerenes for the Control of Burn Infections and Wound Healing. *Curr. Pharm. Biotechnol.* **2022**, *23*, 1483–1496. [CrossRef] [PubMed]
19. Chong, Y.; Ge, C.; Fang, G.; Wu, R.; Zhang, H.; Chai, Z.; Chen, C.; Yin, J.-J. Light-Enhanced Antibacterial Activity of Graphene Oxide, Mainly via Accelerated Electron Transfer. *Environ. Sci. Technol.* **2017**, *51*, 10154–10161. [CrossRef] [PubMed]
20. Kang, S.; Mauter, M.S.; Elimelech, M. Physicochemical Determinants of Multiwalled Carbon Nanotube Bacterial Cytotoxicity. *Environ. Sci. Technol.* **2008**, *42*, 7528–7534. [CrossRef]
21. Haung, C.-F.; Chan, Y.-H.; Chen, L.-K.; Liu, C.-M.; Huang, W.-C.; Ou, S.-F.; Ou, K.-L.; Wang, D.-Y. Preparation, Characterization, and Properties of Anticoagulation and Antibacterial Films of Carbon-Based Nanowires Fabricated on Surfaces of Ti Implants. *J. Electrochem. Soc.* **2013**, *160*, H392–H397. [CrossRef]
22. Lee, E.-S.; Kim, Y.-O.; Ha, Y.-M.; Lim, D.; Hwang, J.Y.; Kim, J.; Park, M.; Cho, J.W.; Jung, Y.C. Antimicrobial Properties of Lignin-Decorated Thin Multi-Walled Carbon Nanotubes in Poly(Vinyl Alcohol) Nanocomposites. *Eur. Polym. J.* **2018**, *105*, 79–84. [CrossRef]
23. Azizi-Lalabadi, M.; Hashemi, H.; Feng, J.; Jafari, S.M. Carbon Nanomaterials against Pathogens; the Antimicrobial Activity of Carbon Nanotubes, Graphene/Graphene Oxide, Fullerenes, and Their Nanocomposites. *Adv. Colloid. Interface Sci.* **2020**, *284*, 102250. [CrossRef]
24. Oyelami, A.O.; Semple, K.T. Impact of Carbon Nanomaterials on Microbial Activity in Soil. *Soil. Biol. Biochem.* **2015**, *86*, 172–180. [CrossRef]
25. Maleki Dizaj, S.; Mennati, A.; Jafari, S.; Khezri, K.; Adibkia, K. Antimicrobial Activity of Carbon-Based Nanoparticles. *Adv. Pharm. Bull.* **2015**, *5*, 19–23. [CrossRef]
26. Kang, S.; Pinault, M.; Pfefferle, L.D.; Elimelech, M. Single-Walled Carbon Nanotubes Exhibit Strong Antimicrobial Activity. *Langmuir* **2007**, *23*, 8670–8673. [CrossRef]
27. Kang, S.; Herzberg, M.; Rodrigues, D.F.; Elimelech, M. Antibacterial Effects of Carbon Nanotubes: Size Does Matter! *Langmuir* **2008**, *24*, 6409–6413. [CrossRef]
28. Chen, H.; Wang, B.; Gao, D.; Guan, M.; Zheng, L.; Ouyang, H.; Chai, Z.; Zhao, Y.; Feng, W. Broad-Spectrum Antibacterial Activity of Carbon Nanotubes to Human Gut Bacteria. *Small* **2013**, *9*, 2735–2746. [CrossRef]
29. Awasthi, S.; Srivastava, A.; Kumar, D.; Pandey, S.K.; Mubarak, N.M.; Dehghani, M.H.; Ansari, K. An Insight into the Toxicological Impacts of Carbon Nanotubes (CNTs) on Human Health: A Review. *Environ. Adv.* **2024**, *18*, 100601. [CrossRef]
30. Kostarelos, K.; Lacerda, L.; Pastorin, G.; Wu, W.; Wieckowski, S.; Luangsivilay, J.; Godefroy, S.; Pantarotto, D.; Briand, J.P.; Muller, S.; et al. Cellular Uptake of Functionalized Carbon Nanotubes Is Independent of Functional Group and Cell Type. *Nat. Nanotechnol.* **2007**, *2*, 108–113. [CrossRef] [PubMed]
31. Noonan, D.M.; Pagani, A.; Pulze, L.; Bruno, A.; Principi, E.; Congiu, T.; Grimaldi, A.; Bassani, B.; De Flora, S.; Gini, E.; et al. Environmental Impact of Multi-Wall Carbon Nanotubes in a Novel Model of Exposure: Systemic Distribution, Macrophage Accumulation, and Amyloid Deposition. *Int. J. Nanomed.* **2015**, *10*, 6133–6145. [CrossRef] [PubMed]
32. Poland, C.A.; Duffin, R.; Kinloch, I.; Maynard, A.; Wallace, W.A.H.; Seaton, A.; Stone, V.; Brown, S.; MacNee, W.; Donaldson, K. Carbon Nanotubes Introduced into the Abdominal Cavity of Mice Show Asbestos-like Pathogenicity in a Pilot Study. *Nat. Nanotechnol.* **2008**, *3*, 423–428. [CrossRef]
33. Ge, C.; Meng, L.; Xu, L.; Bai, R.; Du, J.; Zhang, L.; Li, Y.; Chang, Y.; Zhao, Y.; Chen, C. Acute Pulmonary and Moderate Cardiovascular Responses of Spontaneously Hypertensive Rats after Exposure to Single-Wall Carbon Nanotubes. *Nanotoxicology* **2012**, *6*, 526–542. [CrossRef]
34. Goerzen, D.; Heller, D.A.; Rachel, A. *Meidl Balancing Safety and Innovation: Shaping Responsible Carbon Nanotube Policy*; Rice Universities', Baker Institute for Public Policy: Houston, TX, USA, 2024.
35. Alfei, S.; Schito, G.C. Nanotubes: Carbon-Based Fibers and Bacterial Nano-Conduits Both Arousing a Global Interest and Conflicting Opinions. *Fibers* **2022**, *10*, 75. [CrossRef]
36. Ebbesen, T.W.; Ajayan, P.M. Large-Scale Synthesis of Carbon Nanotubes. *Nature* **1992**, *358*, 220–222. [CrossRef]
37. Eatemadi, A.; Daraee, H.; Karimkhanloo, H.; Kouhi, M.; Zarghami, N.; Akbarzadeh, A.; Abasi, M.; Hanifehpour, Y.; Joo, S.W. Carbon Nanotubes: Properties, Synthesis, Purification, and Medical Applications. *Nanoscale Res. Lett.* **2014**, *9*, 393. [CrossRef]
38. Guo, T.; Nikolaev, P.; Rinzler, A.G.; Tomanek, D.; Colbert, D.T.; Smalley, R.E. Self-Assembly of Tubular Fullerenes. *J. Phys. Chem.* **1995**, *99*, 10694–10697. [CrossRef]
39. Guo, T.; Nikolaev, P.; Thess, A.; Colbert, D.T.; Smalley, R.E. Catalytic Growth of Single-Walled Nanotubes by Laser Vaporization. *Chem. Phys. Lett.* **1995**, *243*, 49–54. [CrossRef]

40. Kumar, M.; Ando, Y. Chemical Vapor Deposition of Carbon Nanotubes: A Review on Growth Mechanism and Mass Production. *J. Nanosci. Nanotechnol.* **2010**, *10*, 3739–3758. [CrossRef] [PubMed]
41. Neupane, S.; Lastres, M.; Chiarella, M.; Li, W.; Su, Q.; Du, G. Synthesis and Field Emission Properties of Vertically Aligned Carbon Nanotube Arrays on Copper. *Carbon* **2012**, *50*, 2641–2650. [CrossRef]
42. Inami, N.; Mohamed, M.A.; Shikoh, E.; Fujiwara, A. Synthesis-Condition Dependence of Carbon Nanotube Growth by Alcohol Catalytic Chemical Vapor Deposition Method. *Sci. Technol. Adv. Mater.* **2007**, *8*, 292–295. [CrossRef]
43. Ishigami, N.; Ago, H.; Imamoto, K.; Tsuji, M.; Iakoubovskii, K.; Minami, N. Crystal Plane Dependent Growth of Aligned Single-Walled Carbon Nanotubes on Sapphire. *J. Am. Chem. Soc.* **2008**, *130*, 9918–9924. [CrossRef]
44. Naha, S.; Puri, I.K. A Model for Catalytic Growth of Carbon Nanotubes. *J. Phys. D Appl. Phys.* **2008**, *41*, 065304. [CrossRef]
45. Banerjee, S.; Naha, S.; Puri, I.K. Molecular Simulation of the Carbon Nanotube Growth Mode during Catalytic Synthesis. *Appl. Phys. Lett.* **2008**, *92*, 233121. [CrossRef]
46. Eftekhari, A.; Jafarkhani, P.; Moztarzadeh, F. High-Yield Synthesis of Carbon Nanotubes Using a Water-Soluble Catalyst Support in Catalytic Chemical Vapor Deposition. *Carbon* **2006**, *44*, 1343–1345. [CrossRef]
47. Ren, Z.F.; Huang, Z.P.; Xu, J.W.; Wang, J.H.; Bush, P.; Siegal, M.P.; Provencio, P.N. Synthesis of Large Arrays of Well-Aligned Carbon Nanotubes on Glass. *Science* **1998**, *282*, 1105–1107. [CrossRef]
48. Hata, K.; Futaba, D.N.; Mizuno, K.; Namai, T.; Yumura, M.; Iijima, S. Water-Assisted Highly Efficient Synthesis of Impurity-Free Single-Walled Carbon Nanotubes. *Science* **2004**, *306*, 1362–1364. [CrossRef]
49. Futaba, D.N.; Hata, K.; Yamada, T.; Hiraoka, T.; Hayamizu, Y.; Kakudate, Y.; Tanaike, O.; Hatori, H.; Yumura, M.; Iijima, S. Shape-Engineerable and Highly Densely Packed Single-Walled Carbon Nanotubes and Their Application as Super-Capacitor Electrodes. *Nat. Mater.* **2006**, *5*, 987–994. [CrossRef]
50. Smiljanic, O.; Stansfield, B.L.; Dodelet, J.-P.; Serventi, A.; Désilets, S. Gas-Phase Synthesis of SWNT by an Atmospheric Pressure Plasma Jet. *Chem. Phys. Lett.* **2002**, *356*, 189–193. [CrossRef]
51. Kim, K.S.; Cota-Sanchez, G.; Kingston, C.T.; Imris, M.; Simard, B.; Soucy, G. Large-Scale Production of Single-Walled Carbon Nanotubes by Induction Thermal Plasma. *J. Phys. D Appl. Phys.* **2007**, *40*, 2375–2387. [CrossRef]
52. Ren, J.; Li, F.-F.; Lau, J.; González-Urbina, L.; Licht, S. One-Pot Synthesis of Carbon Nanofibers from $CO_2$. *Nano Lett.* **2015**, *15*, 6142–6148. [CrossRef] [PubMed]
53. Service, R.F. Conjuring Chemical Cornucopias out of Thin Air. *Science* **2015**, *349*, 1160. [CrossRef]
54. Yuan, L.; Saito, K.; Pan, C.; Williams, F.A.; Gordon, A.S. Nanotubes from Methane Flames. *Chem. Phys. Lett.* **2001**, *340*, 237–241. [CrossRef]
55. Yuan, L.; Saito, K.; Hu, W.; Chen, Z. Ethylene Flame Synthesis of Well-Aligned Multi-Walled Carbon Nanotubes. *Chem. Phys. Lett.* **2001**, *346*, 23–28. [CrossRef]
56. Duan, H.M.; McKinnon, J.T. Nanoclusters Produced in Flames. *J. Phys. Chem.* **1994**, *98*, 12815–12818. [CrossRef]
57. Murr, L.E.; Bang, J.J.; Esquivel, E.V.; Guerrero, P.A.; Lopez, D.A. Carbon Nanotubes, Nanocrystal Forms, and Complex Nanoparticle Aggregates in Common Fuel-Gas Combustion Sources and the Ambient Air. *J. Nanopart. Res.* **2004**, *6*, 241–251. [CrossRef]
58. Vander Wal, R.L. Fe-Catalyzed Single-Walled Carbon Nanotube Synthesis within a Flame Environment. *Combust. Flame* **2002**, *130*, 37–47. [CrossRef]
59. Saveliev, A.V.; Merchan-Merchan, W.; Kennedy, L.A. Metal Catalyzed Synthesis of Carbon Nanostructures in an Opposed Flow Methane Oxygen Flame. *Combust. Flame* **2003**, *135*, 27–33. [CrossRef]
60. Height, M.J.; Howard, J.B.; Tester, J.W.; Vander Sande, J.B. Flame Synthesis of Single-Walled Carbon Nanotubes. *Carbon* **2004**, *42*, 2295–2307. [CrossRef]
61. Sen, S.; Puri, I.K. Flame Synthesis of Carbon Nanofibres and Nanofibre Composites Containing Encapsulated Metal Particles. *Nanotechnology* **2004**, *15*, 264–268. [CrossRef]
62. Naha, S.; Sen, S.; De, A.K.; Puri, I.K. A Detailed Model for the Flame Synthesis of Carbon Nanotubes and Nanofibers. *Proc. Combust. Inst.* **2007**, *31*, 1821–1829. [CrossRef]
63. Jha, R.; Singh, A.; Sharma, P.K.; Fuloria, N.K. Smart carbon nanotubes for drug delivery system: A comprehensive study. *J. Drug Deliv. Sci. Technol.* **2020**, *58*, 101811. [CrossRef]
64. Monthioux, M.; Kuznetsov, V.L. Who Should Be given the Credit for the Discovery of Carbon Nanotubes? *Carbon* **2006**, *44*, 1621–1623. [CrossRef]
65. Chernozatonskii, L.A. Carbon Nanotube Connectors and Planar Jungle Gyms. *Phys. Lett. A* **1992**, *172*, 173–176. [CrossRef]
66. Menon, M.; Srivastava, D. Carbon Nanotube "T Junctions": Nanoscale Metal-Semiconductor-Metal Contact Devices. *Phys. Rev. Lett.* **1997**, *79*, 4453–4456. [CrossRef]
67. Ma, K.L.; Yan, X.H.; Guo, Y.D.; Xiao, Y. Electronic Transport Properties of Junctions between Carbon Nanotubes and Graphene Nanoribbons. *Eur. Phys. J. B* **2011**, *83*, 487–492. [CrossRef]
68. Harris, P.J.F.; Suarez-Martinez, I.; Marks, N.A. The Structure of Junctions between Carbon Nanotubes and Graphene Shells. *Nanoscale* **2016**, *8*, 18849–18854. [CrossRef]

69. Lambin, P.; Vigneron, J.P.; Fonseca, A.; Nagy, J.B.; Lucas, A.A. Atomic Structure and Electronic Properties of a Bent Carbon Nanotube. *Synth. Met.* **1996**, *77*, 249–252. [CrossRef]
70. Dimitrakakis, G.K.; Tylianakis, E.; Froudakis, G.E. Pillared Graphene: A New 3-D Network Nanostructure for Enhanced Hydrogen Storage. *Nano Lett.* **2008**, *8*, 3166–3170. [CrossRef] [PubMed]
71. Lalwani, G.; Kwaczala, A.T.; Kanakia, S.; Patel, S.C.; Judex, S.; Sitharaman, B. Fabrication and Characterization of Three-Dimensional Macroscopic All-Carbon Scaffolds. *Carbon* **2013**, *53*, 90–100. [CrossRef] [PubMed]
72. Lalwani, G.; Gopalan, A.; D'Agati, M.; Srinivas Sankaran, J.; Judex, S.; Qin, Y.-X.; Sitharaman, B. Porous Three-Dimensional Carbon Nanotube Scaffolds for Tissue Engineering. *J. Biomed. Mater. Res. A* **2015**, *103*, 3212–3225. [CrossRef]
73. Noyce, S.G.; Vanfleet, R.R.; Craighead, H.G.; Davis, R.C. High Surface-Area Carbon Microcantilevers. *Nanoscale Adv.* **2019**, *1*, 1148–1154. [CrossRef]
74. Yu, M.-F.; Lourie, O.; Dyer, M.J.; Moloni, K.; Kelly, T.F.; Ruoff, R.S. Strength and Breaking Mechanism of Multiwalled Carbon Nanotubes Under Tensile Load. *Science* **2000**, *287*, 637–640. [CrossRef]
75. Peng, B.; Locascio, M.; Zapol, P.; Li, S.; Mielke, S.L.; Schatz, G.C.; Espinosa, H.D. Measurements of Near-Ultimate Strength for Multiwalled Carbon Nanotubes and Irradiation-Induced Crosslinking Improvements. *Nat. Nanotechnol.* **2008**, *3*, 626–631. [CrossRef]
76. Jensen, K.; Mickelson, W.; Kis, A.; Zettl, A. Buckling and Kinking Force Measurements on Individual Multiwalled Carbon Nanotubes. *Phys. Rev. B* **2007**, *76*, 195436. [CrossRef]
77. Mingo, N.; Stewart, D.A.; Broido, D.A.; Srivastava, D. Phonon Transmission through Defects in Carbon Nanotubes from First Principles. *Phys. Rev. B* **2008**, *77*, 033418. [CrossRef]
78. Kumanek, B.; Janas, D. Thermal Conductivity of Carbon Nanotube Networks: A Review. *J. Mater. Sci.* **2019**, *54*, 7397–7427. [CrossRef]
79. Koziol, K.K.; Janas, D.; Brown, E.; Hao, L. Thermal Properties of Continuously Spun Carbon Nanotube Fibres. *Phys. E Low. Dimens. Syst. Nanostruct.* **2017**, *88*, 104–108. [CrossRef]
80. Hong, S.; Myung, S. A Flexible Approach to Mobility. *Nat. Nanotechnol.* **2007**, *2*, 207–208. [CrossRef] [PubMed]
81. Wang, X.; Li, Q.; Xie, J.; Jin, Z.; Wang, J.; Li, Y.; Jiang, K.; Fan, S. Fabrication of Ultralong and Electrically Uniform Single-Walled Carbon Nanotubes on Clean Substrates. *Nano Lett.* **2009**, *9*, 3137–3141. [CrossRef] [PubMed]
82. Lu, X.; Chen, Z. Curved Pi-Conjugation, Aromaticity, and the Related Chemistry of Small Fullerenes (<$C_{60}$) and Single-Walled Carbon Nanotubes. *Chem. Rev.* **2005**, *105*, 3643–3696. [CrossRef] [PubMed]
83. Anzar, N.; Hasan, R.; Tyagi, M.; Yadav, N.; Narang, J. Carbon Nanotube—A Review on Synthesis, Properties and Plethora of Applications in the Field of Biomedical Science. *Sens. Int.* **2020**, *1*, 100003. [CrossRef]
84. Khan, M.U.; Reddy, K.R.; Snguanwongchai, T.; Haque, E.; Gomes, V.G. Polymer Brush Synthesis on Surface Modified Carbon Nanotubes via in Situ Emulsion Polymerization. *Colloid. Polym. Sci.* **2016**, *294*, 1599–1610. [CrossRef]
85. Che, J.; Çagin, T.; Goddard, W.A. Thermal Conductivity of Carbon Nanotubes. *Nanotechnology* **2000**, *11*, 65–69. [CrossRef]
86. Smart, S.K.; Cassady, A.I.; Lu, G.Q.; Martin, D.J. The Biocompatibility of Carbon Nanotubes. *Carbon* **2006**, *44*, 1034–1047. [CrossRef]
87. Negri, V.; Pacheco-Torres, J.; Calle, D.; López-Larrubia, P. Carbon Nanotubes in Biomedicine. *Top. Curr. Chem.* **2020**, *378*, 15. [CrossRef]
88. Lamberti, M.; Pedata, P.; Sannolo, N.; Porto, S.; De Rosa, A.; Caraglia, M. Carbon Nanotubes: Properties, Biomedical Applications, Advantages and Risks in Patients and Occupationally-Exposed Workers. *Int. J. Immunopathol. Pharmacol.* **2015**, *28*, 4–13. [CrossRef]
89. Kumar, S.; Rani, R.; Dilbaghi, N.; Tankeshwar, K.; Kim, K.-H. Carbon Nanotubes: A Novel Material for Multifaceted Applications in Human Healthcare. *Chem. Soc. Rev.* **2017**, *46*, 158–196. [CrossRef]
90. Teixeira-Santos, R.; Gomes, M.; Gomes, L.C.; Mergulhão, F.J. Antimicrobial and Anti-Adhesive Properties of Carbon Nanotube-Based Surfaces for Medical Applications: A Systematic Review. *iScience* **2021**, *24*, 102001. [CrossRef] [PubMed]
91. Nagai, H.; Toyokuni, S. Differences and Similarities between Carbon Nanotubes and Asbestos Fibers during Mesothelial Carcinogenesis: Shedding Light on Fiber Entry Mechanism. *Cancer Sci.* **2012**, *103*, 1378–1390. [CrossRef] [PubMed]
92. Aslan, S.; Loebick, C.Z.; Kang, S.; Elimelech, M.; Pfefferle, L.D.; Van Tassel, P.R. Antimicrobial Biomaterials Based on Carbon Nanotubes Dispersed in Poly(Lactic-Co-Glycolic Acid). *Nanoscale* **2010**, *2*, 1789. [CrossRef] [PubMed]
93. Olivi, M.; Zanni, E.; De Bellis, G.; Talora, C.; Sarto, M.S.; Palleschi, C.; Flahaut, E.; Monthioux, M.; Rapino, S.; Uccelletti, D.; et al. Inhibition of Microbial Growth by Carbon Nanotube Networks. *Nanoscale* **2013**, *5*, 9023. [CrossRef]
94. Liu, S.; Ng, A.K.; Xu, R.; Wei, J.; Tan, C.M.; Yang, Y.; Chen, Y. Antibacterial Action of Dispersed Single-Walled Carbon Nanotubes on *Escherichia Coli* and *Bacillus Subtilis* Investigated by Atomic Force Microscopy. *Nanoscale* **2010**, *2*, 2744. [CrossRef]
95. Arias, L.R.; Yang, L. Inactivation of Bacterial Pathogens by Carbon Nanotubes in Suspensions. *Langmuir* **2009**, *25*, 3003–3012. [CrossRef]
96. Young, Y.-F.; Lee, H.-J.; Shen, Y.-S.; Tseng, S.-H.; Lee, C.-Y.; Tai, N.-H.; Chang, H.-Y. Toxicity Mechanism of Carbon Nanotubes on *Escherichia coli*. *Mater. Chem. Phys.* **2012**, *134*, 279–286. [CrossRef]

97. Saleemi, M.A.; Fouladi, M.H.; Yong, P.V.C.; Wong, E.H. Elucidation of Antimicrobial Activity of Non-Covalently Dispersed Carbon Nanotubes. *Materials* **2020**, *13*, 1676. [CrossRef]
98. Vecitis, C.D.; Zodrow, K.R.; Kang, S.; Elimelech, M. Electronic-Structure-Dependent Bacterial Cytotoxicity of Single-Walled Carbon Nanotubes. *ACS Nano* **2010**, *4*, 5471–5479. [CrossRef]
99. Liu, S.; Wei, L.; Hao, L.; Fang, N.; Chang, M.W.; Xu, R.; Yang, Y.; Chen, Y. Sharper and Faster "Nano Darts" Kill More Bacteria: A Study of Antibacterial Activity of Individually Dispersed Pristine Single-Walled Carbon Nanotube. *ACS Nano* **2009**, *3*, 3891–3902. [CrossRef]
100. Rodrigues, D.F.; Elimelech, M. Toxic Effects of Single-Walled Carbon Nanotubes in the Development of *E. Coli* Biofilm. *Environ. Sci. Technol.* **2010**, *44*, 4583–4589. [CrossRef] [PubMed]
101. Malek, I.; Schaber, C.F.; Heinlein, T.; Schneider, J.J.; Gorb, S.N.; Schmitz, R.A. Vertically Aligned Multi Walled Carbon Nanotubes Prevent Biofilm Formation of Medically Relevant Bacteria. *J. Mater. Chem. B* **2016**, *4*, 5228–5235. [CrossRef] [PubMed]
102. Yick, S.; Mai-Prochnow, A.; Levchenko, I.; Fang, J.; Bull, M.K.; Bradbury, M.; Murphy, A.B.; Ostrikov, K.K. The Effects of Plasma Treatment on Bacterial Biofilm Formation on Vertically-Aligned Carbon Nanotube Arrays. *RSC Adv.* **2015**, *5*, 5142–5148. [CrossRef]
103. Vassallo, J.; Besinis, A.; Boden, R.; Handy, R.D. The Minimum Inhibitory Concentration (MIC) Assay with *Escherichia Coli*: An Early Tier in the Environmental Hazard Assessment of Nanomaterials? *Ecotoxicol. Environ. Saf.* **2018**, *162*, 633–646. [CrossRef]
104. Hartono, M.R.; Kushmaro, A.; Chen, X.; Marks, R.S. Probing the Toxicity Mechanism of Multiwalled Carbon Nanotubes on Bacteria. *Environ. Sci. Pollut. Res.* **2018**, *25*, 5003–5012. [CrossRef]
105. Zhang, Q.; Nghiem, J.; Silverberg, G.J.; Vecitis, C.D. Semiquantitative Performance and Mechanism Evaluation of Carbon Nanomaterials as Cathode Coatings for Microbial Fouling Reduction. *Appl. Environ. Microbiol.* **2015**, *81*, 4744–4755. [CrossRef]
106. Kim, J.; Shashkov, E.V.; Galanzha, E.I.; Kotagiri, N.; Zharov, V.P. Photothermal Antimicrobial Nanotherapy and Nanodiagnostics with Self-assembling Carbon Nanotube Clusters. *Lasers Surg. Med.* **2007**, *39*, 622–634. [CrossRef]
107. Mohammad, M.R.; Ahmed, D.S.; Mohammed, M.K.A. Synthesis of Ag-Doped $TiO_2$ Nanoparticles Coated with Carbon Nanotubes by the Sol–Gel Method and Their Antibacterial Activities. *J. Sol-Gel Sci. Technol.* **2019**, *90*, 498–509. [CrossRef]
108. Mocan, T.; Matea, C.T.; Pop, T.; Mosteanu, O.; Buzoianu, A.D.; Suciu, S.; Puia, C.; Zdrehus, C.; Iancu, C.; Mocan, L. Carbon Nanotubes as Anti-Bacterial Agents. *Cell. Mol. Life Sci.* **2017**, *74*, 3467–3479. [CrossRef]
109. Huang, Y.-P.; Lin, I.-J.; Chen, C.-C.; Hsu, Y.-C.; Chang, C.-C.; Lee, M.-J. Delivery of Small Interfering RNAs in Human Cervical Cancer Cells by Polyethylenimine-Functionalized Carbon Nanotubes. *Nanoscale Res. Lett.* **2013**, *8*, 267. [CrossRef]
110. Martincic, M.; Tobias, G. Filled Carbon Nanotubes in Biomedical Imaging and Drug Delivery. *Expert. Opin. Drug Deliv.* **2015**, *12*, 563–581. [CrossRef] [PubMed]
111. Zhang, B.; Wang, H.; Shen, S.; She, X.; Shi, W.; Chen, J.; Zhang, Q.; Hu, Y.; Pang, Z.; Jiang, X. Fibrin-Targeting Peptide CREKA-Conjugated Multi-Walled Carbon Nanotubes for Self-Amplified Photothermal Therapy of Tumor. *Biomaterials* **2016**, *79*, 46–55. [CrossRef] [PubMed]
112. Pasquini, L.M.; Hashmi, S.M.; Sommer, T.J.; Elimelech, M.; Zimmerman, J.B. Impact of Surface Functionalization on Bacterial Cytotoxicity of Single-Walled Carbon Nanotubes. *Environ. Sci. Technol.* **2012**, *46*, 6297–6305. [CrossRef] [PubMed]
113. Saleemi, M.A.; Kong, Y.L.; Yong, P.V.C.; Wong, E.H. An Overview of Antimicrobial Properties of Carbon Nanotubes-Based Nanocomposites. *Adv. Pharm. Bull.* **2022**, *12*, 449–465. [CrossRef]
114. Chaudhari, A.A.; Joshi, S.; Vig, K.; Sahu, R.; Dixit, S.; Baganizi, R.; Dennis, V.A.; Singh, S.R.; Pillai, S. A Three-Dimensional Human Skin Model to Evaluate the Inhibition of *Staphylococcus aureus* by Antimicrobial Peptide-Functionalized Silver Carbon Nanotubes. *J. Biomater. Appl.* **2019**, *33*, 924–934. [CrossRef]
115. Prabhu, S.; Poulose, E.K. Silver Nanoparticles: Mechanism of Antimicrobial Action, Synthesis, Medical Applications, and Toxicity Effects. *Int. Nano Lett.* **2012**, *2*, 32. [CrossRef]
116. Zhu, Z.; Wang, Z.; Li, S.; Yuan, X. Antimicrobial Strategies for Urinary Catheters. *J. Biomed. Mater. Res. A* **2019**, *107*, 445–467. [CrossRef]
117. Chaudhari, A.A.; Ashmore, D.; Nath, S.d.; Kate, K.; Dennis, V.; Singh, S.R.; Owen, D.R.; Palazzo, C.; Arnold, R.D.; Miller, M.E.; et al. A Novel Covalent Approach to Bio-Conjugate Silver Coated Single Walled Carbon Nanotubes with Antimicrobial Peptide. *J. Nanobiotechnol.* **2016**, *14*, 58. [CrossRef]
118. Kumar, A.; Dalal, J.; Dahiya, S.; Punia, R.; Sharma, K.D.; Ohlan, A.; Maan, A.S. In Situ Decoration of Silver Nanoparticles on Single-Walled Carbon Nanotubes by Microwave Irradiation for Enhanced and Durable Anti-Bacterial Finishing on Cotton Fabric. *Ceram. Int.* **2019**, *45*, 1011–1019. [CrossRef]
119. Karumuri, A.K.; Oswal, D.P.; Hostetler, H.A.; Mukhopadhyay, S.M. Silver Nanoparticles Supported on Carbon Nanotube Carpets: Influence of Surface Functionalization. *Nanotechnology* **2016**, *27*, 145603. [CrossRef]
120. Barbinta-Patrascu, M.E.; Ungureanu, C.; Iordache, S.M.; Iordache, A.M.; Bunghez, I.-R.; Ghiurea, M.; Badea, N.; Fierascu, R.-C.; Stamatin, I. Eco-Designed Biohybrids Based on Liposomes, Mint–Nanosilver and Carbon Nanotubes for Antioxidant and Antimicrobial Coating. *Mater. Sci. Eng. C* **2014**, *39*, 177–185. [CrossRef] [PubMed]

121. Chang, Y.-N.; Gong, J.-L.; Zeng, G.-M.; Ou, X.-M.; Song, B.; Guo, M.; Zhang, J.; Liu, H.-Y. Antimicrobial Behavior Comparison and Antimicrobial Mechanism of Silver Coated Carbon Nanocomposites. *Process Saf. Environ. Prot.* **2016**, *102*, 596–605. [CrossRef]
122. Park, S.B.; Steadman, C.S.; Chaudhari, A.A.; Pillai, S.R.; Singh, S.R.; Ryan, P.L.; Willard, S.T.; Feugang, J.M. Proteomic Analysis of Antimicrobial Effects of Pegylated Silver Coated Carbon Nanotubes in *Salmonella enterica* serovar Typhimurium. *J. Nanobiotechnol.* **2018**, *16*, 31. [CrossRef] [PubMed]
123. Singh, A.; Goswami, A.; Nain, S. Enhanced Antibacterial Activity and Photo-Remediation of Toxic Dyes Using Ag/SWCNT/PPy Based Nanocomposite with Core–Shell Structure. *Appl. Nanosci.* **2020**, *10*, 2255–2268. [CrossRef]
124. Zhu, Y.; Xu, J.; Wang, Y.; Chen, C.; Gu, H.; Chai, Y.; Wang, Y. Silver Nanoparticles-Decorated and Mesoporous Silica Coated Single-Walled Carbon Nanotubes with an Enhanced Antibacterial Activity for Killing Drug-Resistant Bacteria. *Nano Res.* **2020**, *13*, 389–400. [CrossRef]
125. Yun, H.; Kim, J.D.; Choi, H.C.; Lee, C.W. Antibacterial Activity of CNT-Ag and GO-Ag Nanocomposites Against Gram-Negative and Gram-Positive Bacteria. *Bull. Korean Chem. Soc.* **2013**, *34*, 3261–3264. [CrossRef]
126. Leid, J.G.; Ditto, A.J.; Knapp, A.; Shah, P.N.; Wright, B.D.; Blust, R.; Christensen, L.; Clemons, C.B.; Wilber, J.P.; Young, G.W.; et al. In Vitro Antimicrobial Studies of Silver Carbene Complexes: Activity of Free and Nanoparticle Carbene Formulations against Clinical Isolates of Pathogenic Bacteria. *J. Antimicrob. Chemother.* **2012**, *67*, 138–148. [CrossRef]
127. Nepal, D.; Balasubramanian, S.; Simonian, A.L.; Davis, V.A. Strong Antimicrobial Coatings: Single-Walled Carbon Nanotubes Armored with Biopolymers. *Nano Lett.* **2008**, *8*, 1896–1901. [CrossRef]
128. Upadhyayula, V.K.K.; Gadhamshetty, V. Appreciating the Role of Carbon Nanotube Composites in Preventing Biofouling and Promoting Biofilms on Material Surfaces in Environmental Engineering: A Review. *Biotechnol. Adv.* **2010**, *28*, 802–816. [CrossRef]
129. Levashov, P.A.; Sedov, S.A.; Shipovskov, S.; Belogurova, N.G.; Levashov, A.V. Quantitative Turbidimetric Assay of Enzymatic Gram-Negative Bacteria Lysis. *Anal. Chem.* **2010**, *82*, 2161–2163. [CrossRef]
130. Ahmed, F.; Santos, C.M.; Vergara, R.A.M.V.; Tria, M.C.R.; Advincula, R.; Rodrigues, D.F. Antimicrobial Applications of Electroactive PVK-SWNT Nanocomposites. *Environ. Sci. Technol.* **2012**, *46*, 1804–1810. [CrossRef] [PubMed]
131. Aslan, S.; Deneufchatel, M.; Hashmi, S.; Li, N.; Pfefferle, L.D.; Elimelech, M.; Pauthe, E.; Van Tassel, P.R. Carbon Nanotube-Based Antimicrobial Biomaterials Formed via Layer-by-Layer Assembly with Polypeptides. *J. Colloid. Interface Sci.* **2012**, *388*, 268–273. [CrossRef] [PubMed]
132. Goodwin, D.G.; Marsh, K.M.; Sosa, I.B.; Payne, J.B.; Gorham, J.M.; Bouwer, E.J.; Fairbrother, D.H. Interactions of Microorganisms with Polymer Nanocomposite Surfaces Containing Oxidized Carbon Nanotubes. *Environ. Sci. Technol.* **2015**, *49*, 5484–5492. [CrossRef] [PubMed]
133. Sah, U.; Sharma, K.; Chaudhri, N.; Sankar, M.; Gopinath, P. Antimicrobial Photodynamic Therapy: Single-Walled Carbon Nanotube (SWCNT)-Porphyrin Conjugate for Visible Light Mediated Inactivation of *Staphylococcus aureus*. *Colloids Surf. B Biointerfaces* **2018**, *162*, 108–117. [CrossRef]
134. Tiraferri, A.; Vecitis, C.D.; Elimelech, M. Covalent Binding of Single-Walled Carbon Nanotubes to Polyamide Membranes for Antimicrobial Surface Properties. *ACS Appl. Mater. Interfaces* **2011**, *3*, 2869–2877. [CrossRef]
135. Cajero-Zul, L.R.; López-Dellamary, F.A.; Gómez-Salazar, S.; Vázquez-Lepe, M.; Vera-Graziano, R.; Torres-Vitela, M.R.; Olea-Rodríguez, M.A.; Nuño-Donlucas, S.M. Evaluation of the Resistance to Bacterial Growth of Star-Shaped Poly(ε-Caprolactone)-Co-Poly(Ethylene Glycol) Grafted onto Functionalized Carbon Nanotubes Nanocomposites. *J. Biomater. Sci. Polym. Ed.* **2019**, *30*, 163–189. [CrossRef]
136. Murugan, E.; Vimala, G. Effective Functionalization of Multiwalled Carbon Nanotube with Amphiphilic Poly(Propyleneimine) Dendrimer Carrying Silver Nanoparticles for Better Dispersability and Antimicrobial Activity. *J. Colloid. Interface Sci.* **2011**, *357*, 354–365. [CrossRef]
137. Neelgund, G.M.; Oki, A. Deposition of Silver Nanoparticles on Dendrimer Functionalized Multiwalled Carbon Nanotubes: Synthesis, Characterization and Antimicrobial Activity. *J. Nanosci. Nanotechnol.* **2011**, *11*, 3621–3629. [CrossRef]
138. Neelgund, G.M.; Oki, A.; Luo, Z. Antimicrobial Activity of CdS and $Ag_2S$ Quantum Dots Immobilized on Poly(Amidoamine) Grafted Carbon Nanotubes. *Colloids Surf. B Biointerfaces* **2012**, *100*, 215–221. [CrossRef]
139. Ong, L.C.; Tan, Y.F.; Tan, B.S.; Chung, F.F.; Cheong, S.K.; Leong, C.O. Single-walled carbon nanotubes (SWCNTs) inhibit heat shock protein 90 (HSP90) signaling in human lung fibroblasts and keratinocytes. *Toxicol. Appl. Pharmacol.* **2017**, *15*, 329:347–357. [CrossRef]
140. Zardini, H.Z.; Davarpanah, M.; Shanbedi, M.; Amiri, A.; Maghrebi, M.; Ebrahimi, L. Microbial Toxicity of Ethanolamines—Multiwalled Carbon Nanotubes. *J. Biomed. Mater. Res. A* **2014**, *102*, 1774–1781. [CrossRef] [PubMed]
141. Gilbertson, L.M.; Goodwin, D.G.; Taylor, A.D.; Pfefferle, L.; Zimmerman, J.B. Toward Tailored Functional Design of Multi-Walled Carbon Nanotubes (MWNTs): Electrochemical and Antimicrobial Activity Enhancement via Oxidation and Selective Reduction. *Environ. Sci. Technol.* **2014**, *48*, 5938–5945. [CrossRef] [PubMed]
142. Choi, J.; Seo, Y.; Hwang, J.; Kim, J.; Jeong, Y.; Hwang, M. Antibacterial Activity and Cytotoxicity of Multi-Walled Carbon Nanotubes Decorated with Silver Nanoparticles. *Int. J. Nanomed.* **2014**, *9*, 4621–4629. [CrossRef] [PubMed]

143. Jung, J.H.; Hwang, G.B.; Lee, J.E.; Bae, G.N. Preparation of Airborne Ag/CNT Hybrid Nanoparticles Using an Aerosol Process and Their Application to Antimicrobial Air Filtration. *Langmuir* **2011**, *27*, 10256–10264. [CrossRef]
144. Rusen, E.; Mocanu, A.; Nistor, L.C.; Dinescu, A.; Călinescu, I.; Mustăţea, G.; Voicu, Ş.I.; Andronescu, C.; Diacon, A. Design of Antimicrobial Membrane Based on Polymer Colloids/Multiwall Carbon Nanotubes Hybrid Material with Silver Nanoparticles. *ACS Appl. Mater. Interfaces* **2014**, *6*, 17384–17393. [CrossRef]
145. Mohan, R.; Shanmugharaj, A.M.; Sung Hun, R. An Efficient Growth of Silver and Copper Nanoparticles on Multiwalled Carbon Nanotube with Enhanced Antimicrobial Activity. *J. Biomed. Mater. Res. B Appl. Biomater.* **2011**, *96B*, 119–126. [CrossRef]
146. Sui, M.; Zhang, L.; Sheng, L.; Huang, S.; She, L. Synthesis of ZnO Coated Multi-Walled Carbon Nanotubes and Their Antibacterial Activities. *Sci. Total Environ.* **2013**, *452–453*, 148–154. [CrossRef]
147. Karthika, V.; Arumugam, A. Synthesis and Characterization of MWCNT/$TiO_2$/Au Nanocomposite for Photocatalytic and Antimicrobial Activity. *IET Nanobiotechnol.* **2017**, *11*, 113–118. [CrossRef]
148. Kim, S.H.; Kim, D.Y.; Park, J.S.; Park, M.; Park, M.C. Antimicrobial and Antibiofilm Activities of Urinary Catheter Incorporated with ZnO-Carbon Nanotube. *ACS Appl. Bio Mater.* **2025**, *8*, 1397–1405. [CrossRef]
149. Hirschfeld, J.; Akinoglu, E.M.; Wirtz, D.C.; Hoerauf, A.; Bekeredjian-Ding, I.; Jepsen, S.; Haddouti, E.-M.; Limmer, A.; Giersig, M. Long-Term Release of Antibiotics by Carbon Nanotube-Coated Titanium Alloy Surfaces Diminish Biofilm Formation by Staphylococcus Epidermidis. *Nanomedicine* **2017**, *13*, 1587–1593. [CrossRef]
150. Grover, N.; Borkar, I.V.; Dinu, C.Z.; Kane, R.S.; Dordick, J.S. Laccase- and Chloroperoxidase-Nanotube Paint Composites with Bactericidal and Sporicidal Activity. *Enzym. Microb. Technol.* **2012**, *50*, 271–279. [CrossRef] [PubMed]
151. Venkatesan, J.; Jayakumar, R.; Mohandas, A.; Bhatnagar, I.; Kim, S.-K. Antimicrobial Activity of Chitosan-Carbon Nanotube Hydrogels. *Materials* **2014**, *7*, 3946–3955. [CrossRef] [PubMed]
152. Mohamed, N.A.; Al-Harby, N.F.; Almarshed, M.S. Synthesis and Characterization of Novel Trimellitic Anhydride Isothiocyanate-Cross Linked Chitosan Hydrogels Modified with Multi-Walled Carbon Nanotubes for Enhancement of Antimicrobial Activity. *Int. J. Biol. Macromol.* **2019**, *132*, 416–428. [CrossRef] [PubMed]
153. Ding, L.; Wang, H.; Liu, D.; Zeng, X.-A.; Mao, Y. Bacteria Capture and Inactivation with Functionalized Multi-Walled Carbon Nanotubes (MWCNTs). *J. Nanosci. Nanotechnol.* **2020**, *20*, 2055–2062. [CrossRef]
154. Levi-Polyachenko, N.; Young, C.; MacNeill, C.; Braden, A.; Argenta, L.; Reid, S. Eradicating Group A Streptococcus Bacteria and Biofilms Using Functionalised Multi-Wall Carbon Nanotubes. *Int. J. Hyperth.* **2014**, *30*, 490–501. [CrossRef]
155. Bai, Y.; Park, I.S.; Lee, S.J.; Bae, T.S.; Watari, F.; Uo, M.; Lee, M.H. Aqueous Dispersion of Surfactant-Modified Multiwalled Carbon Nanotubes and Their Application as an Antibacterial Agent. *Carbon* **2011**, *49*, 3663–3671. [CrossRef]
156. Khazaee, M.; Ye, D.; Majumder, A.; Baraban, L.; Opitz, J.; Cuniberti, G. Non-Covalent Modified Multi-Walled Carbon Nanotubes: Dispersion Capabilities and Interactions with Bacteria. *Biomed. Phys. Eng. Express* **2016**, *2*, 055008. [CrossRef]
157. Akhavan, O.; Azimirad, R.; Safa, S. Functionalized Carbon Nanotubes in ZnO Thin Films for Photoinactivation of Bacteria. *Mater. Chem. Phys.* **2011**, *130*, 598–602. [CrossRef]
158. Amiri, A.; Zardini, H.Z.; Shanbedi, M.; Maghrebi, M.; Baniadam, M.; Tolueinia, B. Efficient Method for Functionalization of Carbon Nanotubes by Lysine and Improved Antimicrobial Activity and Water-Dispersion. *Mater. Lett.* **2012**, *72*, 153–156. [CrossRef]
159. Hasnain, M.S.; Nayak, A.K. Regulatory Considerations of Carbon Nanotubes. In *Carbon Nanotubes for Targeted Drug Delivery*; Springer: Cham, Switzerland, 2019; pp. 103–106.
160. Ma-Hock, L.; Treumann, S.; Strauss, V.; Brill, S.; Luizi, F.; Mertler, M.; Wiench, K.; Gamer, A.O.; van Ravenzwaay, B.; Landsiedel, R. Inhalation Toxicity of Multiwall Carbon Nanotubes in Rats Exposed for 3 Months. *Toxicol. Sci.* **2009**, *112*, 468–481. [CrossRef]
161. Wikipedia. Toxicology of Carbon Nanotubes. Available online: https://en.wikipedia.org/wiki/Toxicology_of_carbon_nanomaterials#:~:text=The%20toxicity%20of%20carbon%20nanotubes%20has%20been%20an,in%20evaluating%20the%20toxicity%20of%20this%20heterogeneous%20material (accessed on 7 March 2025).
162. Wang, H.; Wang, J.; Deng, X.; Sun, H.; Shi, Z.; Gu, Z.; Liu, Y.; Zhaoc, Y. Biodistribution of Carbon Single-Wall Carbon Nanotubes in Mice. *J. Nanosci. Nanotechnol.* **2004**, *4*, 1019–1024. [CrossRef] [PubMed]
163. Cheng, J.; Flahaut, E.; Cheng, S.H. Effect of Carbon Nanotubes on Developing Zebrafish (*Danio Rerio*) Embryos. *Environ. Toxicol. Chem.* **2007**, *26*, 708–716. [CrossRef] [PubMed]
164. Li, J.G.; Li, Q.N.; Xu, J.Y.; Cai, X.Q.; Liu, R.L.; Li, Y.J.; Ma, J.F.; Li, W.X. The Pulmonary Toxicity of Multi-Wall Carbon Nanotubes in Mice 30 and 60 Days After Inhalation Exposure. *J. Nanosci. Nanotechnol.* **2009**, *9*, 1384–1387. [CrossRef]
165. Scoville, C.; Cole, R.; Hogg, J.; Farooque, O.; Russel, A. Carbon Nanotubes. Available online: https://docslib.org/doc/13279829/carbon-nanotubes-by-chris-scoville-robin-cole-jason-hogg-omar-farooque-and-archie-russell (accessed on 18 March 2025).
166. Li, Y.; Wang, S.; Lv, Z.; Wang, Z.; Zhao, Y.; Xie, Y.; Xu, Y.; Qian, L.; Yang, Y.; Zhao, Z.; et al. Transforming the Synthesis of Carbon Nanotubes with Machine Learning Models and Automation. *Matter* **2025**, *8*, 101913. [CrossRef]

167. Simon-Deckers, A.; Gouget, B.; Mayne-L'Hermite, M.; Herlin-Boime, N.; Reynaud, C.; Carrière, M. In Vitro Investigation of Oxide Nanoparticle and Carbon Nanotube Toxicity and Intracellular Accumulation in A549 Human Pneumocytes. *Toxicology* **2008**, *253*, 137–146. [CrossRef]
168. Pulskamp, K.; Diabate, S.; Krug, H. Carbon Nanotubes Show No Sign of Acute Toxicity but Induce Intracellular Reactive Oxygen Species in Dependence on Contaminants. *Toxicol. Lett.* **2007**, *168*, 58–74. [CrossRef]
169. Wick, P.; Manser, P.; Limbach, L.; Dettlaffweglikowska, U.; Krumeich, F.; Roth, S.; Stark, W.; Bruinink, A. The Degree and Kind of Agglomeration Affect Carbon Nanotube Cytotoxicity. *Toxicol. Lett.* **2007**, *168*, 121–131. [CrossRef]
170. Monteiro-Riviere, N.A.; Inman, A.O.; Wang, Y.Y.; Nemanich, R.J. Surfactant Effects on Carbon Nanotube Interactions with Human Keratinocytes. *Nanomedicine* **2005**, *1*, 293–299. [CrossRef]
171. Pantarotto, D.; Briand, J.-P.; Prato, M.; Bianco, A. Translocation of Bioactive Peptides across Cell Membranes by Carbon Nanotubes. *Chem. Commun.* **2004**, *1*, 16–17. [CrossRef]
172. Manna, S.K.; Sarkar, S.; Barr, J.; Wise, K.; Barrera, E.V.; Jejelowo, O.; Rice-Ficht, A.C.; Ramesh, G.T. Single-Walled Carbon Nanotube Induces Oxidative Stress and Activates Nuclear Transcription Factor-KB in Human Keratinocytes. *Nano Lett.* **2005**, *5*, 1676–1684. [CrossRef]
173. Shvedova, A.; Castranova, V.; Kisin, E.; Schwegler-Berry, D.; Murray, A.; Gandelsman, V.; Maynard, A.; Baron, P. Exposure to Carbon Nanotube Material: Assessment of Nanotube Cytotoxicity Using Human Keratinocyte Cells. *J. Toxicol. Environ. Health A* **2003**, *66*, 1909–1926. [CrossRef] [PubMed]
174. Shvedova, A.A.; Kisin, E.R.; Mercer, R.; Murray, A.R.; Johnson, V.J.; Potapovich, A.I.; Tyurina, Y.Y.; Gorelik, O.; Arepalli, S.; Schwegler-Berry, D.; et al. Unusual Inflammatory and Fibrogenic Pulmonary Responses to Single-Walled Carbon Nanotubes in Mice. *Am. J. Physiol.-Lung Cell. Mol. Physiol.* **2005**, *289*, L698–L708. [CrossRef] [PubMed]
175. Dumortier, H.; Lacotte, S.; Pastorin, G.; Marega, R.; Wu, W.; Bonifazi, D.; Briand, J.-P.; Prato, M.; Muller, S.; Bianco, A. Functionalized Carbon Nanotubes Are Non-Cytotoxic and Preserve the Functionality of Primary Immune Cells. *Nano Lett.* **2006**, *6*, 1522–1528. [CrossRef] [PubMed]
176. Sayes, C.M.; Liang, F.; Hudson, J.L.; Mendez, J.; Guo, W.; Beach, J.M.; Moore, V.C.; Doyle, C.D.; West, J.L.; Billups, W.E.; et al. Functionalization Density Dependence of Single-Walled Carbon Nanotubes Cytotoxicity in Vitro. *Toxicol. Lett.* **2006**, *161*, 135–142. [CrossRef]
177. Shi Kam, N.W.; Jessop, T.C.; Wender, P.A.; Dai, H. Nanotube Molecular Transporters: Internalization of Carbon Nanotube−Protein Conjugates into Mammalian Cells. *J. Am. Chem. Soc.* **2004**, *126*, 6850–6851. [CrossRef]
178. Fenoglio, I.; Tomatis, M.; Lison, D.; Muller, J.; Fonseca, A.; Nagy, J.B.; Fubini, B. Reactivity of Carbon Nanotubes: Free Radical Generation or Scavenging Activity? *Free Radic. Biol. Med.* **2006**, *40*, 1227–1233. [CrossRef]
179. Sun, Z.; Liu, Z.; Meng, J.; Meng, J.; Duan, J.; Xie, S.; Lu, X.; Zhu, Z.; Wang, C.; Chen, S.; et al. Carbon Nanotubes Enhance Cytotoxicity Mediated by Human Lymphocytes In Vitro. *PLoS ONE* **2011**, *6*, e21073. [CrossRef]
180. Cui, D.; Tian, F.; Ozkan, C.S.; Wang, M.; Gao, H. Effect of Single Wall Carbon Nanotubes on Human HEK293 Cells. *Toxicol. Lett.* **2005**, *155*, 73–85. [CrossRef]
181. Jia, G.; Wang, H.; Yan, L.; Wang, X.; Pei, R.; Yan, T.; Zhao, Y.; Guo, X. Cytotoxicity of Carbon Nanomaterials: Single-Wall Nanotube, Multi-Wall Nanotube, and Fullerene. *Environ. Sci. Technol.* **2005**, *39*, 1378–1383. [CrossRef]
182. Pacurari, M.; Yin, X.J.; Zhao, J.; Ding, M.; Leonard, S.S.; Schwegler-Berry, D.; Ducatman, B.S.; Sbarra, D.; Hoover, M.D.; Castranova, V.; et al. Raw Single-Wall Carbon Nanotubes Induce Oxidative Stress and Activate MAPKs, AP-1, NF-KB, and Akt in Normal and Malignant Human Mesothelial Cells. *Environ. Health Perspect.* **2008**, *116*, 1211–1217. [CrossRef]
183. Kisin, E.R.; Murray, A.R.; Keane, M.J.; Shi, X.-C.; Schwegler-Berry, D.; Gorelik, O.; Arepalli, S.; Castranova, V.; Wallace, W.E.; Kagan, V.E.; et al. Single-Walled Carbon Nanotubes: Geno- and Cytotoxic Effects in Lung Fibroblast V79 Cells. *J. Toxicol. Environ. Health A* **2007**, *70*, 2071–2079. [CrossRef] [PubMed]
184. Francis, A.P.; Devasena, T. Toxicity of Carbon Nanotubes: A Review. *Toxicol. Ind. Health* **2018**, *34*, 200–210. [CrossRef] [PubMed]
185. Muller, J.; Huaux, F.; Moreau, N.; Misson, P.; Heilier, J.-F.; Delos, M.; Arras, M.; Fonseca, A.; Nagy, J.B.; Lison, D. Respiratory Toxicity of Multi-Wall Carbon Nanotubes. *Toxicol. Appl. Pharmacol.* **2005**, *207*, 221–231. [CrossRef] [PubMed]
186. Hirano, S.; Kanno, S.; Furuyama, A. Multi-Walled Carbon Nanotubes Injure the Plasma Membrane of Macrophages. *Toxicol. Appl. Pharmacol.* **2008**, *232*, 244–251. [CrossRef]
187. Monteiro-Riviere, N.A.; Nemanich, R.J.; Inman, A.O.; Wang, Y.Y.; Riviere, J.E. Multi-Walled Carbon Nanotube Interactions with Human Epidermal Keratinocytes. *Toxicol. Lett.* **2005**, *155*, 377–384. [CrossRef]
188. Monteiro-Riviere, N.A.; Inman, A.O. Challenges for Assessing Carbon Nanomaterial Toxicity to the Skin. *Carbon* **2006**, *44*, 1070–1078. [CrossRef]
189. Yang, H.; Liu, C.; Yang, D.; Zhang, H.; Xi, Z. Comparative Study of Cytotoxicity, Oxidative Stress and Genotoxicity Induced by Four Typical Nanomaterials: The Role of Particle Size, Shape and Composition. *J. Appl. Toxicol.* **2009**, *29*, 69–78. [CrossRef]

190. Ding, L.; Stilwell, J.; Zhang, T.; Elboudwarej, O.; Jiang, H.; Selegue, J.P.; Cooke, P.A.; Gray, J.W.; Chen, F.F. Molecular Characterization of the Cytotoxic Mechanism of Multiwall Carbon Nanotubes and Nano-Onions on Human Skin Fibroblast. *Nano Lett.* **2005**, *5*, 2448–2464. [CrossRef]
191. Soto, K.F.; Carrasco, A.; Powell, T.G.; Garza, K.M.; Murr, L.E. Comparative in Vitro Cytotoxicity Assessment of Some Manufacturednanoparticulate Materials Characterized by Transmissionelectron Microscopy. *J. Nanopart. Res.* **2005**, *7*, 145–169. [CrossRef]
192. Murr, L.E.; Garza, K.M.; Soto, K.F.; Carrasco, A.; Powell, T.G.; Ramirez, D.A.; Guerrero, P.A.; Lopez, D.A.; Venzor, J. Cytotoxicity Assessment of Some Carbon Nanotubes and Related Carbon Nanoparticle Aggregates and the Implications for Anthropogenic Carbon Nanotube Aggregates in the Environment. *Int. J. Environ. Res. Public. Health* **2005**, *2*, 31–42. [CrossRef]
193. Bottini, M.; Bruckner, S.; Nika, K.; Bottini, N.; Bellucci, S.; Magrini, A.; Bergamaschi, A.; Mustelin, T. Multi-Walled Carbon Nanotubes Induce T Lymphocyte Apoptosis. *Toxicol. Lett.* **2006**, *160*, 121–126. [CrossRef] [PubMed]
194. Lam, C.-W. Pulmonary Toxicity of Single-Wall Carbon Nanotubes in Mice 7 and 90 Days After Intratracheal Instillation. *Toxicol. Sci.* **2003**, *77*, 126–134. [CrossRef] [PubMed]
195. Warheit, D.B. Comparative Pulmonary Toxicity Assessment of Single-Wall Carbon Nanotubes in Rats. *Toxicol. Sci.* **2003**, *77*, 117–125. [CrossRef] [PubMed]
196. Li, Z.; Hulderman, T.; Salmen, R.; Chapman, R.; Leonard, S.S.; Young, S.-H.; Shvedova, A.; Luster, M.I.; Simeonova, P.P. Cardiovascular Effects of Pulmonary Exposure to Single-Wall Carbon Nanotubes. *Environ. Health Perspect.* **2007**, *115*, 377–382. [CrossRef]
197. Mutlu, G.M.; Budinger, G.R.S.; Green, A.A.; Urich, D.; Soberanes, S.; Chiarella, S.E.; Alheid, G.F.; McCrimmon, D.R.; Szleifer, I.; Hersam, M.C. Biocompatible Nanoscale Dispersion of Single-Walled Carbon Nanotubes Minimizes in Vivo Pulmonary Toxicity. *Nano Lett.* **2010**, *10*, 1664–1670. [CrossRef]
198. Mitchell, L.A.; Lauer, F.T.; Burchiel, S.W.; McDonald, J.D. Mechanisms for How Inhaled Multiwalled Carbon Nanotubes Suppress Systemic Immune Function in Mice. *Nat. Nanotechnol.* **2009**, *4*, 451–456. [CrossRef]
199. Muller, J.; Huaux, F.; Fonseca, A.; Nagy, J.B.; Moreau, N.; Delos, M.; Raymundo-Piñero, E.; Béguin, F.; Kirsch-Volders, M.; Fenoglio, I.; et al. Structural Defects Play a Major Role in the Acute Lung Toxicity of Multiwall Carbon Nanotubes: Toxicological Aspects. *Chem. Res. Toxicol.* **2008**, *21*, 1698–1705. [CrossRef]
200. Donaldson, K.; Aitken, R.; Tran, L.; Stone, V.; Duffin, R.; Forrest, G.; Alexander, A. Carbon Nanotubes: A Review of Their Properties in Relation to Pulmonary Toxicology and Workplace Safety. *Toxicol. Sci.* **2006**, *92*, 5–22. [CrossRef]
201. Inoue, K.; Yanagisawa, R.; Koike, E.; Nishikawa, M.; Takano, H. Repeated Pulmonary Exposure to Single-Walled Carbon Nanotubes Exacerbates Allergic Inflammation of the Airway: Possible Role of Oxidative Stress. *Free Radic. Biol. Med.* **2010**, *48*, 924–934. [CrossRef]
202. Kobayashi, N.; Naya, M.; Ema, M.; Endoh, S.; Maru, J.; Mizuno, K.; Nakanishi, J. Biological Response and Morphological Assessment of Individually Dispersed Multi-Wall Carbon Nanotubes in the Lung after Intratracheal Instillation in Rats. *Toxicology* **2010**, *276*, 143–153. [CrossRef]
203. Morimoto, Y.; Hirohashi, M.; Ogami, A.; Oyabu, T.; Myojo, T.; Todoroki, M.; Yamamoto, M.; Hashiba, M.; Mizuguchi, Y.; Lee, B.W.; et al. Pulmonary Toxicity of Well-Dispersed Multi-Wall Carbon Nanotubes Following Inhalation and Intratracheal Instillation. *Nanotoxicology* **2012**, *6*, 587–599. [CrossRef] [PubMed]
204. Grubek-Jaworska, H.; Nejman, P.; Czumińska, K.; Przybyłowski, T.; Huczko, A.; Lange, H.; Bystrzejewski, M.; Baranowski, P.; Chazan, R. Preliminary Results on the Pathogenic Effects of Intratracheal Exposure to One-Dimensional Nanocarbons. *Carbon* **2006**, *44*, 1057–1063. [CrossRef]
205. Francis, A.P.; Ganapathy, S.; Palla, V.R.; Murthy, P.B.; Ramaprabhu, S.; Devasena, T. One Time Nose-Only Inhalation of MWCNTs: Exploring the Mechanism of Toxicity by Intermittent Sacrifice in Wistar Rats. *Toxicol. Rep.* **2015**, *2*, 111–120. [CrossRef] [PubMed]
206. Simeonova, P.P.; Erdely, A. Engineered Nanoparticle Respiratory Exposure and Potential Risks for Cardiovascular Toxicity: Predictive Tests and Biomarkers. *Inhal. Toxicol.* **2009**, *21*, 68–73. [CrossRef]
207. Pietroiusti, A.; Massimiani, M.; Fenoglio, I.; Colonna, M.; Valentini, F.; Palleschi, G.; Camaioni, A.; Magrini, A.; Siracusa, G.; Bergamaschi, A.; et al. Low Doses of Pristine and Oxidized Single-Wall Carbon Nanotubes Affect Mammalian Embryonic Development. *ACS Nano* **2011**, *5*, 4624–4633. [CrossRef]
208. Bai, Y.; Zhang, Y.; Zhang, J.; Mu, Q.; Zhang, W.; Butch, E.R.; Snyder, S.E.; Yan, B. Repeated Administrations of Carbon Nanotubes in Male Mice Cause Reversible Testis Damage without Affecting Fertility. *Nat. Nanotechnol.* **2010**, *5*, 683–689. [CrossRef]
209. Philbrook, N.A.; Walker, V.K.; Afrooz, A.R.M.N.; Saleh, N.B.; Winn, L.M. Investigating the Effects of Functionalized Carbon Nanotubes on Reproduction and Development in Drosophila Melanogaster and CD-1 Mice. *Reprod. Toxicol.* **2011**, *32*, 442–448. [CrossRef]
210. Møller, P.; Wils, R.S.; Di Ianni, E.; Gutierrez, C.A.T.; Roursgaard, M.; Jacobsen, N.R. Genotoxicity of Multi-Walled Carbon Nanotube Reference Materials in Mammalian Cells and Animals. *Mutat. Res./Rev. Mutat. Res.* **2021**, *788*, 108393. [CrossRef]

211. Sargent, L.M.; Hubbs, A.F.; Young, S.-H.; Kashon, M.L.; Dinu, C.Z.; Salisbury, J.L.; Benkovic, S.A.; Lowry, D.T.; Murray, A.R.; Kisin, E.R.; et al. Single-Walled Carbon Nanotube-Induced Mitotic Disruption. *Mutat. Res./Genet. Toxicol. Environ. Mutagen.* **2012**, *745*, 28–37. [CrossRef]
212. Lash, L.H. Special Issue: Membrane Transporters in Toxicology. *Toxicol. Appl. Pharmacol.* **2005**, *204*, 197. [CrossRef]
213. Murray, A.R.; Kisin, E.; Leonard, S.S.; Young, S.H.; Kommineni, C.; Kagan, V.E.; Castranova, V.; Shvedova, A.A. Oxidative Stress and Inflammatory Response in Dermal Toxicity of Single-Walled Carbon Nanotubes. *Toxicology* **2009**, *257*, 161–171. [CrossRef] [PubMed]
214. Orecchioni, M.; Bedognetti, D.; Sgarrella, F.; Marincola, F.M.; Bianco, A.; Delogu, L.G. Impact of Carbon Nanotubes and Graphene on Immune Cells. *J. Transl. Med.* **2014**, *12*, 138. [CrossRef]
215. Svadlakova, T.; Holmannova, D.; Kolackova, M.; Malkova, A.; Krejsek, J.; Fiala, Z. Immunotoxicity of Carbon-Based Nanomaterials, Starring Phagocytes. *Int. J. Mol. Sci.* **2022**, *23*, 8889. [CrossRef] [PubMed]
216. Kan, H.; Pan, D.; Castranova, V. Engineered Nanoparticle Exposure and Cardiovascular Effects: The Role of a Neuronal-Regulated Pathway. *Inhal. Toxicol.* **2018**, *30*, 335–342. [CrossRef]
217. Alidori, S.; Bowman, R.L.; Yarilin, D.; Romin, Y.; Barlas, A.; Mulvey, J.J.; Fujisawa, S.; Xu, K.; Ruggiero, A.; Riabov, V.; et al. Deconvoluting Hepatic Processing of Carbon Nanotubes. *Nat. Commun.* **2016**, *7*, 12343. [CrossRef]
218. Reddy, A.R.N.; Reddy, Y.N.; Krishna, D.R.; Himabindu, V. Multi Wall Carbon Nanotubes Induce Oxidative Stress and Cytotoxicity in Human Embryonic Kidney (HEK293) Cells. *Toxicology* **2010**, *272*, 11–16. [CrossRef]
219. Facciolà, A.; Visalli, G.; La Maestra, S.; Ceccarelli, M.; D'Aleo, F.; Nunnari, G.; Pellicanò, G.F.; Di Pietro, A. Carbon Nanotubes and Central Nervous System: Environmental Risks, Toxicological Aspects and Future Perspectives. *Environ. Toxicol. Pharmacol.* **2019**, *65*, 23–30. [CrossRef]
220. Zheng, W.; McKinney, W.; Kashon, M.; Salmen, R.; Castranova, V.; Kan, H. The Influence of Inhaled Multi-Walled Carbon Nanotubes on the Autonomic Nervous System. *Part. Fibre Toxicol.* **2015**, *13*, 8. [CrossRef]
221. Aragon, M.J.; Topper, L.; Tyler, C.R.; Sanchez, B.; Zychowski, K.; Young, T.; Herbert, G.; Hall, P.; Erdely, A.; Eye, T.; et al. Serum-Borne Bioactivity Caused by Pulmonary Multiwalled Carbon Nanotubes Induces Neuroinflammation via Blood–Brain Barrier Impairment. *Proc. Natl. Acad. Sci. USA* **2017**, *114*, E1968–E1976. [CrossRef]
222. Gholamine, B.; Karimi, I.; Salimi, A.; Mazdarani, P.; Becker, L.A. Neurobehavioral Toxicity of Carbon Nanotubes in Mice. *Toxicol. Ind. Health* **2017**, *33*, 340–350. [CrossRef]
223. Larner, S.F.; Wang, J.; Goodman, J.; O'Donoghue Altman, M.B.; Xin, M.; Wang, K.K.W. In Vitro Neurotoxicity Resulting from Exposure of Cultured Neural Cells to Several Types of Nanoparticles. *J. Cell Death* **2017**, *10*, 117967071769452. [CrossRef] [PubMed]
224. Chen, H.; Zheng, X.; Nicholas, J.; Humes, S.T.; Loeb, J.C.; Robinson, S.E.; Bisesi, J.H.; Das, D.; Saleh, N.B.; Castleman, W.L.; et al. Single-Walled Carbon Nanotubes Modulate Pulmonary Immune Responses and Increase Pandemic Influenza a Virus Titers in Mice. *Virol. J.* **2017**, *14*, 242. [CrossRef] [PubMed]
225. Park, E.-J.; Choi, J.; Kim, J.-H.; Lee, B.-S.; Yoon, C.; Jeong, U.; Kim, Y. Subchronic Immunotoxicity and Screening of Reproductive Toxicity and Developmental Immunotoxicity Following Single Instillation of HIPCO-Single-Walled Carbon Nanotubes: Purity-Based Comparison. *Nanotoxicology* **2016**, *10*, 1188–1202. [CrossRef] [PubMed]
226. Khang, D.; Lee, S.; Kim, S.-H. High Dispersity of Carbon Nanotubes Diminishes Immunotoxicity in Spleen. *Int. J. Nanomed.* **2015**, *10*, 2697–2710. [CrossRef]
227. Qi, W.; Bi, J.; Zhang, X.; Wang, J.; Wang, J.; Liu, P.; Li, Z.; Wu, W. Damaging Effects of Multi-Walled Carbon Nanotubes on Pregnant Mice with Different Pregnancy Times. *Sci. Rep.* **2014**, *4*, 4352. [CrossRef]
228. Cheng, J.; Cheng, S.H. Influence of Carbon Nanotube Length on Toxicity to Zebrafish Embryos. *Int. J. Nanomed.* **2012**, *7*, 3731–3739. [CrossRef]
229. Huang, X.; Zhang, F.; Sun, X.; Choi, K.-Y.; Niu, G.; Zhang, G.; Guo, J.; Lee, S.; Chen, X. The Genotype-Dependent Influence of Functionalized Multiwalled Carbon Nanotubes on Fetal Development. *Biomaterials* **2014**, *35*, 856–865. [CrossRef]
230. Zhu, B.; Zhu, S.; Li, J.; Hui, X.; Wang, G.-X. The Developmental Toxicity, Bioaccumulation and Distribution of Oxidized Single Walled Carbon Nanotubes in *Artemia salina*. *Toxicol. Res.* **2018**, *7*, 897–906. [CrossRef]
231. Lindberg, H.K.; Falck, G.C.-M.; Singh, R.; Suhonen, S.; Järventaus, H.; Vanhala, E.; Catalán, J.; Farmer, P.B.; Savolainen, K.M.; Norppa, H. Genotoxicity of Short Single-Wall and Multi-Wall Carbon Nanotubes in Human Bronchial Epithelial and Mesothelial Cells in Vitro. *Toxicology* **2013**, *313*, 24–37. [CrossRef]
232. Lettiero, B.; Andersen, A.J.; Hunter, A.C.; Moghimi, S.M. Complement System and the Brain: Selected Pathologies and Avenues toward Engineering of Neurological Nanomedicines. *J. Control. Release* **2012**, *161*, 283–289. [CrossRef]
233. Moghimi, S.M.; Peer, D.; Langer, R. Reshaping the Future of Nanopharmaceuticals: *Ad iudicium*. *ACS Nano* **2011**, *5*, 8454–8458. [CrossRef] [PubMed]

234. Liu, Z.; Davis, C.; Cai, W.; He, L.; Chen, X.; Dai, H. Circulation and Long-Term Fate of Functionalized, Biocompatible Single-Walled Carbon Nanotubes in Mice Probed by Raman Spectroscopy. *Proc. Natl. Acad. Sci. USA* **2008**, *105*, 1410–1415. [CrossRef] [PubMed]
235. Yang, S.; Fernando, K.A.S.; Liu, J.; Wang, J.; Sun, H.; Liu, Y.; Chen, M.; Huang, Y.; Wang, X.; Wang, H.; et al. Covalently PEGylated Carbon Nanotubes with Stealth Character In Vivo. *Small* **2008**, *4*, 940–944. [CrossRef] [PubMed]
236. Pondman, K.M.; Pednekar, L.; Paudyal, B.; Tsolaki, A.G.; Kouser, L.; Khan, H.A.; Shamji, M.H.; ten Haken, B.; Stenbeck, G.; Sim, R.B.; et al. Innate Immune Humoral Factors, C1q and Factor H, with Differential Pattern Recognition Properties, Alter Macrophage Response to Carbon Nanotubes. *Nanomedicine* **2015**, *11*, 2109–2118. [CrossRef]
237. Silva, R.M.; Doudrick, K.; Franzi, L.M.; TeeSy, C.; Anderson, D.S.; Wu, Z.; Mitra, S.; Vu, V.; Dutrow, G.; Evans, J.E.; et al. Instillation *versus* Inhalation of Multiwalled Carbon Nanotubes: Exposure-Related Health Effects, Clearance, and the Role of Particle Characteristics. *ACS Nano* **2014**, *8*, 8911–8931. [CrossRef]
238. Liu, Y.; Zhao, Y.; Sun, B.; Chen, C. Understanding the Toxicity of Carbon Nanotubes. *Acc. Chem. Res.* **2013**, *46*, 702–713. [CrossRef]
239. Zhu, Y.; Li, W.; Li, Q.; Li, Y.; Li, Y.; Zhang, X.; Huang, Q. Effects of Serum Proteins on Intracellular Uptake and Cytotoxicity of Carbon Nanoparticles. *Carbon* **2009**, *47*, 1351–1358. [CrossRef]
240. Dong, X.; Sun, Z.; Wang, X.; Leng, X. An Innovative MWCNTs/DOX/TC Nanosystem for Chemo-Photothermal Combination Therapy of Cancer. *Nanomedicine* **2017**, *13*, 2271–2280. [CrossRef]
241. Yan, Y.; Wang, R.; Hu, Y.; Sun, R.; Song, T.; Shi, X.; Yin, S. Stacking of Doxorubicin on Folic Acid-Targeted Multiwalled Carbon Nanotubes for in Vivo Chemotherapy of Tumors. *Drug Deliv.* **2018**, *25*, 1607–1616. [CrossRef]
242. Awasthi, S.; De, S.; Pandey, S.K. Surface Grafting of Carbon Nanostructures. In *Handbook of Functionalized Carbon Nanostructures*; Springer International Publishing: Cham, Switzerland, 2024; pp. 1–45.
243. Rele, S.; Thakur, C.K.; Khan, F.; Baral, B.; Saini, V.; Karthikeyan, C.; Moorthy, N.S.H.N.; Jha, H.C. Curcumin Coating: A Novel Solution to Mitigate Inherent Carbon Nanotube Toxicity. *J. Mater. Sci. Mater. Med.* **2024**, *35*, 24. [CrossRef]
244. Wu, L.; Man, C.; Wang, H.; Lu, X.; Ma, Q.; Cai, Y.; Ma, W. PEGylated Multi-Walled Carbon Nanotubes for Encapsulation and Sustained Release of Oxaliplatin. *Pharm. Res.* **2013**, *30*, 412–423. [CrossRef] [PubMed]
245. Costa, P.M.; Bourgognon, M.; Wang, J.T.-W.; Al-Jamal, K.T. Functionalised Carbon Nanotubes: From Intracellular Uptake and Cell-Related Toxicity to Systemic Brain Delivery. *J. Control. Release* **2016**, *241*, 200–219. [CrossRef] [PubMed]
246. Sohaebuddin, S.K.; Thevenot, P.T.; Baker, D.; Eaton, J.W.; Tang, L. Nanomaterial Cytotoxicity Is Composition, Size, and Cell Type Dependent. *Part. Fibre Toxicol.* **2010**, *7*, 22. [CrossRef] [PubMed]
247. Heras, A.; Colina, A.; López-Palacios, J.; Ayala, P.; Sainio, J.; Ruiz, V.; Kauppinen, E.I. Electrochemical Purification of Carbon Nanotube Electrodes. *Electrochem. Commun.* **2009**, *11*, 1535–1538. [CrossRef]
248. Mercier, G.; Hérold, C.; Marêché, J.-F.; Cahen, S.; Gleize, J.; Ghanbaja, J.; Lamura, G.; Bellouard, C.; Vigolo, B. Selective Removal of Metal Impurities from Single Walled Carbon Nanotube Samples. *New J. Chem.* **2013**, *37*, 790. [CrossRef]
249. Pełech, I.; Narkiewicz, U.; Kaczmarek, A.; Jędrzejewska, A.; Pełech, R. Removal of Metal Particles from Carbon Nanotubes Using Conventional and Microwave Methods. *Sep. Purif. Technol.* **2014**, *136*, 105–110. [CrossRef]
250. Madani, S.Y.; Mandel, A.; Seifalian, A.M. A Concise Review of Carbon Nanotube's Toxicology. *Nano Rev.* **2013**, *4*, 21521. [CrossRef]
251. Sallam, A.A.; Ahmed, M.M.; El-Magd, M.A.; Magdy, A.; Ghamry, H.I.; Alshahrani, M.Y.; Abou El-Fotoh, M.F. Quercetin-Ameliorated, Multi-Walled Carbon Nanotubes-Induced Immunotoxic, Inflammatory, and Oxidative Effects in Mice. *Molecules* **2022**, *27*, 2117. [CrossRef]
252. Zhang, D.; Deng, X.; Ji, Z.; Shen, X.; Dong, L.; Wu, M.; Gu, T.; Liu, Y. Long-Term Hepatotoxicity of Polyethylene-Glycol Functionalized Multi-Walled Carbon Nanotubes in Mice. *Nanotechnology* **2010**, *21*, 175101. [CrossRef]
253. Pérez-Luna, V.; Moreno-Aguilar, C.; Arauz-Lara, J.L.; Aranda-Espinoza, S.; Quintana, M. Interactions of Functionalized Multi-Wall Carbon Nanotubes with Giant Phospholipid Vesicles as Model Cellular Membrane System. *Sci. Rep.* **2018**, *8*, 17998. [CrossRef]
254. Adenuga, A.A.; Truong, L.; Tanguay, R.L.; Remcho, V.T. Preparation of Water Soluble Carbon Nanotubes and Assessment of Their Biological Activity in Embryonic Zebrafish. *Int. J. Biomed. Nanosci. Nanotechnol.* **2013**, *3*, 38. [CrossRef] [PubMed]
255. Lay, C.L.; Liu, H.Q.; Tan, H.R.; Liu, Y. Delivery of Paclitaxel by Physically Loading onto Poly(Ethylene Glycol) (PEG)-Graftcarbon Nanotubes for Potent Cancer Therapeutics. *Nanotechnology* **2010**, *21*, 065101. [CrossRef] [PubMed]
256. Naqvi, S.T.R.; Rasheed, T.; Hussain, D.; Najam ul Haq, M.; Majeed, S.; Shafi, S.; Ahmed, N.; Nawaz, R. Modification Strategies for Improving the Solubility/Dispersion of Carbon Nanotubes. *J. Mol. Liq.* **2020**, *297*, 111919. [CrossRef]
257. Saleemi, M.A.; Hosseini Fouladi, M.; Yong, P.V.C.; Chinna, K.; Palanisamy, N.K.; Wong, E.H. Toxicity of Carbon Nanotubes: Molecular Mechanisms, Signaling Cascades, and Remedies in Biomedical Applications. *Chem. Res. Toxicol.* **2021**, *34*, 24–46. [CrossRef]
258. Barosova, H.; Karakocak, B.B.; Septiadi, D.; Petri-Fink, A.; Stone, V.; Rothen-Rutishauser, B. An In Vitro Lung System to Assess the Proinflammatory Hazard of Carbon Nanotube Aerosols. *Int. J. Mol. Sci.* **2020**, *21*, 5335. [CrossRef]
259. Hou, P.-X.; Liu, C.; Cheng, H.-M. Purification of Carbon Nanotubes. *Carbon* **2008**, *46*, 2003–2025. [CrossRef]
260. Yang, M.; Zhang, M. Biodegradation of Carbon Nanotubes by Macrophages. *Front. Mater.* **2019**, *6*, 225. [CrossRef]

261. Yang, H.; Yang, T.; Heng, C.; Zhou, Y.; Jiang, Z.; Qian, X.; Du, L.; Mao, S.; Yin, X.; Lu, Q. Quercetin Improves Nonalcoholic Fatty Liver by Ameliorating Inflammation, Oxidative Stress, and Lipid Metabolism in *Db/Db* Mice. *Phytother. Res.* **2019**, *33*, 3140–3152. [CrossRef]
262. Di Petrillo, A.; Orrù, G.; Fais, A.; Fantini, M.C. Quercetin and Its Derivates as Antiviral Potentials: A Comprehensive Review. *Phytother. Res.* **2022**, *36*, 266–278. [CrossRef]
263. Nakanishi, J.; Morimoto, Y.; Ogura, I.; Kobayashi, N.; Naya, M.; Ema, M.; Endoh, S.; Shimada, M.; Ogami, A.; Myojyo, T.; et al. Risk Assessment of the Carbon Nanotube Group. *Risk Anal.* **2015**, *35*, 1940–1956. [CrossRef]
264. Kumar Babele, P.; Kumar Verma, M.; Kant Bhatia, R. Carbon Nanotubes: A Review on Risks Assessment, Mechanism of Toxicity and Future Directives to Prevent Health Implication. *BIOCELL* **2021**, *45*, 267–279. [CrossRef]
265. Duraia, E.M.; Opoku, M.; Beall, G.W. Efficient Eco-Friendly Synthesis of Carbon Nanotubes over Graphite Nanosheets from Yellow Corn: A One-Step Green Approach. *Sci. Rep.* **2024**, *14*, 16405. [CrossRef] [PubMed]
266. Fisher, C.; Rider, A.E.; Jun Han, Z.; Kumar, S.; Levchenko, I.; Ostrikov, K. Applications and Nanotoxicity of Carbon Nanotubes and Graphene in Biomedicine. *J. Nanomater.* **2012**, *2012*, 315185. [CrossRef]
267. Gendron, D.; Bubak, G. Carbon Nanotubes and Graphene Materials as Xenobiotics in Living Systems: Is There a Consensus on Their Safety? *J. Xenobiot.* **2023**, *13*, 740–760. [CrossRef]
268. Zumwalde, R.D. *Current Intelligence Bulletin 65: Occupational Exposure to Carbon Nanotubes and Nanofibers*; The National Institute for Occupational Safety and Health (NIOSH): Atlanta, GA, USA, 2013.
269. European Union Eur-Lex. Available online: https://eur-lex.europa.eu/legal-content/EN/TXT/?uri=OJ:JOC_2022_229_R_0001 (accessed on 13 March 2025).
270. EU-OSHA—The Swedish Work Environment Authority. *Carbon Nanotubes—Exposure, Toxicology and Protective Measures in the Work Environment*; EU-OSHA—The Swedish Work Environment Authority: Solna, Sweden, 2011.
271. Schulte, P.A.; Murashov, V.; Zumwalde, R.; Kuempel, E.D.; Geraci, C.L. Occupational Exposure Limits for Nanomaterials: State of the Art. *J. Nanopart. Res.* **2010**, *12*, 1971–1987. [CrossRef]
272. Kerr, R.; McHugh, M.; McCrory, M. HSE Management Standards and Stress-Related Work Outcomes. *Occup. Med.* **2009**, *59*, 574–579. [CrossRef]

*nanomaterials*

MDPI

*Review*

# Architecting Durability: Synergies in Assembly, Self-Repair, and Advanced Characterization of Carbon Nanotube Materials

**Monika R. Snowdon [1,*], Shasvat Rathod [2], Robert L. F. Liang [1,2] and Marina Freire-Gormaly [2,*]**

[1] Waterloo Institute for Nanotechnology, University of Waterloo, Waterloo, ON N2L 3G1, Canada; robliang@yorku.ca

[2] Lassonde School of Engineering, York University, North York, ON M3J 2S5, Canada; shasvat.rathod@uwaterloo.ca

* Correspondence: monika.snowdon@uwaterloo.ca (M.R.S.); marina.freire-gormaly@lassonde.yorku.ca (M.F.-G.)

**Abstract**

Carbon nanotubes (CNTs) have remarkable mechanical, electrical, and thermal properties, making them highly attractive as foundational elements for advanced materials. However, translating their nanoscale potential into macroscale reliability and longevity requires a holistic design approach that integrates precise architectural control with robust damage mitigation strategies. This review presents a synergistic perspective on enhancing the durability of CNT-based systems by critically examining the interplay between molecular assembly, self-repair mechanisms, and the advanced characterization techniques required for their validation. We first establish how foundational architectural control—achieved through strategies like chemical functionalization, field-directed alignment, and dispersion—governs the ultimate performance of CNT materials. A significant focus is placed on advanced functionalization, such as fluorination, and its verification using high-powered spectroscopic tools, including X-ray photoelectron spectroscopy (XPS) and near-edge X-ray absorption fine structure (NEXAFS) spectroscopy. Subsequently, this manuscript delves into the mechanisms of self-repair, systematically analyzing both the intrinsic capacity of the carbon lattice to heal atomic-level defects and the extrinsic strategies that incorporate engineered healing agents into composites. This discussion is uniquely supplemented by an exploration of the experimental techniques, such as electron energy loss spectroscopy (EELS) and Auger electron spectroscopy (AES), that provide crucial evidence for irradiation-induced healing dynamics. Finally, we argue that a "characterization gap" has limited the field's progress and highlight the critical role of techniques like in situ Raman spectroscopy for quantitatively monitoring healing efficiency at the molecular level. By identifying current challenges and future research frontiers, this review underscores that the creation of truly durable materials depends on an integrated understanding of how to build, repair, and precisely measure CNT-based systems.

**Keywords:** carbon nanotubes (CNTs); assembly; self-repair; durability; composites; advanced characterization; functionalization; spectroscopy

## 1. Introduction

### 1.1. The Durability Imperative in Advanced Carbon Materials

Carbon nanotubes (CNTs), one-dimensional allotropes of carbon, have been the subject of intense scientific and technological interest due to their extraordinary combination of properties. These cylindrical nanostructures, which can be conceptualized as rolled-up sheets of graphene, exist as either single-walled (SWCNTs) or multi-walled (MWCNTs) as per Figure 1 and exhibit exceptionally high tensile strength, stiffness, electrical conductivity, and thermal conductivity [1]. These attributes make them ideal candidates for reinforcing polymers, metals, and ceramics, leading to the development of high-performance composites for demanding applications in aerospace, automotive, electronics, and energy storage [2–4]. However, the promise of these nanoscale properties is often unfulfilled in macroscopic applications due to a fundamental challenge: the accumulation of damage during service. Mechanical fatigue, environmental degradation, and operational stress can introduce defects, microcracks, and delamination, which compromise the structural integrity and functional performance of the material, ultimately leading to premature failure [4,5]. This reality has given rise to a "durability imperative" in materials science. It is no longer sufficient to design materials with high initial strength; they must be engineered to resist, tolerate, and even reverse damage over their intended lifespan. To this end, researchers have increasingly turned to nature for inspiration, where biological systems exhibit remarkable capabilities for autonomous damage repair, such as the healing of wounds or the regeneration of bone [2]. The translation of these biological principles to synthetic materials has led to the burgeoning field of self-repair, which aims to create materials that can autonomously detect and heal damage, thereby restoring their original properties and extending their service life. In the context of CNT-based systems, self-repair is not merely an intriguing feature but a critical design principle necessary to unlock their full potential for creating truly resilient and reliable next-generation materials.

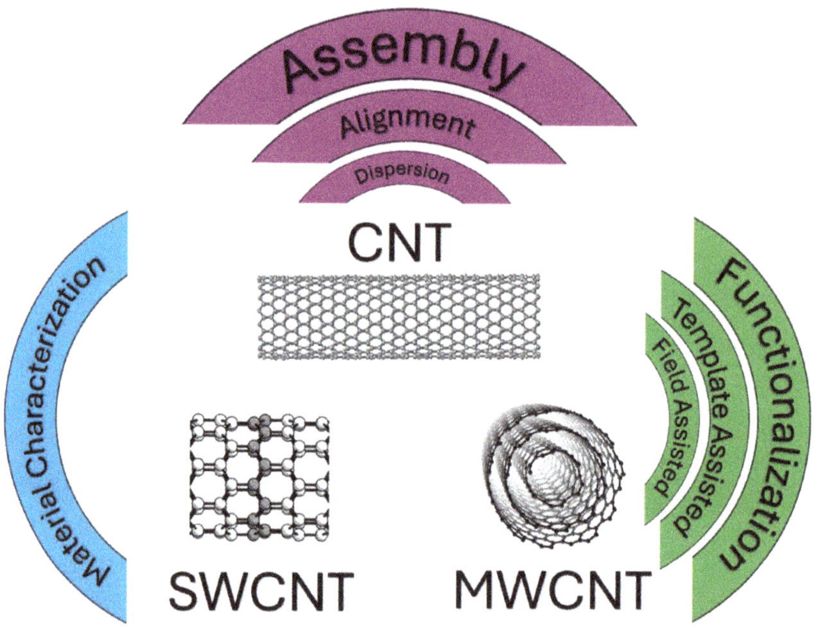


**Figure 1.** A schematic illustrating the fundamental concepts in engineering carbon nanotube (CNT) materials. The central image displays the basic molecular structure of a CNT, with its single-walled (SWCNT) and multi-walled (MWCNT) variants shown below. This is framed by the key engineering strategies discussed in this review: assembly (encompassing dispersion and alignment) and functionalization, which are critically validated through advanced material characterization.

### 1.2. State of the Art

The concepts of CNT assembly and self-healing materials are not new, and numerous reviews have capably summarized the progress in these individual areas [6]. The existing literature typically addresses topics such as assembly techniques (e.g., alignment, dispersion), self-healing mechanisms (e.g., microencapsulation, reversible bonds), and characterization methods as distinct and separate subjects. While valuable, this compartmentalized approach often overlooks the profound and synergistic relationships between these domains. The effectiveness of a self-repair mechanism is not an independent variable; it is fundamentally dependent on the initial material architecture established during assembly and the interfacial chemistry engineered through functionalization. A poorly dispersed network of CNTs, for instance, cannot provide the uniform conductive pathways necessary for efficient, thermally activated healing.

The novelty and innovation of this review lie in its integrated, synergistic framework. It posits that the design of durable CNT materials must be approached as a cohesive process, linking architecture to function. More critically, this review addresses a significant characterization gap that has emerged in the field. The maturation of self-healing materials research demands a shift from purely phenomenological observations of repair (e.g., a crack appears closed under a microscope) to a rigorous, quantitative understanding of the underlying molecular and atomic-level reconstruction. Conventional characterization methods, such as mechanical testing and standard microscopy, reveal the outcome of healing but are often insufficient to elucidate the process—the chemical rebonding, interfacial reconstruction, and atomic rearrangements that constitute true repair.

Therefore, this review provides a unique contribution by being the first to comprehensively bridge the design of self-healing CNT materials with the advanced analytical toolkit required to validate their function at a fundamental level. It will build a logical argument that flows from foundational chemical functionalization and physical assembly, through a detailed examination of intrinsic and extrinsic repair mechanisms, to a critical discussion of the high-powered spectroscopic techniques (e.g., XPS, NEXAFS, EELS, Auger, and in situ Raman) that are essential for probing and proving the efficacy of these processes. By framing the discussion in this manner, this work transforms from a descriptive summary into a prescriptive guide for the rigorous design and analysis of the next generation of durable, intelligent materials.

## 2. Foundational Architectures: Controlled Assembly and Functionalization of CNTs

The macroscopic performance of any CNT-based material is dictated by the precise arrangement and interaction of its nanoscale constituents. This section reviews the critical parameters of CNT assembly—alignment, dispersion, and density—and introduces advanced functionalization as a foundational strategy to enable this architectural control.

### 2.1. Impact of Alignment on Mechanical, Electrical, and Thermal Properties

The alignment of CNTs within a matrix is a pivotal factor in transitioning their anisotropic nanoscale properties to the macroscale. In terms of electrical properties, aligned CNTs create highly efficient, direct pathways for electron transport. This dramatically lowers the electrical percolation threshold—the minimum concentration of conductive filler required to form a connected network [7]. For example, studies on CNT–epoxy composites have shown that aligning the nanotubes can reduce the percolation threshold by more than an order of magnitude, from approximately 0.034 vol% for randomly oriented CNTs to as low as 0.0031 vol% when measured parallel to the alignment direction [8]. This structural precision is a significant advancement for applications in miniaturized electronics, as it allows for the same or better conductivity to be achieved with over 90% less CNT content [8,9].

This architectural control simultaneously enhances mechanical robustness. When aligned, CNTs act as reinforcing fibers, effectively bearing a significant portion of any applied load and hindering the propagation of cracks [10,11]. This leads to measurable improvements in both the elastic modulus and fracture toughness of the composite. The aligned nanotubes force cracks to deviate and follow more tortuous paths, dissipating more energy and increasing the material's resistance to failure [12]. Similarly, thermal conductivity is greatly improved along the direction of alignment, as the ordered structure provides efficient pathways for phonon transport, which is critical for thermal management applications requiring effective heat dissipation [12,13]. Conversely, misalignment leads to increased phonon scattering at nanotube–nanotube interfaces, reducing overall thermal conductivity (see Figure 2) [14].

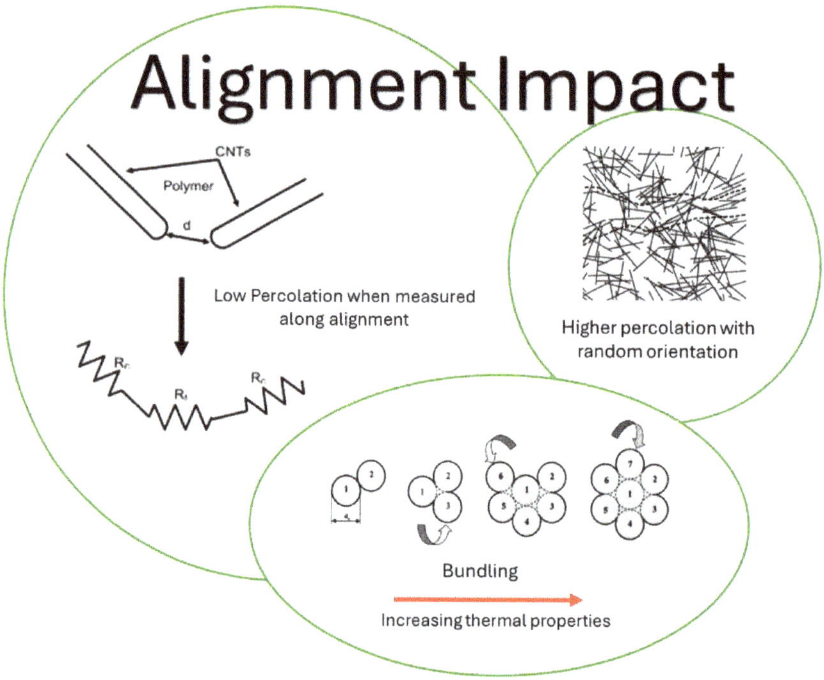


**Figure 2.** Illustration of alignment impact on CNT materials. Adapted from [12,13].

### 2.2. *Influence of Dispersion on Material Characteristics*

Achieving a uniform dispersion of CNTs within a matrix material is fundamental to harnessing their full potential. Due to strong van der Waals forces and $\pi-\pi$ stacking interactions, CNTs have a natural tendency to agglomerate into bundles [15]. These agglomerates act as defects within the composite, creating stress concentration points that can initiate premature failure under a mechanical load and disrupting the formation of continuous networks required for electrical and thermal conductivity [12]. Poor dispersion has been directly linked to reduced mechanical strength, toughness, and conductivity in composites [16,17]. For instance, while well-dispersed CNTs can form a reinforcing meshwork in materials like cement mortar, inadequate dispersion can lead to increased porosity and the creation of weak zones, ultimately diminishing the material's mechanical integrity [7,18]. A homogeneous distribution ensures effective load transfer from the matrix to the high-strength nanotubes and allows for the formation of an efficient, interconnected conductive network at lower CNT concentrations [15].

### 2.3. Effects of Density on Material Performance

The density of CNTs, both in their individual form and in bulk assemblies, is another key parameter governing material performance [19]. The inherently low density of individual CNTs is a primary reason for their exceptionally high specific strength (strength-to-weight ratio), which far surpasses that of traditional materials like high-carbon steel [1]. This characteristic is particularly advantageous in weight-sensitive industries such as aerospace and automotive. At the bulk level, the packing density of CNT assemblies significantly influences their collective properties [20]. For instance, research has shown that the specific thermal conductivity of CNT films and fibers tends to decrease as the bulk density increases, highlighting the complex interplay between individual tube properties and their collective arrangement. Advanced synthesis techniques, such as plasma-enhanced chemical vapor deposition (PECVD), allow for precise control over CNT density by modulating catalyst density and growth conditions [21,22]. This tunability is essential for optimizing performance in specific applications, such as thermal interface materials that require high conductivity at specific densities or field emitters where the density of emission sites is a critical parameter [17,23,24].

### 2.4. Challenges in Achieving Uniform and Scalable Assembly

Despite significant progress, achieving uniform and scalable assembly of CNTs remains a major hurdle. The primary challenge stems from the strong intermolecular forces that cause CNTs to aggregate, making homogeneous dispersion difficult to achieve and maintain [1]. Overcoming these forces is a prerequisite for creating high-performance composites. Furthermore, many laboratory-scale techniques that produce highly ordered CNT structures, such as chemical vapor deposition (CVD) and electrophoresis, face significant challenges when scaling up for industrial production. These methods often require stringent control over process parameters that are difficult to maintain over large areas, rendering them costly and limiting their throughput [25]. The inherent variability in CNTs produced by common synthesis methods, which yield a mixture of diameters and chiralities, further complicates uniform assembly and leads to inconsistencies in the final material's properties [1]. Finally, residual catalyst nanoparticles and amorphous carbon impurities from the synthesis process can contaminate the CNTs, negatively affecting their properties and requiring purification steps that can themselves introduce structural damage [26].

### 2.5. Advanced Functionalization for Enhanced Interfacial Engineering

The challenges in achieving uniform dispersion and assembly are fundamentally rooted in the poor interfacial compatibility between the pristine, hydrophobic surfaces of CNTs and many host matrices, particularly polymers. Chemical functionalization is a powerful strategy to overcome this barrier by chemically modifying the CNT surface to improve its compatibility, thereby enabling more effective assembly [27]. This process is not merely an optional enhancement but often a necessary prerequisite for creating the well-defined architectures discussed previously.

#### 2.5.1. Covalent Functionalization: The Case of Fluorination

Among various covalent functionalization strategies, direct fluorination has emerged as a well-established and highly effective method for altering the surface properties of CNTs [28]. The process involves treating CNTs with gaseous fluorine, often at elevated temperatures, to form covalent C-F bonds on the nanotube sidewalls. This reaction changes the hybridization of the carbon atoms at the bonding sites from graphitic $sp^2$ to diamond-like $sp^3$. The degree of fluorination can be controlled by tuning reaction conditions such as temperature and fluorine concentration, allowing for the synthesis of materials with a desired stoichiometry, such as CFx, where x can be tuned up to ~0.5 while preserving the core tubular structure of the nanotube [18,28,29].

The covalent attachment of fluorine atoms dramatically alters the properties of the CNTs. The introduction of polar C-F bonds significantly increases the hydrophilicity of the nanotubes, rendering them soluble or dispersible in various solvents, including alcohols and water, which is a major advantage of solution-based processing and integration into polymer matrices [28]. This enhanced compatibility directly addresses the challenge of agglomeration, facilitating the creation of the uniform dispersions that are critical for achieving optimal mechanical and electrical properties in the final composite.

#### 2.5.2. Spectroscopic Verification of Surface Chemistry: XPS, NEXAFS, and SXE Analysis

Verifying the success and nature of the chemical modification is a critical step in any functionalization process. A suite of high-powered, surface-sensitive spectroscopic techniques is essential for this task, providing detailed information on the elemental composition and electronic structure of the modified CNTs.

X-ray Photoelectron Spectroscopy (XPS) is a cornerstone technique for surface chemical analysis. By irradiating a sample with X-rays and analyzing the kinetic energy of the emitted photoelectrons, XPS provides quantitative information about the elemental composition and the chemical bonding environment of the near-surface region (~10 nm depth) [30]. In the case of fluorinated CNTs, XPS can unambiguously confirm the presence of fluorine and, through high-resolution analysis of the C 1s and F 1s core-level spectra, can identify the formation of covalent C-F bonds and distinguish them from non-covalently adsorbed fluorine species [28,31].

Near-Edge X-ray Absorption Fine Structure (NEXAFS) Spectroscopy provides complementary information by probing the unoccupied electronic states of the material. This technique is particularly sensitive to the local bonding environment and hybridization of carbon atoms. In pristine CNTs, the NEXAFS spectrum is characterized by a sharp peak corresponding to transitions into unoccupied $\pi*$ orbitals, a hallmark of $sp^2$ hybridization. Upon covalent fluorination, the intensity of this $\pi*$ peak decreases, and new features corresponding to $\sigma*$ orbitals associated with $sp^3$ C-F bonds appear [32]. Studies on fluorinated MWCNTs using NEXAFS have confirmed the formation of covalent C-F bonds and demonstrated that, under optimal conditions, the fluorination process can be uniform throughout the probing depth of the technique (~15 nm) [31,33].

Soft X-ray Emission (SXE) Spectroscopy completes the electronic structural characterization by probing the occupied electronic states (the valence band). When used in conjunction with NEXAFS, which probes the unoccupied states (the conduction band), SXE provides a comprehensive picture of how functionalization modifies the entire electronic band structure of the CNTs. The successful application of these techniques is detailed in the literature, for example, in the works of Krestinin et al. and Brzhezinskaya et al., which provide foundational studies on the production and advanced spectroscopic characterization of fluorinated single-walled and multi-walled CNTs, respectively [28,31].

## 3. Mechanisms and Analysis of Self-Repair in CNT-Based Systems

The ability of a material to autonomously heal damage is a transformative property that can dramatically enhance its durability and reliability. This chapter explores the fundamental mechanisms of self-repair in the context of CNT-based materials, beginning with the inherent repair capabilities of the carbon lattice itself, followed by engineered extrinsic strategies. Critically, it also details the advanced analytical tools required to observe and quantify these healing processes at the molecular level [34,35].

### 3.1. Intrinsic Repair: Atomic-Level Healing in Carbon Lattices

Intrinsic self-repair refers to the inherent ability of a material's structure to heal defects without the need for external healing agents. In CNTs, this capability is rooted in the dynamic nature of the carbon lattice under certain conditions.

#### 3.1.1. Defect Annihilation and Bond Reconfiguration

The carbon atoms in a CNT lattice are not static. When provided with sufficient activation energy, they can rearrange to form more stable configurations. One of the most well-known mechanisms for this is the Stone–Wales transformation, a 90-degree bond rotation that can create or, more importantly for healing, annihilate topological defects like pentagon–heptagon pairs within the hexagonal carbon lattice. Beyond topological defects, the lattice can also heal vacancies and other structural imperfections. This process is driven by external stimuli that provide the necessary energy for atomic migration and bond reconfiguration. High temperatures can supply the thermal energy for atoms to move and fill vacancies, effectively annealing out defects. Similarly, irradiation with electron or ion beams can both induce defect formation and, under controlled conditions, facilitate healing by promoting atomic rearrangement. The local chemical environment can also play a role, for example, by providing atoms to fill vacancies or by altering the energy barriers for atomic rearrangements. However, these intrinsic mechanisms are typically effective only for atomic scale damage and have limitations when faced with larger cracks or harsh operating conditions.

#### 3.1.2. Experimental Probes of Irradiation-Induced Healing: Insights from EELS and Auger Spectroscopy

While theoretical simulations provide valuable insights into atomic-level healing, experimental verification is crucial. Ion irradiation serves as a powerful tool for this purpose, acting dually as a method to controllably introduce defects for study and as an energy source to potentially trigger healing processes. The atomic-scale structural and electronic changes resulting from this process can be probed with high-resolution spectroscopic techniques. Electron energy loss spectroscopy (EELS), typically performed in a scanning transmission electron microscope (STEM), offers the ability to acquire chemical and electronic information with sub-nanometer spatial resolution [36,37]. EELS spectra from carbonaceous materials exhibit distinct features, including a $\pi-\pi*$ plasmon peak at lower energy losses, which is characteristic of sp2-bonded carbon, and a higher-energy $\pi+\sigma$ plasmon peak [38]. The fine structure of the carbon K-edge ionization loss provides detailed information on the unoccupied density of states, allowing for the quantification of the $sp^2/sp^3$ bonding ratio [38]. Ion irradiation induces damage that disrupts the $sp^2$ lattice, leading to a decrease in the energy and intensity of the $\pi$ plasmon and changes in the C K-edge fine structure [38,39]. By monitoring these spectral features as a function of irradiation dose and other conditions (e.g., temperature), EELS can provide direct experimental evidence of the dynamics of defect formation and subsequent atomic-level healing or structural rearrangement. Auger electron spectroscopy (AES) is a complementary, highly surface-sensitive technique that provides information on the elemental composition and chemical bonding environment of the top few atomic layers of a material. The kinetic energy and, more importantly, the shape of the C KVV Auger transition are highly sensitive to the local chemical environment and hybridization of the carbon atoms. The C KVV lineshape serves as a "fingerprint" that can distinguish between graphitic ($sp^2$), diamond-like ($sp^3$), and amorphous carbon [39,40]. Consequently, AES can be used to characterize the nature of the CNT surface before and after ion irradiation. Changes in the C KVV lineshape can reveal the transformation from a well-ordered graphitic structure to a more disordered or $sp^3$-rich state upon damage, and any subsequent recovery of the graphitic signature would provide evidence of intrinsic healing [41]. These experimental approaches, as detailed in studies such as those investigating $\pi$-plasmons in ion-irradiated nanotubes, provide the crucial empirical data needed to validate and understand the fundamental mechanisms of intrinsic repair [39].

### 3.2. *Extrinsic Repair: Engineering Healing into the Composite*

Extrinsic self-repair involves incorporating external healing functionalities into a material that does not possess an inherent healing capacity. This is typically achieved by embedding reservoirs of healing agents within the composite matrix.

#### 3.2.1. Reservoir-Based Healing: Microcapsules and Vascular Networks

The most common extrinsic approach is microencapsulation [1]. In this method, a liquid healing agent, such as the monomer dicyclopentadiene (DCPD), is encapsulated within small, brittle-shelled microcapsules such as urea-formaldehyde. These microcapsules are then dispersed throughout the composite matrix along with a corresponding catalyst, including Grubbs' catalyst [13,42,43]. When a crack propagates through the material, it ruptures the embedded microcapsules, releasing the healing agent into the crack plane via capillary action [44]. The monomer then comes into contact with the dispersed catalyst, initiating a polymerization reaction that bonds the crack faces together, thereby restoring structural integrity [42]. For healing chemistries that require two separate components (e.g., epoxy and a hardener), a dual-microcapsule system can be employed, where each component is encapsulated separately. A more advanced, bio-inspired strategy utilizes vascular networks. These systems consist of interconnected microchannels or hollow fibers embedded within the composite, mimicking biological circulatory systems [45]. These networks can be filled with a healing agent, which is delivered to a damaged site when the network is breached by a crack. The primary advantage of vascular networks over microcapsules is their potential for repeated healing. While microcapsules are a finite resource that is depleted after a single rupture event, vascular networks can be refilled from an external reservoir, allowing for multiple healing cycles at the same damage location [45].

#### 3.2.2. The Multifaceted Role of CNTs in Matrix Reconstruction

Carbon nanotubes can be integrated into extrinsic self-healing systems to play several crucial roles that enhance the repair process and the properties of the healed material [45]. Their contribution goes beyond simple reinforcement and can be tailored to enable more advanced, stimulus-responsive healing mechanisms. First, CNTs serve as potent mechanical reinforcement for the healed region [3,46]. When dispersed within the healing agent (e.g., co-encapsulated within microcapsules), they are delivered to the crack plane along with the monomer. After polymerization, the CNTs are embedded within the newly formed polymer, where they can bridge the crack faces and improve the load transfer across the repaired interface, leading to a higher recovery of mechanical properties like strength and toughness [3,46]. Second, the hollow core of CNTs can be utilized as nanoscale healing agent reservoirs. In this concept, the nanotubes themselves are filled with a healing agent and dispersed in the matrix. Damage that causes the nanotubes to rupture would then trigger the release of the encapsulated agent to initiate repair. Third, and perhaps most significantly, a percolated network of CNTs within a composite matrix can enable stimulus-responsive healing via Joule heating [47,48]. When an electric current is passed through the conductive CNT network, resistance heating (Joule heating) generates thermal energy. This can be used to trigger self-repair in several ways [47,48]. In a thermoplastic matrix, the localized heat can melt the polymer in the vicinity of the crack, allowing it to flow, fill the void, and solidify upon cooling [47,48]. In thermoset matrices incorporating reversible chemical bonds (e.g., Diels–Alder adducts), the heat can provide the activation energy needed to break and reform these bonds across the crack interface, effectively re-welding the material. Finally, the high surface area and unique electronic properties of CNTs give them the potential to act as catalysts for the polymerization of certain healing agents, potentially eliminating the need for a separate, dispersed catalyst and simplifying the material system.

### 3.3. A Modern Toolkit for Characterizing Self-Repair Efficiency

Quantifying the efficiency of a self-repair process requires a multi-faceted characterization approach that can assess the restoration of properties from the macroscopic to the molecular scale.

#### 3.3.1. Mechanical and Morphological Restoration

The most common method for evaluating healing efficiency is through destructive mechanical testing. Techniques such as tensile tests, fracture toughness tests, and fatigue tests are used to compare the mechanical properties of a pristine material with those of a sample that has been damaged and subsequently healed [49,50]. The healing efficiency is typically reported as the percentage of the original property that is recovered [3,46,49,50]. These tests provide a critical measure of the functional restoration of the material. Complementing this, microscopy techniques like scanning electron microscopy (SEM) and atomic force microscopy (AFM) are used for visual assessment. These methods can directly image the damaged area before and after healing, providing qualitative and quantitative information on crack closure and the morphology of the healed region [49,50].

#### 3.3.2. In Situ Monitoring of Healing Dynamics with Raman Spectroscopy

Raman spectroscopy is a non-destructive optical technique that provides detailed information about the vibrational modes of molecules, making it exceptionally well-suited for studying the molecular structure of carbon materials [8,17,51–54]. The Raman spectrum of CNTs is characterized by several key features: the G-band (~1582 $cm^{-1}$), which arises from the in-plane bond stretching of $sp^2$ carbon atoms and is indicative of the graphitic nature of the nanotube; the D-band (~1350 $cm^{-1}$), which is activated by defects, disorder, or the edges of the graphene lattice; and the 2D (or G′) band (~2700 $cm^{-1}$), which is an overtone of the D-band [51–54]. The intensity ratio of the D-band to the G-band (ID/IG) is a widely used and well-established metric for the density of defects in the carbon lattice. This relationship provides a powerful avenue for quantitatively monitoring self-healing. Damage to a CNT-based composite, such as a crack, inherently introduces disorder and breaks bonds, leading to a localized increase in the ID/IG ratio [8,17,51–54]. A successful intrinsic healing event, which involves the reconstruction of the graphitic lattice, should reduce this disorder, resulting in a decrease in the ID/IG ratio. This allows for the development of a quantitative, spectroscopic measure of healing efficiency. By using Raman mapping to scan the damaged area before and after a healing cycle, one can generate a spatially resolved map of the defect density [51–54]. The reduction in the integrated ID/IG ratio across the healed region provides direct, molecular-level evidence of structural repair. For extrinsic systems where healing involves polymerization, Raman spectroscopy can be used in situ to monitor the reaction kinetics by tracking the disappearance of monomer-specific vibrational peaks and the appearance of polymer peaks [8,17].

#### 3.3.3. Interfacial Chemistry in Repaired Zones via High-Powered Spectroscopy

While mechanical tests and Raman spectroscopy can confirm that healing has occurred, a deeper understanding requires probing the chemical nature of the newly formed interface within the repaired zone. The advanced spectroscopic techniques discussed earlier for functionalization analysis are also invaluable for post-mortem analysis of healed materials. After a sample has been damaged and healed, techniques like NEXAFS, EELS, AES, and SXE can be employed to analyze the chemical composition and bonding across the healed crack plane [37]. For instance, these methods can verify that an extrinsic healing agent has successfully formed covalent bonds with the original matrix material, or they can characterize the electronic structure of an intrinsically healed region to confirm the restoration of the graphitic lattice. This level of detailed chemical information is critical for optimizing healing chemistries and understanding failure mechanisms in healed materials, as demonstrated in studies of polymer–nanocarbon composite interfaces [55].

To synthesize and clarify the roles of these advanced techniques, Table 1 provides a comparative overview of their capabilities in the context of self-healing analysis.

**Table 1.** Advanced spectroscopic techniques for characterization of self-healing in CNT composites.

| Technique | Acronym | Information Provided | Application in Self-Healing Analysis | Advantages/Limitations |
|---|---|---|---|---|
| Raman Spectroscopy | - | Vibrational modes, molecular structure, defect density (ID/IG ratio), stress/strain state. | In situ monitoring of polymerization kinetics; quantitative mapping of defect healing; assessing stress transfer at healed interface. | Non-destructive, high spatial resolution/Signal can be weak, with fluorescence interference. |
| X-ray Photoelectron Spectroscopy | XPS | Elemental composition, chemical state, C-F/C-O/C-N bonding environments. | Verifying CNT functionalization; characterizing surface chemistry of fractured and healed surfaces. | Quantitative, high chemical specificity/Surface sensitive (~10 nm), requires vacuum. |
| Near-Edge X-ray Absorption Fine Structure | NEXAFS | Unoccupied electronic states, bond hybridization (sp2/sp3), molecular orientation. | Confirming covalent bond formation during functionalization or healing; probing polymer chain alignment at interfaces. | Sensitive to bonding and orientation/Requires synchrotron source. |
| Electron Energy Loss Spectroscopy | EELS | Elemental composition, electronic structure (plasmons, C K-edge), bonding state. | Atomic-resolution analysis of defects induced by irradiation; characterizing bonding across the healed interface in a TEM. | Very high spatial resolution/Requires electron-transparent samples, potential beam damage. |
| Auger Electron Spectroscopy | AES | Elemental composition, chemical state (from C KVV lineshape). | Characterizing the "fingerprint" of carbon bonding (graphitic vs. disordered) at surfaces before/after damage and healing. | Extremely surface sensitive (~1–3 nm)/Can be destructive, requires vacuum. |
| Soft X-ray Emission Spectroscopy | SXE | Occupied electronic states (valence band). | Provides a complete picture of electronic structural changes when combined with NEXAFS. | Complements NEXAFS/Requires synchrotron source. |

## 4. Synergistic Effects of Molecular Assembly and Self-Repair: Influence of CNT Assembly

The efficiency and effectiveness of a self-repair mechanism are not isolated properties but are intimately linked to the underlying architecture of the CNT network within the material. The deliberate control over CNT assembly, as discussed in Section 2, creates foundational structures that can be leveraged to enhance the transport of healing agents, the accessibility of damage sites, and the overall durability of the composite.

### 4.1. Impact of CNT Alignment on Healing Agent Transport and Diffusion

The assembly of CNTs into ordered architectures, particularly through alignment, offers a promising strategy for enhancing the efficiency of self-repair mechanisms in advanced materials. A key advantage of aligned CNT networks is their capacity to serve as nanoscale conduits for the transport and diffusion of healing agents to sites of material damage. Research in water purification has shown that aligned and encapsulated CNTs with open ends facilitate the flow of water molecules through their interiors [56]. This inherent nanoscale fluid transport capability suggests that aligned CNTs can similarly function as efficient pathways for the delivery of liquid healing agents within a material matrix. Directional alignment enables a more targeted and rapid delivery of these agents compared to the slower, isotropic diffusion characteristic of bulk material matrices. Furthermore, the diameter of CNTs can be tailored to control the transport of specific healing agents based on molecular size, offering an additional degree of control over the self-repair process. The role of CNT alignment in guiding biological entities is also well-documented. For instance, aligned CNT yarn-patterned substrates have been shown to act as topographic scaffolds that direct neurite outgrowth. This demonstrates that the directional guidance provided by aligned CNTs extends beyond fluid transport to influence the movement and growth of micro- and nanoscale entities [57]. In the context of self-repair, this suggests that aligned CNTs might not only facilitate the transport of healing agents but could also guide their subsequent reaction or polymerization at the damage site, leading to a more structured and effective repair. The foundational principle is that uniaxial alignment, whether achieved by hot rubbing, blade coating, or gas flow, enhances directional transport properties parallel to the orientation. The development of advanced deposition techniques, such as those utilizing a vapor-mediated Marangoni effect to create locally aligned networks, further expands the toolkit for creating these functional architectures [58,59]. Moreover, the effectiveness of these aligned networks is predicated on starting with high-purity, single-chirality nanotubes, which can be isolated using advanced, scalable sorting methods like salt-driven aqueous two-phase extraction [60]. This demonstrates that the directional guidance provided by aligned CNTs extends beyond fluid transport to influence the movement and growth of micro- and nanoscale entities. In the context of self-repair, this suggests that aligned CNTs might not only facilitate the transport of healing agents but could also guide their subsequent reaction or polymerization at the damage site, leading to a more structured and effective repair. The mechanical properties of CNT-based materials are strongly influenced by the degree of CNT alignment.

For CNT fibers, tensile strength is closely related to the orientation of constituent CNTs. Achieving a high degree of alignment—often through processes such as wet-spinning followed by drafting and thermal annealing—is critical for maximizing mechanical performance [61]. Similarly, for efficient healing agent transport, a well-aligned and densely packed network of CNTs is essential to provide continuous and unobstructed pathways to the damage. Imperfect alignment can result in tortuous routes and reduced transport efficiency [61]. CNT alignment also plays a critical role in enhancing the efficiency of stimulus-based self-healing mechanisms, particularly those involving Joule heating [48]. By creating efficient conductive pathways, aligned CNTs allow for localized heat generation when an electric field is applied. This heat can activate thermally responsive self-healing mechanisms in the surrounding material. Alignment ensures that electrical current flows more readily through the network, leading to targeted and effective delivery of thermal energy to the damage site compared to randomly dispersed CNT networks [48].

The consistent improvement in performance observed with aligned CNTs across various applications—from fluid transport and biological guidance to enhanced electrical and thermal properties—highlights the fundamental importance of controlling CNT orientation for optimizing material functionalities relevant to self-repair. This control can influence both the direct transport of healing agents through CNT lumens and the material's response to external stimuli used to activate healing mechanisms.

### 4.2. Role of CNT Dispersion in Damage Site Accessibility for Self-Repair Mechanisms

As discussed in Section 2, beyond alignment, the dispersion of CNTs within a material matrix is another critical factor influencing self-repair efficiency, particularly with respect to the accessibility of damage sites for healing agents and mechanisms [62]. Microencapsulation strategies, in which CNTs are suspended within a self-healing monomer, illustrate the importance of dispersion. Uniform distribution of CNTs within the healing agent reservoir ensures that nanoparticles are effectively delivered to the damage site upon capsule rupture. This intimate mixing allows CNTs to reinforce the repaired area or facilitate healing reactions at the crack interface. Poor dispersion can lead to CNT agglomeration, resulting in uneven distribution within the released healing agent and localized property enhancement rather than uniform improvement across the healed region [62]. Research has directly linked the quality of CNT dispersion to the effectiveness of self-healing mechanisms, especially those involving Joule heating [48,62]. A homogeneous dispersion of CNTs leads to a more uniform conductive network throughout the material, which is essential for generating heat evenly across the damaged area when an electric current is applied. This ensures that the entire crack interface experiences the necessary temperature for efficient healing. For example, in composites designed for thermal healing, well-dispersed CNTs act as efficient heat conductors that enable the rapid transfer of thermal energy to a healing agent like polycaprolactone (PCL), accelerating its transport into damaged areas [63].

Conversely, poor dispersion can result in localized CNT concentrations, leading to uneven heating and potentially incomplete or inefficient repair in areas with fewer nanotubes. The importance of dispersion is also highlighted in repair techniques for composite laminates, where functionalized CNTs with improved dispersibility can be transported deep into microcracks via the capillary action of a solvent carrier, ensuring the reinforcing agent reaches the damage site [64]. The dispersibility of CNTs is also influenced by their physical characteristics, such as length. Shorter CNTs tend to disperse more readily in a matrix compared to longer ones, which are more prone to entanglement and aggregation due to increased van der Waals forces. While longer CNTs may be desirable for reinforcement due to their higher aspect ratio, improved dispersion of shorter CNTs enhances damage site accessibility by ensuring a more homogeneous distribution throughout the material. The impact of CNT dispersion on mechanical properties further underscores its importance for self-repair. Properly dispersed CNTs improve load transfer efficiency within a composite material, whereas poor dispersion leading to agglomerated clusters can restrict stress distribution and weaken mechanical properties. Since self-repair aims to restore mechanical integrity, good CNT dispersion is crucial for successful repair. Agglomerated CNTs can act as stress concentration points, potentially leading to re-cracking or incomplete restoration of strength after healing [62]. Achieving uniform and stable CNT dispersion often requires tailored strategies based on the properties of both CNTs and the matrix material. Surface functionalization of CNTs can improve their compatibility with polymer matrices, enhancing dispersibility in aqueous or organic solvents. Specialized assembly techniques, such as electrostatic self-assembly, have been used to promote uniform CNT distribution in materials like cement paste [65]. These approaches highlight the need for careful consideration of interfacial interactions between CNTs and the surrounding matrix to achieve optimal dispersion for effective self-repair.

### *4.3. Demonstrations of Enhanced Durability Through Combined Approaches*

The literature provides numerous compelling case studies that demonstrate the successful integration of carbon nanotube assembly with various self-repair mechanisms, leading to significant enhancements in material durability, lifespan, and healing efficiency. These studies collectively highlight the versatility of CNTs and the synergistic effects achieved by precisely controlling their architecture within different material systems. Table 2 provides a summary of representative case studies.

**Table 2.** Representative studies of CNT assembly and self-repair.

| Reference | Material | CNT Assembly Method | Mechanism | Key |
|---|---|---|---|---|
| Zamal et al., 2020 [66] | Polymer composite | Dispersion in microcapsules | Polymers | Up to 80% restoration of mechanical and electrical properties. |
| Gómez-Sánchez et al., 2024 [50] | Epoxy | Homogeneous dispersion | Joule heating of CNT network | Efficient crack recovery with homogeneous resistive heating, healing over 90% of material. |
| Ouyang et al., 2022 [67] | Carbon fiber polymer | Network in interleaves | Joule heating at interlayer | Delamination repair. |
| Sambandan, 2012 [68] | Integrated circuits | Dispersion in fluid | Electric-field-induced aggregation | Automated open fault repair triggered by the electric field across the gap. |
| Wang et al., 2020 [69] | Liquid crystal | Paving aligned sheets | Joule heating of aligned CNT sheets | Healing of scratches with almost no loss of mechanical properties. |
| Lanzara et al., 2009 [35] | Polymer composite | Nanoreservoirs | Healing agent release from ruptured CNTs | Potential for tougher and automated self-healing materials. |
| Pu et al., 2018 [70] | Polyurethane composite colydimethylsiloxane (PDMS) | Dispersion | Joule heating via retro-Diels–Alder, polymerization | Crack diagnosing and rapid, repeatable self-healing by electricity or NIR light. Healing time reduced from 24 h to 10 min. |
| Zhao et al., 2023 [47] | Fiber-reinforced thermoplastic | Embedded film | Joule heating of CNT film | 87% self-healing efficiency in terms of flexural strength after low-energy impact. |
| Gong et al., 2021 [71] | Asphalt binder | Dispersion | Enhanced thermal conductivity for healing | Up to 652% enhancement of mean square displacement at optimal temperatures. |

These studies illustrate a variety of approaches, including the use of dispersed CNTs to enable Joule heating for thermally activated healing [50], the assembly of CNTs into specific architectures like interleaves for targeted repair in composites [67], and the utilization of the unique structure of CNTs for healing agent storage and delivery [35]. Furthermore, the dynamic assembly of dispersed CNTs in response to an electric field offers a novel solution for self-repair in electronic circuits [68]. The consistent theme across these studies is that by carefully controlling the assembly of CNTs within a material, it is possible to significantly enhance its ability to autonomously or through external stimuli repair damage, leading to improved durability and an extended lifespan.

## 5. Challenges and Future Opportunities in Self-Healing CNT Materials

### 5.1. Advanced Characterization Techniques for Studying Self-Healing Processes at Different Scales

While mechanical characterization techniques effectively quantify the recovery of mechanical properties, specialized methods are needed to assess the restoration of other crucial functionalities, such as electrical and thermal conductivity, as discussed in Section 3.3, are paramount in many CNT applications. Developing techniques to measure these properties in situ during healing and after repeated cycles would offer a more complete evaluation of the self-healing efficiency. Complementing these experimental methods, advanced computational approaches like ReaxFF molecular dynamics and density functional theory (DFT) are critical for investigating microscopic mechanisms that are difficult to observe directly [72]. Such simulations can provide atomic-level insights into the structural evolution of CNTs at elevated temperatures, including defect formation, bond rearrangement, and the transformation of end-cap structures, which are fundamental to understanding intrinsic repair and material stability during thermal healing cycles [72].

Characterizing the interface between CNTs and the matrix remains a significant challenge. Advanced techniques like high-resolution electron microscopy with energy-dispersive X-ray spectroscopy (EDS) or focused ion beam (FIB) milling combined with SEM could provide more detailed information about the interfacial structure and bonding, and how these are affected by damage and repair [73].

Finally, the establishment of standardized protocols and metrics for characterizing self-healing in CNT materials is crucial for ensuring reproducibility and comparability across different research studies. This includes defining clear methods for inducing damage, quantifying the extent of healing (e.g., recovery of mechanical, electrical, and thermal properties), and assessing the durability of the self-healing capability over multiple cycles [73].

### 5.2. Investigating the Long-Term Durability and Performance After Repeated Self-Repair Cycles

The long-term performance of self-healing CNT materials requires careful consideration of the healing efficiency over numerous cycles. While initial healing might be highly effective, the ability to maintain this level of repair over prolonged use is crucial. Recent studies have demonstrated a remarkable performance over multiple cycles [74,75]. For instance, a CNT/epoxy composite toughened with polycaprolactone (PCL) not only recovered its properties but also exhibited tensile and flexural strength increases of over 15% and 25%, respectively, after a second and third repair cycle under 40% prefabricated damage, with healing efficiencies reaching as high as 116% [63]. Similarly, repair of delaminated carbon/glass composites using resin pre-coating with aminated CNTs restored up to 30% of the compressive strength and significantly increased tensile (19%) and flexural (29%) loads compared to the damaged state [64]. Factors such as the completeness of the repair at the molecular level, demonstrated by the filling of 2–4 μm cracks, and the long-term stability of the doping in functional composites, which has been shown to be stable for over 60 h in some systems, are critical indicators of durability that must be assessed [63,64]. Studies have demonstrated the potential for repeated healing in composite materials. For instance, fiberglass composites with a 3D vascular system have shown autonomous and repeated healing over multiple cycles with high efficiency. Similarly, continuous carbon fiber-reinforced composites with a thermoplastic matrix have been healed numerous times, although with a slight decrease in healing efficiency in subsequent cycles [42]. Epoxy matrix composites have also exhibited self-repair capabilities, but some have shown reduced efficiency after multiple repair cycles [76]. Intrinsic self-healing mechanisms, relying on reversible interactions within the material, offer the potential for numerous healing events, as shown in Table 2. In contrast, extrinsic mechanisms that depend on encapsulated healing agents are limited by the amount of available agent.

The long-term performance of self-healing CNT materials requires careful consideration of the healing efficiency over numerous cycles. While initial healing might be highly effective, the ability to maintain this level of repair over prolonged use is crucial. Factors such as the completeness of the repair at the molecular level, the potential for fatigue of the self-healing mechanism, and the accumulation of byproducts from the repair process can all influence the long-term durability. Self-healing CNT materials intended for use in harsh environments, such as deep space or oceans, will need to demonstrate robustness and the ability to maintain their self-healing capabilities under extreme conditions [25]. Studies on CNT composite elastomers with dynamic bonds have shown promise for long-term stability in applications like electrophysiological signal acquisition [46]. Additionally, certain CNT composites have exhibited good bending durability over a large number of cycles, indicating their potential for long-term performance under mechanical stress [77].

The long-term effectiveness of self-healing CNT materials is a critical area for future research. Evaluating the healing efficiency over many cycles, understanding the degradation mechanisms of the self-healing capability, and assessing performance under realistic operating conditions are essential steps towards realizing the full potential of these materials for extended service life and enhanced reliability. Potential applications of these CNT materials are outlined in Table 3 and Figure 3.

**Table 3.** Examples of applications of self-healing CNT materials.

| Application Area | Application Examples | Mechanisms Involved | Reference |
|---|---|---|---|
| Energy Storage | Electrodes in batteries and supercapacitors (flexible, solid-state) | CNT network rearrangement in self-healing polymer matrix | Ibrahim et al., 2025 [49] |
| Energy Storage | Joule heating for localized repair in epoxy/CNT systems | Thermal activation by the Joule effect | Gómez-Sánchez et al., 2024 [50] |
| Sensors | Strain and pressure sensors, electronic skin | CNT network sensing and matrix repair; dynamic cross-linking and hydrogen bonding in hydrogels | Del Bosque et al., 2022 [78] |
| Biomedical | Protective coatings for implants | Extrinsic self-healing using encapsulated agents | Choi et al., 2023 [79] |
| Biomedical | Drug delivery systems | Potential enhancement of delivery vehicle durability | Zhang et al., 2010 [80] |
| Aerospace/Automotive | Self-sensing structural components | Potential integration of self-healing for damage repair | Dong et al., 2021 [81] |

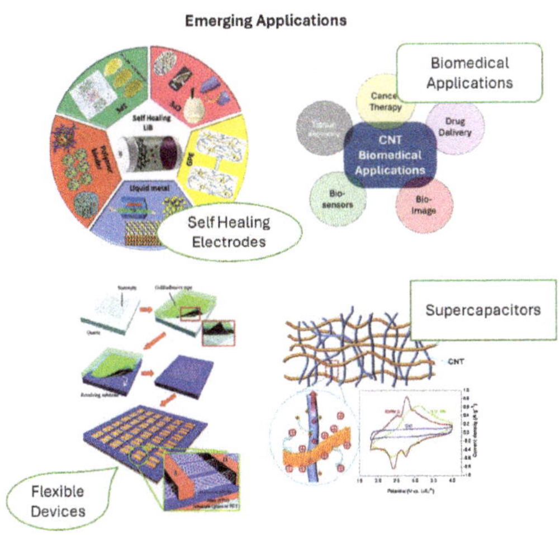


**Figure 3.** Illustrations for application of CNT self-healing materials, adapted from [82–85].

## 6. Conclusions

In conclusion, the strategic integration of controlled carbon nanotube (CNT) assembly with self-repair mechanisms represents a frontier in materials science, offering a clear pathway toward creating materials with unprecedented durability and reliability. This review has established that the precise architectural control of CNTs, encompassing their alignment, dispersion, and functionalization, is not merely a method for property enhancement but a foundational prerequisite for the successful implementation of efficient self-repair processes. The synergy between the initial material design and its subsequent ability to heal damage is a central theme that must guide future research and development. Significant progress has been made in developing a diverse array of assembly techniques and both intrinsic and extrinsic self-healing strategies. However, persistent challenges related to the scalability and cost-effectiveness of these methods, as well as achieving perfect, uniform CNT arrangements, currently limit their widespread industrial adoption. Overcoming these engineering hurdles is critical for translating laboratory innovations into real-world applications. The future of self-healing materials hinges on moving beyond phenomenological observations to a deep, quantitative understanding of the molecular-level repair processes. This necessitates the broader adoption and development of advanced, in situ characterization techniques, such as high-powered spectroscopy (XPS, NEXAFS, EELS, AES) and quantitative Raman mapping. These tools are essential for validating healing mechanisms, optimizing material chemistry, and providing the rigorous data needed to build predictive models for material lifetime and performance. Future research should be directed toward creating multifunctional materials that combine self-healing with other intelligent capabilities, such as self-sensing for damage detection or integrated energy storage. By fostering interdisciplinary collaboration and leveraging the unique versatility of CNTs to enable and enhance self-repair, the scientific community can unlock the full potential of these systems. Ultimately, the synergy between CNT assembly and self-repair, validated by advanced characterization, holds immense promise for creating a new generation of innovative, adaptive materials capable of transforming sectors ranging from aerospace and energy to medicine and consumer electronics.

**Author Contributions:** M.R.S. performed the literature review and writing. S.R. performed writing and made the figures. R.F.L.L. performed writing and editing. M.F.-G. performed editing and provided funding. All authors have read and agreed to the published version of the manuscript.

**Funding:** The authors gratefully acknowledge the financial support of the Natural Sciences and Engineering Research Council of Canada (NSERC) through the Discovery Grants Program [Grant No. RGPIN-1512].

**Acknowledgments:** During the preparation of this manuscript/study, the author(s) used Gemini 2.2 to edit grammar and overall English. The authors have reviewed and edited the output and take full responsibility for the content of this publication.

**Conflicts of Interest:** The authors declare no conflicts of interest.

## References

1. Hughes, K.J.; Iyer, K.A.; Bird, R.E.; Ivanov, J.; Banerjee, S.; Georges, G.; Zhou, Q.A. Review of Carbon Nanotube Research and Development: Materials and Emerging Applications. *ACS Appl. Nano Mater.* **2024**, *7*, 18695–18713. [CrossRef]
2. Sultana, N.; Dewey, H.M.; Arellano, A.G.; Budhathoki-Uprety, J. Understanding the Molecular Assemblies of Single Walled Carbon Nanotubes and Tailoring their Photoluminescence for the Next-Generation Optical Nanosensors. *Chem. Mater.* **2024**, *36*, 4034–4053. [CrossRef]
3. Shabnum, S.S.; Siranjeevi, R.; Raj, C.K.; Nivetha, P.; Benazir, K. A Comprehensive Review on Recent Progress in Carbon Nanotubes for Biomedical Application. *Environ. Qual. Manag.* **2025**, *34*, e70040. [CrossRef]

4. Wulf, V.; Bisker, G. Integrating Single-Walled Carbon Nanotubes into Supramolecular Assemblies: From Basic Interactions to Emerging Applications. *ACS Nano* **2024**, *18*, 29380–29393. [CrossRef]
5. Snowdon, M.R.; Sukhanova, E.V.; Popov, Z.I.; Li, S.; Nurdiwijayanto, L.; Taniguchi, T.; Ishihara, S.; Tanaka, T.; Kataura, H.; Tsukagoshi, K.; et al. Iptycene-Assisted Alignment of Chirality-Sorted SWCNTs for Field-Effect Transistors. *ACS Appl. Nano Mater.* **2025**, *8*, 944–951. [CrossRef]
6. Kausar, A.; Ahmad, I.; Maaza, M.; Bocchetta, P. Self-Healing Nanocomposites—Advancements and Aerospace Applications. *J. Compos. Sci.* **2023**, *7*, 148. [CrossRef]
7. Hu, Z.; Hong, H. Review on Material Performance of Carbon Nanotube-Modified Polymeric Nanocomposites. *Recent Prog. Mater.* **2023**, *5*, 031. [CrossRef]
8. Khan, S.U.; Pothnis, J.R.; Kim, J.-K. Effects of carbon nanotube alignment on electrical and mechanical properties of epoxy nanocomposites. *Compos. Part A Appl. Sci. Manuf.* **2013**, *49*, 26–34. [CrossRef]
9. Souier, T.; Santos, S.; Al Ghaferi, A.; Stefancich, M.; Chiesa, M. Enhanced electrical properties of vertically aligned carbon nanotube-epoxy nanocomposites with high packing density. *Nanoscale Res. Lett.* **2012**, *7*, 630. [CrossRef] [PubMed]
10. Kulakov, V.; Aniskevich, A.; Ivanov, S.; Poltimae, T.; Starkova, O. Effective electrical conductivity of carbon nanotube–epoxy nanocomposites. *J. Compos. Mater.* **2017**, *51*, 2979–2988. [CrossRef]
11. Chen, J. Effects of Carbon Nanotube Structure, Purity, and Alignment on the Heat Conduction Properties of Carbon Films and Fibers. *Online J. Mater. Sci.* **2022**, *1*, 1–11. [CrossRef]
12. Pop, E.; Mann, D.; Wang, Q.; Goodson, K.; Dai, H. Thermal Conductance of an Individual Single-Wall Carbon Nanotube above Room Temperature. *Nano Lett.* **2006**, *6*, 96–100. [CrossRef]
13. Jakubinek, M.B.; Johnson, M.B.; White, M.A.; Jayasinghe, C.; Li, G.; Cho, W.; Schulz, M.J.; Shanov, V. Thermal and electrical conductivity of array-spun multi-walled carbon nanotube yarns. *Carbon* **2012**, *50*, 244–248. [CrossRef]
14. Mannino, G.; Condorelli, M.; Compagnini, G.; Faraci, G. Raman amplification for trapped radiation in crystalline single Si nanoparticle. *Sci. Rep.* **2023**, *13*, 1014. [CrossRef]
15. Zhang, D.; Huang, Y.; Chia, L. Effects of carbon nanotube (CNT) geometries on the dispersion characterizations and adhesion properties of CNT reinforced epoxy composites. *Compos. Struct.* **2022**, *296*, 115942. [CrossRef]
16. Siddique, R.; Mehta, A. Effect of carbon nanotubes on properties of cement mortars. *Constr. Build. Mater.* **2014**, *50*, 116–129. [CrossRef]
17. Feijoo, I.; Pena, G.; Cristóbal, M.J.; Cabeza, M.; Rey, P. Effect of Carbon Nanotube Content and Mechanical Milling Conditions on the Manufacture of AA7075/MWCNT Composites. *Metals* **2022**, *12*, 1020. [CrossRef]
18. Maune, H.T.; Han, S.-p.; Barish, R.D.; Bockrath, M.; Iii, W.A.G.; Rothemund, P.W.K.; Winfree, E. Self-assembly of carbon nanotubes into two-dimensional geometries using DNA origami templates. *Nat. Nanotechnol.* **2010**, *5*, 61–66. [CrossRef]
19. Nanda, J.; Maranville, C.; Bollin, S.C.; Sawall, D.; Ohtani, H.; Remillard, J.T.; Ginder, J.M. Thermal Conductivity of Single-Wall Carbon Nanotube Dispersions: Role of Interfacial Effects. *J. Phys. Chem. C* **2008**, *112*, 654–658. [CrossRef]
20. Kotagiri, N.; Kim, J.W. Stealth nanotubes: Strategies of shielding carbon nanotubes to evade opsonization and improve biodistribution. *Int. J. Nanomed.* **2014**, *9* (Suppl. 1), 85–105. [CrossRef]
21. Tayyab, S.; Apponi, A.; Betti, M.G.; Blundo, E.; Cavoto, G.; Frisenda, R.; Jiménez-Arévalo, N.; Mariani, C.; Pandolfi, F.; Polimeni, A.; et al. Spectromicroscopy Study of Induced Defects in Ion-Bombarded Highly Aligned Carbon Nanotubes. *Nanomaterials* **2024**, *14*, 77. [CrossRef]
22. De Luca, P.; Siciliano, C.; B.Nagy, J.; Macario, A. The role of carbon nanotubes in the reactions of heterogeneous catalysis. *Chem. Eng. Res. Des.* **2023**, *197*, 74–84. [CrossRef]
23. Spièce, J.; Lulla, K.; Picquendaele, P.d.C.d.; Divay, L.; Bezencenet, O.; Hackens, B.; Gehring, P.; Robson, A.J.; Evangeli, C.; Kolosov, O.V. Nanoscale Heat Transport of Vertically Aligned Carbon Nanotube Bundles for Thermal Management Applications. *ACS Appl. Mater. Interfaces* **2024**, *16*, 45671–45677. [CrossRef]
24. Zhao, C.; Zhang, P.; Shi, R.; Xu, Y.; Zhang, L.; Fang, R.; Zhao, T.; Qi, S.; Jiang, L.; Liu, M. Super-tough and strong nanocomposite fibers by flow-induced alignment of carbon nanotubes on grooved hydrogel surfaces. *Sci. China Mater.* **2019**, *62*, 1332–1340. [CrossRef]
25. Zecchi, S.; Cristoforo, G.; Piatti, E.; Torsello, D.; Ghigo, G.; Tagliaferro, A.; Rosso, C.; Bartoli, M. A Concise Review of Recent Advancements in Carbon Nanotubes for Aerospace Applications. *Micromachines* **2025**, *16*, 53. [CrossRef]
26. Mohd Nurazzi, N.; Asyraf, M.R.M.; Khalina, A.; Abdullah, N.; Sabaruddin, F.A.; Kamarudin, S.H.; Ahmad, S.; Mahat, A.M.; Lee, C.L.; Aisyah, H.A.; et al. Fabrication, Functionalization, and Application of Carbon Nanotube-Reinforced Polymer Composite: An Overview. *Polymers* **2021**, *13*, 1047. [CrossRef] [PubMed]
27. Jeong, D.-R.; Yoon, S. Characterization of thin-walled self-healing microcapsules reinforced with multi-walled carbon nanotubes. *Adv. Compos. Mater.* **2024**, 1–18. [CrossRef]

28. Krestinin, A.V.; Kharitonov, A.P.; Shul'ga, Y.M.; Zhigalina, O.M.; Knerel'man, E.I.; Dubois, M.; Brzhezinskaya, M.M.; Vinogradov, A.S.; Preobrazhenskii, A.B.; Zvereva, G.I.; et al. Fabrication and characterization of fluorinated single-walled carbon nanotubes. *Nanotechnologies Russ.* **2009**, *4*, 60–78. [CrossRef]
29. Tulevski, G.S.; Hannon, J.; Afzali, A.; Chen, Z.; Avouris, P.; Kagan, C.R. Chemically Assisted Directed Assembly of Carbon Nanotubes for the Fabrication of Large-Scale Device Arrays. *J. Am. Chem. Soc.* **2007**, *129*, 11964–11968. [CrossRef]
30. Speranza, G. Characterization of Carbon Nanostructures by Photoelectron Spectroscopies. *Materials* **2022**, *15*, 4434. [CrossRef] [PubMed]
31. Brzhezinskaya, M.M.; Vinogradov, N.A.; Zimina, A.; Muradyan, V.E.; Shul'ga, Y.M.; Vinogradov, A.S. Characterization of fluorinated multiwalled carbon nanotubes with X-ray absorption, photoelectron and emission spectroscopies. *Appl. Phys. A* **2009**, *94*, 445–448. [CrossRef]
32. Song, T.; Jiang, B.; Li, Y.; Ji, Z.; Zhou, H.; Jiang, D.; Seok, I.; Murugadoss, V.; Wen, N.; Colorado, H. Self-healing Materials: A Review of Recent Developments. *ES Mater. Manuf.* **2021**, *14*, 1–19. [CrossRef]
33. Jamil, M.S.S.; Endot, N.A. Influence of Fluorine Substituents on the Electronic Properties of Selenium-N-Heterocyclic Carbene Compounds. *Molecules* **2020**, *25*, 5161. [CrossRef] [PubMed]
34. Kausar, A. Self-healing polymer/carbon nanotube nanocomposite: A review. *J. Plast. Film. Sheeting* **2021**, *37*, 160–181. [CrossRef]
35. Lanzara, G.; Yoon, Y.; Liu, H.; Peng, S.; Lee, W. Carbon nanotube reservoirs for self-healing materials. *Nanotechnology* **2009**, *20*, 335704. [CrossRef]
36. Lazar, S.; Tiemeijer, P.; Schnohr, C.S.; Meledina, M.; Patzig, C.; Höche, T.; Longo, P.; Freitag, B. Enabling electron-energy-loss spectroscopy at very high energy losses: An opportunity to obtain X-ray absorption spectroscopy--like information using an electron microscope. *Phys. Rev. Appl.* **2025**, *23*, 054095. [CrossRef]
37. Silva, S.R.P.; Stolojan, V. Electron energy loss spectroscopy of carbonaceous materials. *Thin Solid Film.* **2005**, *488*, 283–290. [CrossRef]
38. Zhang, Y.; Chen, L.; Xu, Z.; Li, Y.; Shan, M.; Liu, L.; Guo, Q.; Chen, G.; Wang, Z.; Wang, C. Effects of ion irradiation on carbon nanotubes: A review. *Int. J. Mater. Prod. Technol.* **2012**, *45*, 1–30. [CrossRef]
39. Brzhezinskaya, M.M.; Baitinger, E.M.; Shnitov, V.V. π-plasmons in ion-irradiated multiwall carbon nanotubes. *Phys. B Condens. Matter* **2004**, *348*, 95–100. [CrossRef]
40. Korusenko, P.M.; Nesov, S.N.; Povoroznyuk, S.N.; Bolotov, V.V.; Knyazev, E.V.; Pushkarev, A.I.; Smirnov, D. Modifying the structure of multiwalled carbon nanotubes with continuous and pulsed ion beams. *Phys. Solid State* **2018**, *60*, 2616–2622. [CrossRef]
41. Bolotov, V.; Korusenko, P.; Nesov, S.; Povoroznyuk, S. Effect of pulsed ion irradiation on the electronic structure of multi-walled carbon nanotubes. *Phys. Solid State* **2014**, *56*, 835–838. [CrossRef]
42. Patrick, J.F.; Hart, K.R.; Krull, B.P.; Diesendruck, C.E.; Moore, J.S.; White, S.R.; Sottos, N.R. Continuous Self-Healing Life Cycle in Vascularized Structural Composites. *Adv. Mater.* **2014**, *26*, 4302–4308. [CrossRef]
43. Kessler, M.R.; Sottos, N.R.; White, S.R. Self-healing structural composite materials. *Compos. Part A Appl. Sci. Manuf.* **2003**, *34*, 743–753. [CrossRef]
44. White, S.R.; Sottos, N.R.; Geubelle, P.H.; Moore, J.S.; Kessler, M.R.; Sriram, S.; Brown, E.N.; Viswanathan, S. Autonomic healing of polymer composites. *Nature* **2001**, *409*, 794–797. [CrossRef]
45. Wang, C.; Boulatov, R. Autonomic Self-Healing of Polymers: Mechanisms, Applications, and Challenges. *Molecules* **2025**, *30*, 469. [CrossRef]
46. Tang, M.; Lei, K.; Zhao, X.; Hu, X.; He, Q.; Zhang, K.; Ma, X.; Ni, H.; Shu, Y.; Li, Z. Self-healing and highly adhesive conductive polydimethylsiloxane-based elastomers for chronic epilepsy monitoring. *Nanoscale* **2025**, *17*, 8624–8633. [CrossRef]
47. Zhao, Y.; Liu, Q.; Li, R.; Lomov, S.V.; Abaimov, S.G.; Xiong, K.; Zhang, H.; Chen, C.; Wu, Q. Self-sensing and self-healing innovative fiber-reinforced thermoplastic composite embedded with CNT film. *J. Intell. Mater. Syst. Struct.* **2023**, *34*, 1561–1571. [CrossRef]
48. Jang, S.H.; Kim, D.; Park, Y.L. Accelerated Curing and Enhanced Material Properties of Conductive Polymer Nanocomposites by Joule Heating. *Materials* **2018**, *11*, 1775. [CrossRef]
49. Ibrahim, O.O.; Oluwadunsin, J.D.; Antwi, M.; Mekunye, F.; Oluwatobi, M.A.; Olorunfemi, A.D.; Obanla, O.R.; Abdul-Rahman, M.I.; Babalola, K.O.; Babalola, O. The Evolution of Self-Healing Electrodes: A Critical Review of Nanomaterial Contributions. *Am. J. Nanosci.* **2025**, *9*, 8–31. [CrossRef]
50. Gómez-Sánchez, J.; Fernández Sánchez-Romate, X.X.; Jiménez-Suárez, A.; Prolongo, S.G. Self-Healing Activation by Conventional Resistive Heating through the Addition of Carbon Nanotubes in Epoxy Systems Based on Covalent Adaptable Networks. *ACS Appl. Polym. Mater.* **2024**, *6*, 1106–1115. [CrossRef]
51. Krasnikov, D.V.; Bokova-Sirosh, S.N.; Tsendsuren, T.-O.; Romanenko, A.I.; Obraztsova, E.D.; Volodin, V.A.; Kuznetsov, V.L. Influence of the Growth Temperature on the Defective Structure of the Multi-Walled Carbon Nanotubes. *Phys. Status Solidi* **2018**, *255*, 1700255. [CrossRef]

52. Jackson, E.M.; Laibinis, P.E.; Collins, W.E.; Ueda, A.; Wingard, C.D.; Penn, B. Development and thermal properties of carbon nanotube-polymer composites. *Compos. Part B Eng.* **2016**, *89*, 362–373. [CrossRef]
53. Lutz, T.; Donovan, K.J. Macroscopic scale separation of metallic and semiconducting nanotubes by dielectrophoresis. *Carbon* **2005**, *43*, 2508–2513. [CrossRef]
54. Krupke, R.; Hennrich, F.; Löhneysen, H.; Kappes, M.M. Separation of Metallic from Semiconducting Single-Walled Carbon Nanotubes. *Science* **2003**, *301*, 344–347. [CrossRef]
55. Pozdnyakov, A.O.; Brzhezinskaya, M.M.; Vinogradov, A.S.; Friedrich, K. NEXAFS Spectra of Polymer-nanocarbon Composites. *Fuller. Nanotub. Carbon Nanostruct.* **2008**, *16*, 471–474. [CrossRef]
56. Arora, B.; Attri, P. Carbon nanotubes (CNTs): A potential nanomaterial for water purification. *J. Compos. Sci.* **2020**, *4*, 135. [CrossRef]
57. Fan, L.; Feng, C.; Zhao, W.; Qian, L.; Wang, Y.; Li, Y. Directional neurite outgrowth on superaligned carbon nanotube yarn patterned substrate. *Nano Lett.* **2012**, *12*, 3668–3673. [CrossRef] [PubMed]
58. Rodriguez-Martinez, X.; Hawkey, A.; Bohnet, K.; Lindenthal, S.; Zaumseil, J. Composites of Isotropic and Aligned Semiconducting Single-Walled Carbon Nanotubes with Conjugated Polymers for Air-Processed Thermoelectrics. *J. Mater. Chem. A* **2025**, *13*, 28129–28139. [CrossRef]
59. Li, H.; Zheng, M.; Fagan, J.A. Precise Partitioning of Metallic Single-Wall Carbon Nanotubes and Enantiomers through Aqueous Two-Phase Extraction. *ACS Nano* **2025**, *19*, 14137–14149. [CrossRef] [PubMed]
60. Lyu, M.; Li, C.; Liu, Y.; Li, Y.; Zheng, M. A salt-driven mechanism for precise chirality sorting of carbon nanotubes. *Sci. Adv.* **2025**, *11*, eadx3958. [CrossRef] [PubMed]
61. Tang, Z.; Huang, Q.; Liu, Y.; Chen, Y.; Guo, B.; Zhang, L. Uniaxial Stretching-Induced Alignment of Carbon Nanotubes in Cross-Linked Elastomer Enabled by Dynamic Cross-Link Reshuffling. *ACS Macro Lett.* **2019**, *8*, 1575–1581. [CrossRef] [PubMed]
62. Naveen, V.; Deshpande, A.P.; Raja, S. Self-healing microcapsules encapsulated with carbon nanotubes for improved thermal and electrical properties. *RSC Adv.* **2020**, *10*, 33178–33188. [CrossRef]
63. Xian, G.; Qi, X.; Shi, J.; Tian, J.; Xiao, H. Toughened and self-healing carbon nanotube/epoxy resin composites modified with polycaprolactone filler for coatings, adhesives and FRP. *J. Build. Eng.* **2025**, *111*, 113207. [CrossRef]
64. Duan, L.; Shen, W.; Chen, L.; Zhu, L. Resin precoating solutions containing aminated carbon nanotubes for the improvement of mechanical characteristics and repair behavior of carbon/glass composite laminates. *J. Compos. Mater.* **2025**, 00219983251341617. [CrossRef]
65. Wang, L.; Ge, Z.; Zhang, N.; Feng, Y.; Ling, Y.; Zhang, H. Electrostatic self-assembly effect of Fe3O4 nanoparticles on performance of carbon nanotubes in cement-based materials. *Nanotechnol. Rev.* **2024**, *13*, 20230209. [CrossRef]
66. Zamal, H.H.; Barba, D.; Aïssa, B.; Haddad, E.; Rosei, F. Recovery of electro-mechanical properties inside self-healing composites through microencapsulation of carbon nanotubes. *Sci. Rep.* **2020**, *10*, 2973. [CrossRef]
67. Ouyang, Q.; Liu, L.; Wu, Z. Electrothermally Self-Healing Delamination Cracks in Carbon/Epoxy Composites Using Sandwich and Tough Carbon Nanotube/Copolymer Interleaves. *Polymers* **2022**, *14*, 4313. [CrossRef]
68. Sambandan, S. Self-Repair in Circuits—Automating Open Fault Repair in Integrated Circuits Using Field-Induced Aggregation of Carbon Nanotubes. *IEEE Trans. Electron Devices* **2012**, *59*, 1773–1779. [CrossRef]
69. Wang, H.; Yang, Y.; Zhang, M.; Wang, Q.; Xia, K.; Yin, Z.; Wei, Y.; Ji, Y.; Zhang, Y. Electricity-Triggered Self-Healing of Conductive and Thermostable Vitrimer Enabled by Paving Aligned Carbon Nanotubes. *ACS Appl. Mater. Interfaces* **2020**, *12*, 14315–14322. [CrossRef]
70. Pu, W.; Fu, D.; Wang, Z.; Gan, X.; Lu, X.; Yang, L.; Xia, H. Realizing Crack Diagnosing and Self-Healing by Electricity with a Dynamic Crosslinked Flexible Polyurethane Composite. *Adv. Sci.* **2018**, *5*, 1800101. [CrossRef]
71. Gong, Y.; Xu, J.; Yan, E.-H.; Cai, J.-H. The Self-Healing Performance of Carbon-Based Nanomaterials Modified Asphalt Binders Based on Molecular Dynamics Simulations. *Front. Mater.* **2021**, *7*, 599551. [CrossRef]
72. Hou, F.-C.; Wu, X.-H.; Lin, Z.-C.; Chang, J.-B.; Ma, X.; Fang, T.; Sun, J.; Song, L. Formation behavior of single-walled carbon nanotube end-caps at high temperatures: A combined ReaxFF MD with DFT study. *Comput. Mater. Sci.* **2025**, *258*, 114025. [CrossRef]
73. Sabet, M. Advanced Functionalization Strategies for Carbon Nanotube Polymer Composites: Achieving Superior Dispersion and Compatibility. *Polym.-Plast. Technol. Mater.* **2025**, *64*, 465–494. [CrossRef]
74. Cui, K.; Chang, J.; Feo, L.; Chow, C.L.; Lau, D. Developments and Applications of Carbon Nanotube Reinforced Cement-Based Composites as Functional Building Materials. *Front. Mater.* **2022**, *9*, 861646. [CrossRef]
75. Paladugu, S.R.M.; Sreekanth, P.S.R.; Sahu, S.K.; Naresh, K.; Karthick, S.A.; Venkateshwaran, N.; Ramoni, M.; Mensah, R.A.; Das, O.; Shanmugam, R. A Comprehensive Review of Self-Healing Polymer, Metal, and Ceramic Matrix Composites and Their Modeling Aspects for Aerospace Applications. *Materials* **2022**, *15*, 8521. [CrossRef]
76. Zhu, Y.-F.; Ma, C.; Zhang, W.; Zhang, R.-P.; Koratkar, N.; Liang, J. Alignment of multiwalled carbon nanotubes in bulk epoxy composites via electric field. *J. Appl. Phys.* **2009**, *105*, 054319. [CrossRef]

77. Joh, H.; Anoop, G.; Jo, S.; Seol, W.; Park, S.M.; Eom, K.; Lee, H.J.; Jo, J.Y. Enhanced Flexible Thermoelectric Performance in Polymer–CNT Composites via Carbon Nanotube Alignment. *ACS Appl. Energy Mater.* **2025**, *8*, 3178–3184. [CrossRef]
78. Del Bosque, A.; Calderón-Villajos, R.; Sánchez, M.; Ureña, A. Multifunctional Carbon Nanotubes-Reinforced Surlyn Nanocomposites: A Study of Strain-Sensing and Self-Healing Capabilities. *Nanomaterials* **2022**, *12*, 2878. [CrossRef]
79. Choi, K.; Noh, A.; Kim, J.; Hong, P.H.; Ko, M.J.; Hong, S.W. Properties and Applications of Self-Healing Polymeric Materials: A Review. *Polymers* **2023**, *15*, 4408. [CrossRef]
80. Zhang, Y.; Bai, Y.; Yan, B. Functionalized carbon nanotubes for potential medicinal applications. *Drug Discov. Today* **2010**, *15*, 428–435. [CrossRef] [PubMed]
81. Dong, Z.-J.; Sun, B.; Zhu, H.; Yuan, G.-M.; Li, B.-L.; Guo, J.-G.; Li, X.-K.; Cong, Y.; Zhang, J. A review of aligned carbon nanotube arrays and carbon/carbon composites: Fabrication, thermal conduction properties and applications in thermal management. *New Carbon Mater.* **2021**, *36*, 873–892. [CrossRef]
82. Ezeigwe, E.R.; Dong, L.; Manjunatha, R.; Tan, M.; Yan, W.; Zhang, J. A review of self-healing electrode and electrolyte materials and their mitigating degradation of Lithium batteries. *Nano Energy* **2021**, *84*, 105907. [CrossRef]
83. Chen, Z.; Augustyn, V.; Wen, J.; Zhang, Y.; Shen, M.; Dunn, B.; Lu, Y. High-performance supercapacitors based on intertwined CNT/$V_2O_5$ nanowire nanocomposites. *Adv. Mater.* **2011**, *23*, 791–795. [CrossRef]
84. Ishikawa, F.N.; Chang, H.K.; Ryu, K.; Chen, P.C.; Badmaev, A.; Gomez De Arco, L.; Shen, G.; Zhou, C. Transparent electronics based on transfer printed aligned carbon nanotubes on rigid and flexible substrates. *ACS Nano* **2009**, *3*, 73–79. [CrossRef] [PubMed]
85. Murjani, B.O.; Kadu, P.S.; Bansod, M.; Vaidya, S.S.; Yadav, M.D. Carbon nanotubes in biomedical applications: Current status, promises, and challenges. *Carbon Lett.* **2022**, *32*, 1207–1226. [CrossRef] [PubMed]

*nanomaterials*

MDPI

*Review*

# Machine Learning as a "Catalyst" for Advancements in Carbon Nanotube Research

**Guohai Chen [1,*] and Dai-Ming Tang [2,3,*]**

[1] Nano Carbon Device Research Center, National Institute of Advanced Industrial Science and Technology (AIST), Tsukuba Central 5, 1-1-1 Higashi, Tsukuba 305-8565, Japan
[2] Research Center for Materials Nanoarchitectonics (MANA), National Institute for Materials Science (NIMS), Tsukuba 305-0044, Japan
[3] Institute of Pure and Applied Sciences, University of Tsukuba, Tsukuba 305-8571, Japan
* Correspondence: guohai-chen@aist.go.jp (G.C.); tang.daiming@nims.go.jp (D.-M.T.)

**Abstract:** The synthesis, characterization, and application of carbon nanotubes (CNTs) have long posed significant challenges due to the inherent multiple complexity nature involved in their production, processing, and analysis. Recent advancements in machine learning (ML) have provided researchers with novel and powerful tools to address these challenges. This review explores the role of ML in the field of CNT research, focusing on how ML has enhanced CNT research by (1) revolutionizing CNT synthesis through the optimization of complex multivariable systems, enabling autonomous synthesis systems, and reducing reliance on conventional trial-and-error approaches; (2) improving the accuracy and efficiency of CNT characterizations; and (3) accelerating the development of CNT applications across several fields such as electronics, composites, and biomedical fields. This review concludes by offering perspectives on the future potential of integrating ML further into CNT research, highlighting its role in driving the field forward.

**Keywords:** carbon nanotube; machine learning; data-driven; deep learning; artificial intelligence; chemical vapor deposition; catalyst; synthesis; characterization; application

## 1. Introduction

### 1.1. Machine Learning

Recently, machine learning (ML), a new paradigm, has demonstrated great potential as a tool for materials research [1–3]. ML is a subset of artificial intelligence (AI) concerned with the development and study of algorithms and statistical models that can learn from data, generalize to unseen data, and thus perform tasks without explicit instructions. Instead of following predetermined instructions, ML systems learn from data to recognize patterns, make decisions, and improve their performance over time [4–10].

ML has shown a wide range of applications across various domains, such as materials research, healthcare, finance, retail, autonomous vehicles, language processing, etc. Specifically, in the field of materials science, ML has been applied to various aspects and demonstrated great success [11]. Some of them are briefly discussed here (Figure 1).

(1) Materials discovery. ML has significantly accelerated the discovery of new materials, showing great advantages compared with traditional approaches, which have been slow and labor-intensive, involving extensive experimentation and trial-and-error processes [12,13]. ML offers a data-driven alternative, enabling the prediction of new materials with desired properties. With the assistance of ML, high-throughput screening allows thousands or even millions of potential compounds to be evaluated computationally using trained ML models before selecting a few candidates for real experimental testing. Additionally, inverse design becomes feasible, where researchers begin with the desired material properties and use ML algorithms to identify the structures or compositions that can achieve those properties. For instance, Rao et al. used an iterative scheme that combines

ML, density functional theory, experiments, and thermodynamic calculation to find two new invar alloys with extremely low thermal expansion out of millions of candidates. The alloys are both compositionally complex, high-entropy materials, thus demonstrating the power of this approach for materials discovery [14].

(2) Process optimization. ML is also revolutionizing the optimization of manufacturing/synthesizing processes in materials science. These processes often involve multiple complex variables, such as temperature, pressure, gas, and time, which must be carefully controlled to achieve optimal results; however, the variable space is far too vast to exhaustively explore experimentally. ML enables more efficient optimization by analyzing complex datasets and identifying the most critical parameters. For example, in the field of chemical vapor deposition (CVD) synthesis of single-walled carbon nanotubes (SWCNTs), Rao et al. utilized ML to find optimal synthesis conditions within one hundred experiments to selectively grow SWCNTs within a narrow diameter range [15]. Lin et al. predicted growth conditions toward addressing the synthetic trade-off between the crystallinity and growth efficiency of SWCNT forests by training an ML regression model and validating it in less than 50 tests [16]. ML has significantly advanced process optimization, especially for complex multivariable systems.

(3) Materials characterization. ML is being integrated into the characterization of materials, where it can reveal hidden patterns in complex datasets and automate the analysis of experimental results [17,18]. ML, particularly deep learning techniques such as convolutional neural networks (CNNs), is used to analyze microscopy images to automatically classify microstructural features in the images, such as defects, phase boundaries, grain structures, particles, etc. ML is also used to analyze data from various spectroscopic techniques, including Raman [19,20], infrared [21,22], and X-ray photoelectron spectroscopy (XPS) [23–26]. ML can deconvolute complex spectra, identify chemical species, and quantify material composition. For example, Dee et al. performed quantitative analysis of catalyst nanoparticles for CNT synthesis using high-resolution, high-rate video capture of environmental transmission electron microscopy experimentation coupled with automated image processing, involving computer vision ML algorithms and CNNs [27]. Automation using ML accelerates the analysis process and provides more consistent and objective results [28].

(4) Property prediction. Predicting the properties of materials, such as mechanical, thermal, electronic, and optical properties, based on their composition and structure is one of the key applications of ML in materials science. This is particularly useful for materials where experimental property measurement is challenging or time-consuming. For instance, Hajilounezhad et al. predicted mechanical properties, including the stiffness and buckling properties of CNT forests, using an image-based ML classifier model, showing an accuracy of >91% [29].

(5) Design of experiments (DoE). ML greatly enhances the DoE by enabling more efficient exploration of the experimental parameter space. Traditionally, DoE involves systematically varying a few parameters to identify optimal conditions. However, this approach can be limited when dealing with high-dimensional spaces with many variables. Leveraging ML analysis alongside the DoE approach allows for using limited resources, particularly time, more efficiently and increases the likelihood of achieving a true optimum. For example, Cao et al. successfully optimized the materials and devices of organic photovoltaics via the combination of ML and DoE [30].

In summary, ML is profoundly impacting various aspects of materials science, from the discovery of new materials and the optimization of manufacturing/synthesizing processes to the prediction of material properties and the automation of materials characterization. These advancements are accelerating the pace of innovation, reducing the time and cost associated with materials development, and enabling the design of materials with unprecedented precision and desired functionality. As ML techniques continue to evolve, their integration into materials science will likely become even more pervasive, driving further breakthroughs and opening up new possibilities in this dynamic field.

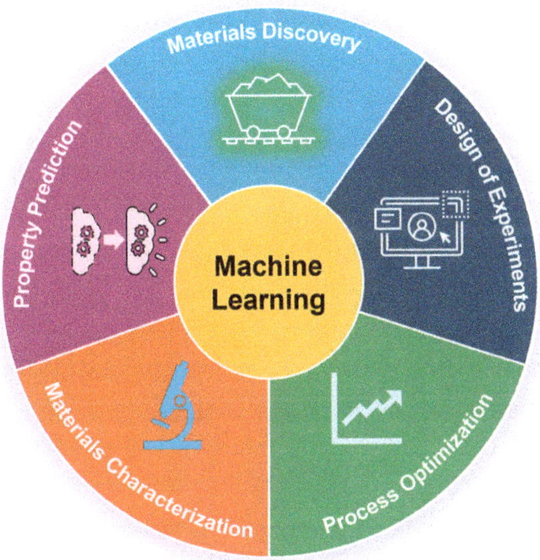


**Figure 1.** Machine learning-assisted materials research in various aspects.

## 1.2. Carbon Nanotubes

CNTs represent a sub-realm of materials science and a cross-discipline field that intersects with chemistry and applied physics. CNTs, a specific allotrope of carbon, are cylindrical structures characterized by diameters in the nanometer range and lengths that can span from nanometers to centimeters. Due to their unique one-directional structures, CNTs exhibit exceptional mechanical, electrical, and thermal properties. Specifically, CNTs possess remarkable tensile strength, excellent electrical conductivity, and superior thermal stability. These qualities make CNTs highly promising for a broad spectrum of applications, including electronics, composites, energy storage, and biomedical engineering. In electronics, CNTs are being explored for their potential in field-effect transistors [31,32], touch panels [33], and field emitters [34,35] due to their high conductivity and electron mobility. In energy-related applications, CNTs have demonstrated significant promise in supercapacitors [36,37], hydrogen storage [38,39], and lithium-ion batteries [40,41], where their high surface area and electrical properties improve energy density and charge capacity. In the biomedical field, CNTs are being researched for biosensors/bioelectrodes [42,43] and drug delivery systems [44], taking advantage of their biocompatibility and structural versatility. Additionally, through-silicon-via interposers [45] and CNT-based dry adhesives [46,47] are being developed for use in advanced manufacturing and semiconductor technologies.

ML has also proven to be highly beneficial across numerous facets of research and development of CNTs, given the nature of the involvement of a vast array of intricate and multifaceted parameters. As a special one-dimensional molecule with a unique tubular structure, controlling the atomic structures, characterizing the distinct chirality, and establishing the relationship between properties, structures, and device performance for CNTs are significantly more challenging. The complexity of CNT synthesis, which includes factors such as catalyst composition/structure, temperature, synthesis ambient, carbon feedstock, growth enhancers, and reaction conditions, presents additional significant challenges. ML can address them by identifying patterns and optimizing processes that would be difficult or time-consuming to achieve through traditional experimentation methods. Moreover, ML has facilitated the interpretation of large datasets from CNT characterizations, enabling researchers to uncover subtle correlations and predict properties with greater accuracy. This ability to manage and analyze complex datasets has proven indispensable in advancing

CNT applications, where precise control over structural properties can lead to innovations in fields ranging from electronics to materials science. Consequently, the role of ML plays in CNT research continues to grow, offering the potential to unlock new possibilities and drive further progress in this cutting-edge area of nanotechnology.

This short review briefly highlights the significant contributions of the ML approach to CNT synthesis, characterizations, and applications, demonstrating how ML has substantially advanced CNT research. Finally, the review concludes with a perspective, offering insights into potential future developments and directions for deeper integrating ML into CNT research to drive the field forward.

## 2. Machine Learning-Assisted CNT Synthesis

Synthesis of CNTs involves complex multivariable parameters. A typical CVD synthesis process of CNTs includes catalyst preparation, catalyst nanoparticle formation, and CNT growth, targeting the desired CNT structural properties (Figure 2a) [48]. As shown in Figure 2b, all these processes are highly sensitive to a vast range of parameters including, but not limited to, catalyst type (composition: commonly used elements such as Fe, Co, Ni, or their alloys; structure: single-layer or multilayer; preparation method: sputtering, evaporation, casting, spin coating, etc.), temperature, pressure, carbon feedstock (type and concentration), growth enhancer (type and concentration), ambient conditions (carrier gas type and concentration), time, etc. [49–51]. These parameters must be meticulously controlled to achieve CNTs with desired structural properties [52], typically including length [53,54], crystallinity [55,56], diameter [57,58], wall number [59–61], surface area [62,63], alignment [64,65], density [66,67], etc. Traditionally, optimizing these parameters has relied heavily on trial-and-error approaches, which are time-consuming and labor-intensive, and may not fully capture the complex interactions between these variables. ML has emerged as a powerful tool to address these challenges, offering data-driven strategies for efficiently optimizing CNT synthesis parameters and even enabling the development of autonomous synthesis systems.

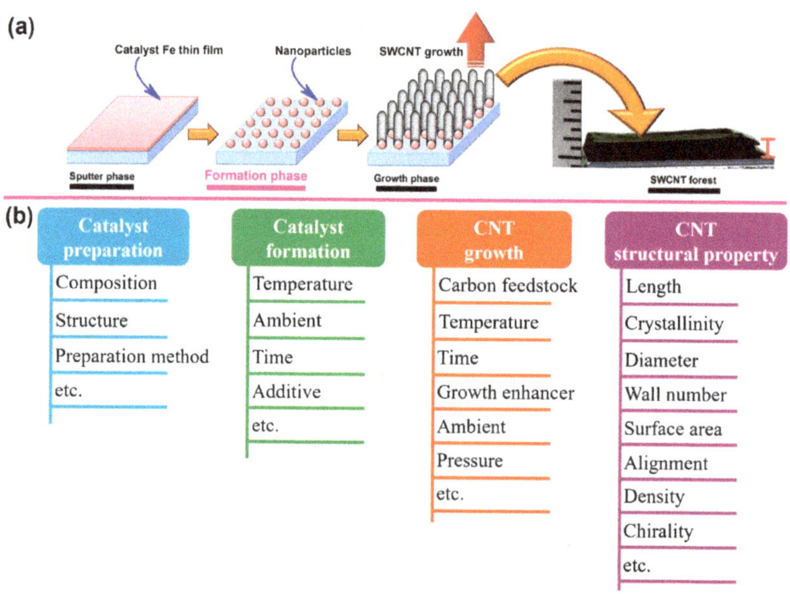


**Figure 2.** (**a**) A typical CNT CVD synthesis process. Reprinted with permission from Reference [48]. (**b**) Complex multivariable parameters in CNT synthesis process.

### 2.1. Data-Driven Optimization of CNT Synthesis

Data-driven optimization generally refers to the use of ML algorithms to analyze large datasets generated from CNT synthesis experiments and focuses on optimizing the multitude of variables involved in the production process. By identifying patterns and correlations within the dataset, ML algorithms can model the complex relationships between synthesis parameters and the resulting properties of CNTs. These models can predict how changes in synthesis parameters will affect the properties of the resulting CNTs, facilitating the identification of optimal synthesis conditions, significantly reducing the need for trial-and-error approaches, and accelerating the discovery of targeted CNTs with desired structural properties. This approach can also provide deeper insights into the underlying mechanisms governing CNT growth.

In CNT synthesis, the parameter space is vast and multidimensional, as shown in Figure 2b. These parameters may interact in nonlinear ways to influence the structural properties of CNTs. ML algorithms such as support vector machines (SVMs), random forests (RFs), XGBoost, multilayer perceptron (MLP), Bayesian optimization, and artificial neural network (ANN) are particularly effective in exploring this parameter space. They can identify complex patterns and correlations that might be missed by traditional statistical methods. These models have been previously applied in material synthesis and showed excellent performance [16,68–73]. In the field of CNT synthesis, successful applications of data-driven optimizations have also been demonstrated [15,16,29,74–78].

For instance, Lin et al. demonstrated a route for addressing the synthetic trade-off between CNT forest crystallinity and growth efficiency through the assistance of data-driven optimization [16]. Specifically, synthetic trade-offs in the synthesis of SWCNT forests are very challenging as growing certain desired properties can often come at the expense of other desirable characteristics. One example is the one between crystallinity and growth efficiency, resulting in the difficulty of achieving both high crystallinity and tall forests simultaneously (Figure 3a). To address this, Lin et al. used a data-driven approach to train a machine learning regression model (XGBoost) using nine input feature descriptors with a set of 585 experimental synthesis data (Figure 3b). Subsequently, 16,000 exploratory "virtual" experiments were performed by the trained model to examine potential routes toward addressing the current crystallinity—height (growth efficiency) trade-off limitation, and possible growth conditions were predicted (Figure 3c). Finally, validation experiments showed very good agreement with the predictions, highlighting the effectiveness and accuracy of the predictive capability of the ML model, which achieved improved results in less than 50 validation tests. Importantly and fundamentally, the trained model revealed the surprising importance of the nature of the carbon feedstock, as a route for indirectly maintaining a high level of crystallinity at a high growth efficiency through the maintenance of the activity of the catalysts to overcome the trade-off. This data-driven approach represents a significant advance in complex multivariable CNT synthesis systems.

Another example is that Ji et al. used ML for high-throughput screening of the efficient growth of high-quality SWCNTs [76]. A database of 1280 experiments was built with four input parameters of the thickness of the cobalt (Co) catalyst film, growth temperature, reduction time, and carbon precursor flow rate and one output of the Raman $I_G/I_D$ ratio. ML models including linear regression (LR), random forest regression (RFR), support vector regression (SVR), and ANN were trained, and RFR was finally chosen due to its highest prediction performance (Figure 3e). The trained model was used to further explore the parameter space for the optimum growth conditions for high-quality SWCNTs with 38,016 combinations of growth parameters (Figure 3f–h). Finally, the predicted optimum growth parameters were validated to achieve a high $I_G/I_D$ value of 138 in the conventional CVD growth process. The efficient modeling, predicting, and learning ability of the combined high-throughput and ML method show great potential to speed up the controlled synthesis of SWCNTs with specific structures and properties.

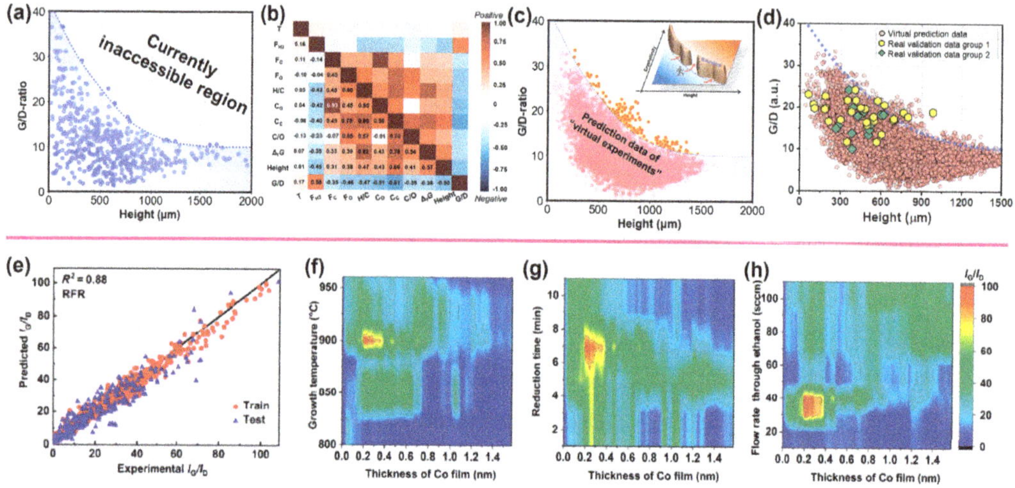


**Figure 3.** (**a**–**d**) Data-driven assisted SWCNT forest synthesis overcoming the synthetic trade-off. Reprinted with permission from Reference [16]. (**a**) Trade-off between G/D ratio and CNT height. (**b**) Heat map of Pearson correlation coefficient matrix. (**c**) Plot of virtual experiment data with prediction points beyond the boundary. (**d**) Validation results. (**e**–**h**) ML-assisted high-throughput screening for high-quality SWCNTs. Reprinted with permission from Reference [76]. (**e**) $I_G/I_D$ from experiments and predictions. Dependence of predicted $I_G/I_D$ on Co film thickness with (**f**) growth temperature, (**g**) reduction time, and (**h**) flow rate through ethanol as secondary variables.

In the data-driven approach, the size of the dataset plays a crucial role in determining the performance and reliability of ML models. The general thinking is that larger datasets are beneficial for training ML models because they can provide more information for the model to learn from, reducing overfitting and leading to more accurate and robust predictions. However, they also require more computational resources. Sufficient unbiased data are crucial for successful ML training and validation [79]. The appropriate dataset size depends on several factors, including the problem type, the level of complexity, the number of features, data quality, and error tolerance. This makes it a case-by-case scenario. The most widely used rule-of-thumb is that the dataset size should be at least 10 times the number of weights in the model. However, other guidelines also exist, such as requiring 50 to 1000 times the number of prediction classes or 10 to 100 times the number of features [80]. Krasnikov et al. examined the implementation of ML techniques and discussed features of the optimal dataset size and density for aerosol synthesis of SWCNTs with a complex carbon source [77]. They originally employed a dataset of 369 points comprising four inputs and three target parameters and assessed the performance of six ML methods. They showed that even a dataset of 250 points with an inhomogeneous distribution of input parameters is sufficient to reach an acceptable performance of the ANN model, wherein the error is most likely to arise from experimental inaccuracy and hidden uncontrolled variables. Therefore, careful consideration of dataset quality and input feature selection, rather than just dataset size, seems to be more important.

In summary, the use of ML in the data-driven optimization of CNT synthesis parameters not only reduces the time and resources required for CNT synthesis experimentations but also enhances the reproducibility of results. By providing a deeper understanding of the synthesis process, ML enables researchers to design experiments more strategically, leading to the discovery of new synthesis pathways and the production of CNTs with tailored properties.

### 2.2. Autonomous Synthesis of CNTs

Autonomous synthesis represents a paradigm shift in materials science, where ML and automation technologies are integrated to enable fully automated production/synthesis processes. In the context of CNT synthesis, this involves the development of systems that can autonomously adjust synthesis parameters in real time based on continuous feedback from ML algorithms, ensuring the production of CNTs with desired, precise, and consistent properties. This concept goes beyond traditional methods that rely on manual adjustments and trial-and-error experimentation. Instead, autonomous synthesis systems can dynamically adjust synthesis parameters on the fly to optimize the quality and yield of CNTs, significantly improving efficiency and reproducibility.

Due to the multivariable nature of CNT synthesis, a closed-loop autonomous synthesis system is uniquely suitable for optimizing the synthesis process continuously and autonomously by integrating real-time data monitoring, ML algorithms, and automated process controls. As conceptually diagrammed in Figure 4a, a simple system generally starts with a CVD reactor to perform the synthesis process and collect the growth parameters. The system also needs to be equipped with in situ monitoring and evaluation units, such as Raman spectroscopy, thermogravimetric analysis (TGA), and mass spectrometry, to provide real-time monitoring of the process. Then, all these data are sent into an ML control unit to analyze the data and predict the optimal synthesis conditions. Finally, the decisions of new synthesis conditions are set and fed back into the CVD process.

Autonomous synthesis of CNTs has been successfully demonstrated in several experimental setups [81–84]. For instance, a proof of concept was exemplified by Maruyama et al. with an adaptive rapid experimentation and in situ spectroscopy (ARES) system (Figure 4b). This system was developed to increase the experimentation rate by 100-fold, to 100 runs per day, with results analyzed in situ and in real time via Raman spectroscopy. This system was used to synthesize CNTs on the surface of micropillars, and linear regression modeling was used to map regions of selectivity toward SWCNT and multiwall CNT (MWCNT) growth in the complex parameter space of 534 CVD experiments [81]. Recently, the same group demonstrated a closed-loop system using Bayesian optimization (BO) as an efficient and robust ML algorithm integrated with the ARES setup (Figure 4b). The CNT growth rates were extracted from the ARES and fed into the BO algorithm. The algorithm then generated a new set of conditions, run by ARES, and the new output data were sent to the BO planner to update the existing dataset and plan a new experiment. This closed-loop system significantly improved the predictive power, showed good consistency performance, exploited a complex parameter space, and achieved a 5-fold faster experimentation speed than before [83].

More recently, Zhang's group introduced an AI-driven platform, Carbon Copilot (CARCO), which integrates transformer-based language models tailored for carbon materials, robotic CVD, and data-driven ML models. Using CARCO, a catalyst discovery was demonstrated by predicting a superior Titanium–Platinum bimetallic catalyst for high-density horizontally aligned CNT (HACNT) array synthesis, validated through over 500 experiments. With the assistance of millions of virtual experiments, an unprecedented 56.25% precision in synthesizing HACNT arrays with predetermined densities was achieved within just 43 days [84]. This work exemplifies a great advance towards the integration of AI/ML with human expertise to overcome the limitations of traditional experimental approaches.

In summary, despite some challenges, the implementation of autonomous synthesis systems could revolutionize CNT research by drastically reducing the time and cost associated with traditional synthesis methods. This approach not only enhances the scalability of CNT production but also allows for the precise control of CNT properties, which is crucial for their integration into commercial applications such as electronics, composites, and biomedical devices.

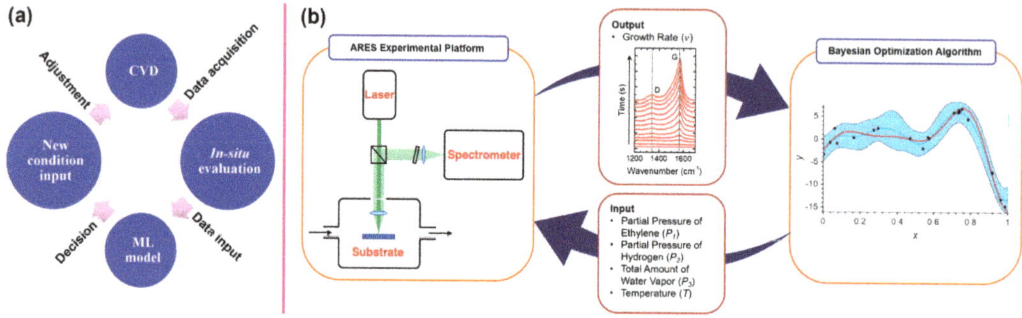


**Figure 4.** (**a**) A conceptual diagram of a closed-loop autonomous synthesis system. (**b**) A closed-loop autonomous synthesis system with adaptive rapid experimentation and in situ spectroscopy (ARES) and Bayesian optimization ML algorithm. Reprinted with permission from Reference [83].

## 3. Machine Learning-Assisted CNT Characterization

The most essential question in materials science including CNT research is to establish the relation of the microstructures with synthesis conditions and physical and chemical properties. Materials characterization is performed to extract information about the structure, including imaging in real space and spectroscopy in energy space. By establishing the relationship between structure and properties, characterization is essential to understand growth mechanisms and design microstructures for desired properties.

For CNTs, structural information extracted from characterization includes purity, morphology, diameter, length, conductivity type, distribution of chirality, and so on. Transmission electron microscopy (TEM) was the key for the discovery of CNTs [85], and for the precise characterization of individual CNTs, including the chiral indices [86]. On the other hand, Raman spectroscopy has been widely used for quick statistical analysis of CNT samples, including purity, distribution of diameter, and chirality [87,88].

In conventional materials characterizations, images and spectra are acquired and analyzed manually one by one. It is challenging to reach the optimum acquisition parameters because of the trade-offs in temporal resolution, spatial resolution, field of view, data volume, acquisition speed, stability, irradiation damage, and signal-to-noise ratio (SNR). For CNTs, from images in real space, diameters and lengths are measured. From diffraction patterns in reciprocal space, crystalline structures could be analyzed by measuring the distances and angles of diffraction spots. And from spectra, peak positions are compared with existing spectra in databases to determine the substances. The conventional analysis process is time-consuming and biased by human factors.

In recent years, ML has been applied to materials characterizations to automate the analysis process and to recognize the hidden patterns in data [89–94]. It has been demonstrated that ML can enhance efficiency, increase accuracy, and most importantly, handle large datasets through high-throughput analysis. In this section, ML-assisted CNT characterization will be introduced with examples of rapid identification of Raman spectroscopy and automated chirality determination from TEM Images.

### 3.1. Machine Learning for Rapid Identification of Raman Spectroscopy

Raman spectroscopy is based on the inelastic scattering of photons with phonons. It is widely used to analyze vibrational modes and to identify molecular structures with the fingerprints in the position, width, and shape of the peaks. Raman spectroscopy has been one of the most important methods for the characterization of CNTs [88]. The purity and quality of CNTs could be extracted from the ratio of G and D bands. In addition, diameter and chirality distribution could be obtained from the radial breathing modes (RBMs) [87,95]. One of the challenges for Raman spectroscopy is the low SNR, since the inelastic Raman scattering cross-section is much smaller than the cross-section for elastic Rayleigh scattering. In recent years, ML algorithms have been applied for Raman spectroscopy-based classification and recognition [19,76,96,97].

For example, Zhang et al. utilized deep learning to automate and accelerate the analysis of Raman spectra for identifying structural features of suspended CNTs, including position, quantity, and metallicity (Figure 5a–f) [96]. Their model was built on a CNN with four convolutional layers, a softmax layer, and a cross-entropy layer for defining the loss function. The database consisted of 62,130 Raman spectra of suspended CNTs, categorized as semiconducting CNTs (S-CNTs), metallic CNTs (M-CNTs), and empty trenches (no CNTs). The CNN was trained on this dataset and validated on a separate set of 48,887 spectra. After training, softmax thresholding was applied to refine classification, reduce false positives, and improve accuracy. The model achieved 90% classification accuracy for spectra with the SNR as low as 0.9, with accuracy increasing to 98% for higher SNR values. The accuracy improved with longer integration times (>30 ms) and higher laser power, peaking at 99% at 1 mW laser power. This deep learning-assisted rapid identification method for CNT structures via Raman spectroscopy can be integrated into industrial production lines for real-time monitoring of CNT growth.

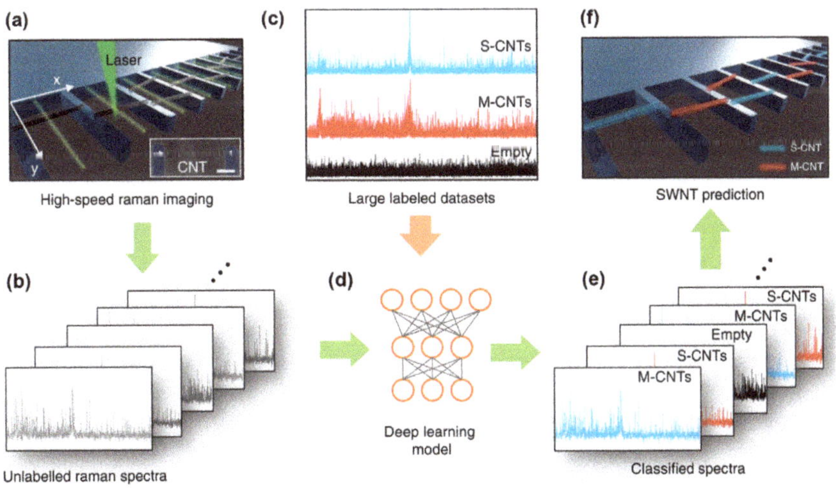


**Figure 5.** Deep learning-assisted rapid identification of Raman spectra from suspended CNTs. (**a**) Raman imaging. (**b**) Data collection. (**c**) Database labeling. (**d**) CNN deep learning model. (**e**) Training and classification. (**f**) CNT identification. Reprinted with permission from Reference [96].

In another example, Kajendirarajah et al. demonstrated a rapid method for diagnosing and analyzing tip-enhanced Raman spectroscopy (TERS) mappings of individual CNTs using deep learning neural networks [97]. In this study, TERS maps were obtained from SWCNTs, where the high spatial resolution and spectral selectivity of TERS were used to distinguish between semiconductive and metallic SWCNTs, identify bundles, and locate defect areas. Two multi-purpose ANNs were developed. ANN Model 1 classified the Raman spectra from TERS experiments as either background or CNT. ANN Model 2 efficiently filtered the three vibrational modes of CNTs and accurately identified their presence in the spectra. Together, ANN Models 1 and 2 generated TERS maps with enhanced contrast and provided clear identification of the vibrational modes. The combined approach also facilitated the enhanced visualization of CNT defect regions. The deep learning ANNs achieved 98% accuracy with ANN Model 1 and 96% accuracy with ANN Model 2. This method was able to diagnose unseen raw TERS hyperspectral data in just 4 to 6 h for a spectral cube containing 8000–10,000 spectra with 1600 points each.

In summary, the utilization of ML combined with Raman spectroscopy has enabled the fast and accurate identification and classification of CNTs, resulting in a powerful and comprehensive analysis tool.

### 3.2. Machine Learning for Automated Chirality Determination from TEM Images

TEM is a key instrument for characterizing nanomaterials such as CNTs. It combines multiple functions, ranging from imaging with atomic resolution, chemical analysis of individual atoms, and mapping of local physical properties to fabrication with nanometer or even atomic precision.

For CNTs, it is quite a challenge to operate a modern TEM to precisely characterize the atomic structures because of the small atomic mass, small diameter, and complex circular geometry. It takes years to train a professional microscopist to master a variety of skills, including high-resolution TEM (HRTEM) imaging, nano-beam electron diffraction for chiral indices, STEM imaging, and 4D-STEM for probing local structures and properties.

In recent years, ML methods have been applied to electron microscopy, ranging from data modeling and data analysis to atomic fabrication [27,98–102]. For example, Lin et al. quantified the fluctuations in catalyst carbon content by using an automated, atomic-scale structural analysis of the time-resolved ETEM images. Such fluctuations in the composition of catalysts were correlated with the SWCNT growth rate [28]. Förster et al. developed a deep learning approach for determining the chiral indices of CNTs from HRTEM images (Figure 6a) [101]. Since the chiral indices entirely determine the atomic structures and electronic and optical properties, it is critical to identify the chiral indices of individual CNTs and more important for the statistical distribution of chirality in a CNT sample. Electron diffraction is the most precise method to determine CNTs' chiral indices; however, individual CNTs as long as 100 nm are required. In this work, Förster decided to use HRTEM images to determine the chirality of CNTs, since sub-angstrom resolution is achievable for modern Cs-corrected TEMs.

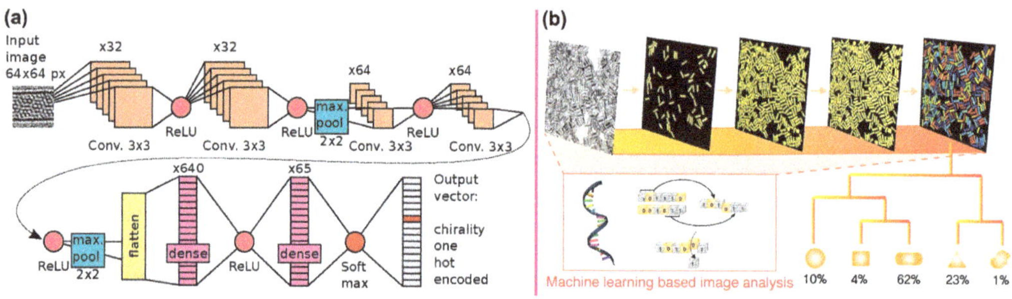


**Figure 6.** (**a**) CNN architecture for determining CNT chiral indices from HRTEM images. Reprinted with permission from Reference [101]. (**b**) Statistical characterization of nanoparticle morphologies through ML-assisted TEM image analysis. Reprinted with permission from Reference [102].

Due to the lack of enough high-quality experimental images of CNTs, a database was constructed by high-throughput atomistic structural simulation and followed by TEM simulations to generate HRTEM images. To increase the diversity of the data and generality of the model, variability was included, such as position, orientation, magnification, defocus, aberration, noise, and so on. In total, all possible 261 chiral indices for the CNTs with a diameter of 0.48 nm to 2.30 nm were considered. And for each chirality, 5000 images were simulated to general a database of 1.3 million images. A CNN based on LeNet-5 was used to model and classify simulated TEM images to identify the chiral indices. To enhance the modeling precision, two CNNs were used for the diameter and chirality, respectively. The first CNN could determine the diameter with an accuracy of 99%. And an overall accuracy of 90.5% was found for the second CNN in classifying chiral indices. The CNN-based chirality classification system was evaluated by using experimental HRTEM images. When images of sufficiently high quality were used, 71% of the results were consistent for classification conducted manually and automatically.

In this pioneering work, ML has demonstrated enhanced efficiency to identify the chiral indices, which typically takes 15–30 min to analyze one picture manually. And the model showed high robustness for the nanotubes with defects. This method opens the door for high-throughput analysis of HRTEM images to gain insight into the distribution of chirality to understand the growth mechanism of CNTs.

As is well known, the structures of CNTs are predominantly determined by the characteristics of the catalyst. Therefore, analyzing catalyst nanoparticles is essential to understanding the CNT growth mechanisms. However, due to their dynamic nature and nanoscale size, quantitatively studying nanoparticle morphologies and conducting statistical analyses on them is extremely challenging. Recently, Lee et al. developed a mass-throughput method for analyzing nanoparticle morphologies by applying a genetic algorithm to an image analysis technique (Figure 6b) [102]. This approach enabled the analysis of over 150,000 nanoparticles with a high precision of 99.75% and a low false discovery rate of 0.25%. The study also introduced clustering techniques to group nanoparticles with similar morphological shapes for extensive statistical analysis. It was determined that analyzing at least 1500 nanoparticles is required to represent the total nanoparticle population at a 95% credible interval. Additionally, the number of TEM images and the average number of nanoparticles per image should be considered to ensure a representative sample. The statistical distribution of polydisperse nanoparticles was also found to be critical in accurately estimating their optical properties.

In summary, combining ML with TEM techniques not only enhances the efficiency and accuracy of CNT analysis but also enables precise characterization of catalyst nanoparticle morphology. This integration provides a powerful tool to significantly advance the field of CNT research.

## 4. Machine Learning-Assisted CNT Applications

Besides the application in CNT synthesis, ML also plays a crucial role in enhancing the CNT applications across various fields to improve their integration into devices and materials and predict their performance. This section explores how ML aids CNT applications in electronics and sensors, composite materials, and the biomedical field.

### *4.1. CNT Applications in Electronics and Sensors*

CNTs possess exceptional electrical properties, such as high conductivity, ballistic transport, and excellent current-carrying capacity, making them ideal for electronic devices and sensors [31]. However, integrating CNTs into practical devices requires precise control over their properties, processing, and placement, which can be challenging due to their nanoscale dimensions and variability in structural properties. ML can help overcome the challenges by providing data-driven insights and optimization strategies [103–109].

For example, Tadokoro et al. developed an AI-assisted method that significantly enhanced the efficiency of fabricating CNT-based nanocantilevers (Figure 7a) compared to traditional methods like CVD and dielectrophoresis [105]. These conventional methods typically involve the manual and time-consuming placement of additional electrodes and careful observation of individually grown CNTs. In their approach, a deep neural network was trained to recognize randomly positioned single CNTs, measure their locations, and determine the precise point where an electrode should be clamped to form the nanocantilever. This AI-driven system drastically improved the recognition and measurement process, reducing the time from 12 h with manual processing to just 2 s.

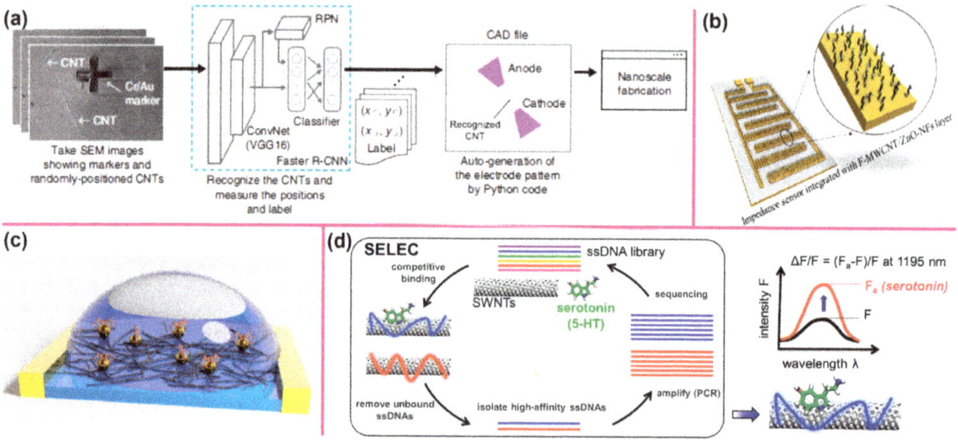


**Figure 7.** (**a**) AI-assisted framework for fabricating nanocantilevers using a neural network to recognize CNTs. Reprinted with permission from Reference [105]. (**b**) $NH_4^+$-selective impedance sensors fabricated by embedding F-MWCNT/ZnO-NF active layers on interdigitated arrays. Reprinted with permission from Reference [106]. (**c**) Schematic diagram for the composition of sensing material and liquid-gated CNT FET. Reprinted with permission from Reference [107]. (**d**) ML-assisted approach to identify DNA sequences in DNA−SWCNT conjugates with high serotonin response. Reprinted with permission from Reference [109].

Another example is that Aliyana et al. utilized an ML approach to accurately correlate the impedance variations in zinc oxide/MWCNT nanocomposite (F-MWCNT/ZnO-NFs) to $NH_4^+$ ion concentrations [106]. The proposed $NH_4^+$ sensor along with the decision-making ML model can identify and operate at specific operating frequencies to continuously collect the most relevant information from a system (Figure 7b). In addition, Bian et al. utilized ML techniques (LR, RF) to create a calibration method for $Hg^{2+}$ sensors based on CNT field-effect transistor (FET) devices (Figure 7c) to solve sensor response saturation [107]. Such applications of ML techniques to investigate which features in the FET signal maximally correlate with concentration changes provide valuable insight into the CNT sensing mechanism and assist in the rational design of future nanosensors. Furthermore, Fan et al. proposed an efficient framework for optimizing the design of CNT FET through the integration of device physics, ML, and multi-objective optimization to expedite the early-stage development of advanced transistors [108]. Moreover, Kelich et al. trained ML models (CNN, SVN) to predict DNA sequences with strong optical responses to the neurotransmitter serotonin (Figure 7d). They discovered five DNA-SWCNT sensors with a higher fluorescence intensity response than those using only a manual screening method [109].

In summary, these studies collectively highlight the transformative potential of ML approaches in advancing the development of CNT applications in electronics and sensors. These examples underscore the potential of ML to not only solve specific technical challenges but also to revolutionize the entire process of CNT-based device development. By automating complex analyses, optimizing sensor design, and providing rapid, data-driven insights, ML approaches are poised to play a pivotal role in accelerating the innovation and deployment of CNT technologies in various electronic and sensing applications.

### *4.2. CNT Applications in Composite Materials*

CNTs are widely used in composite materials to enhance their mechanical, thermal, and electrical properties [45]. However, achieving uniform dispersion and proper alignment of CNTs within the composite matrix is challenging due to the strong van der Waals forces between the nanotubes and their tendency to agglomerate. ML provides powerful tools to optimize these processes and predict the resulting composite properties, thereby enhancing the performance of CNT-based composites across various applications [110–120].

For instance, Yu et al. utilized ML techniques to explore the structure–property linkages of CNT-reinforced AlSi10Mg nanocomposites [110]. The proposed processing framework pipeline starts with SEM image processing of cellular microstructural features, followed by training of ML models (AdaBoost, gradient tree boosting, K-nearest neighbors, decision tree, and extra trees regressors), and property prediction (Figure 8a). The developed models demonstrated satisfactory performance, with the extra tree regression model predicting hardness with a 2.47% error and the decision tree regression model predicting relative mass density with a 1.42% error. This framework can be applied to process optimization and mechanical property manipulation for designing new alloys or composites.

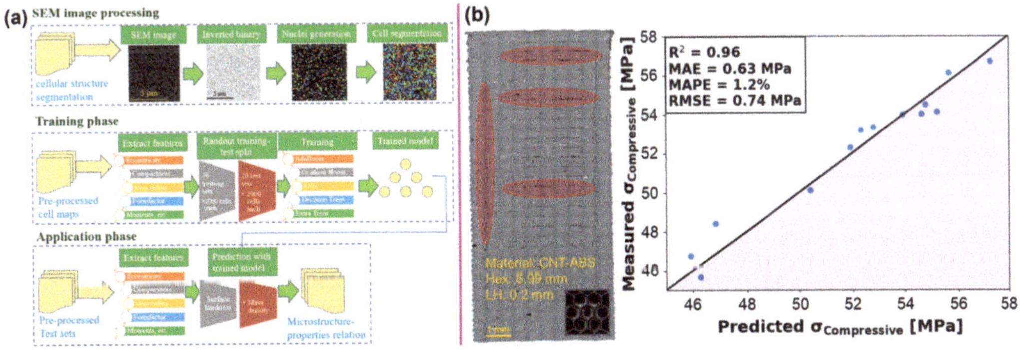


**Figure 8.** (**a**) Schematic of the proposed processing framework pipeline for AlSi10Mg nanocomposites' microstructure–property linkages. Reprinted with permission from Reference [110]. (**b**) Optical microscopy cross-sections of CNT-reinforced ABS honeycomb composite and measured versus predicted ultimate compressive strength. Reprinted with permission from Reference [111].

In another example, Ranaiefar et al. investigated structures and mechanical property predictions of CNT-reinforced acrylonitrile butadiene styrene (ABS) composites fabricated by 3D printing using an ML approach [111]. Multiple regression algorithms were evaluated, including ridge regressor, linear regression, k-neighbors regressor, gradient boosting regressor, random forest regressor, extra trees regressor, decision tree regressor, and lasso regression. The predictive classification and regression supervised ML models supported the experimental results with 0.92 accuracy and a 0.96 coefficient of determination (Figure 8b). This approach utilizing ML techniques can help inform and guide the design of 3D-printed structures for targeted performance.

Furthermore, Jalal et al. demonstrated a concept of 'Big Data' analytics in composite structure with focusing on functionally graded CNT-reinforced composites (FG-CNTRCs) using a mesh-free method and an optimized neural network (ONN) approach to study the effect of structural parameters on vibrational frequence [112]. They built big data containing 15,625 entries with six parameters and developed an optimized ONN model for predictive modeling. The ONN model demonstrated computation speeds thousands of times faster than the mesh-free method while keeping a simulation error as low as 1%, suggesting it is an efficient and reliable approach for handling big data in the optimization and design of composites.

Moreover, different ML techniques have been utilized to study various CNT-based composites for property improvements either in combination with experimental investigations, such as CNT-reinforced cementitious composites [113–115] and CNT-reinforced polymeric composites [116–119], or alongside computer simulations [120].

In summary, ML is revolutionizing the development and application of CNT-based composite materials by providing powerful tools for optimizing fabrication processes, predicting composite properties, and guiding the design of composites with tailored performance characteristics. By leveraging large datasets and advanced ML algorithms, researchers can overcome the challenges associated with CNT dispersion and alignment, enabling the production of high-performance composites for a wide range of applications. As ML technologies continue to evolve, their integration with composite material research and manufacturing processes will likely lead to even greater advancements in this field.

### *4.3. CNT Applications in the Biomedical Field*

CNTs also possess unique properties, such as high surface area, biocompatibility, and the ability to penetrate cell membranes, making them ideal for various biomedical applications [42,121]. ML enhances these applications by providing tools for optimizing the design of CNT functionalization, predicting their interactions with biological molecules, enabling high-throughput selection of targeted DNA sequences, accelerating the discovery of effective biomolecular constructs, and even predicting their genotoxicity [122–126].

For instance, Ouassil et al. developed a random forest classifier (RFC, Figure 9a) to investigate the relationship between a protein's amino acid sequence and a protein's binding propensity to SWCNTs [122]. The classifier aimed to predict which protein–SWCNT interactions to expect in biological environments and to predict high-affinity protein binders to SWCNTs, along with protein features associated with such binding affinity to improve the process of protein–nanoparticle construct design (Figure 9b). The model was validated with a new set of proteins by performing quantitative protein adsorption experiments. The ML classifier can serve as a useful tool for understanding how protein sequences influence nanotube binding.

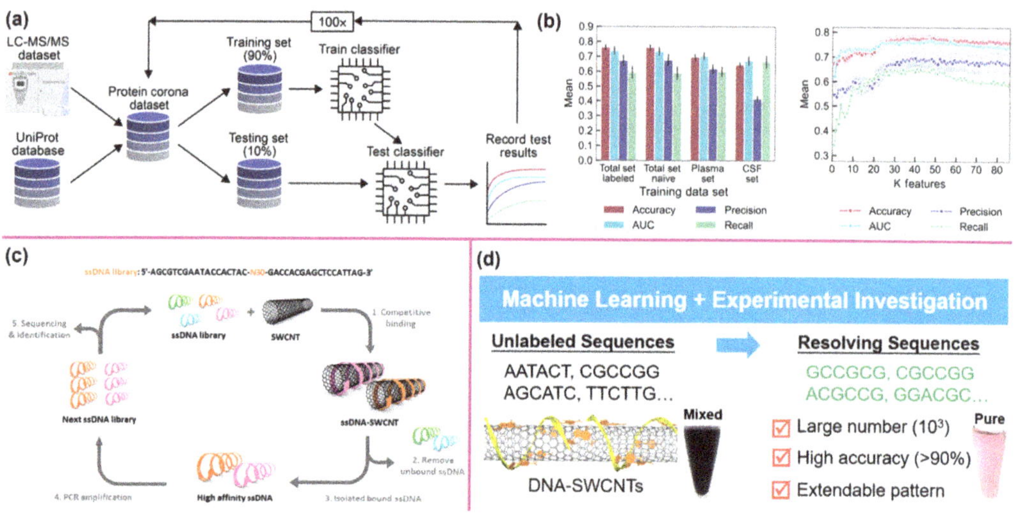


**Figure 9.** (**a**) RFC workflow and (**b**) classifier performance on different biofluid training datasets and varied protein feature inputs. Reprinted with permission from Reference [122]. (**c**) Selection of high-affinity ssDNA sequences on SWCNT surfaces. Reprinted with permission from Reference [123]. (**d**) ML-assisted systematic search of DNA sequences. Reprinted with permission from Reference [124].

Additionally, the use of single-stranded DNA (ssDNA) as a wrapping surfactant for SWCNTs is a promising approach to functionalize and modify the SWCNT surface, enabling precise control over SWCNT alignment. This approach has shown great potential in biomedical applications, including drug delivery, gene therapy, and biosensors. Lee et al. employed high-throughput systematic selection of high-affinity ssDNA sequences using ML models (RF, MLP, CNN, etc.) from a vast random library (Figure 9c). The model accurately distinguished high-affinity ssDNA sequences and provided predictive capabilities for binding affinity, supporting the design of tailored DNA-SWCNT constructs. The stability of ssDNA conformations on SWCNTs were validated by molecular dynamics simulations [123].

Furthermore, the prerequisite of utilizing DNA in sequence-dependent applications, such as biomedical sensors, is to search specific sequences, which represents a significant challenge due to the countless possible sequence combinations. Lin et al. demonstrated an ML-assisted experimental method for systematically searching DNA sequences based on sequence-dependent recognition between DNA and SWCNTs (Figure 9d). Compared with empirical search methods, their approach significantly improved both the number of resolvable sequences and the success rate of finding them, from $\sim 10^2$ to $\sim 10^3$ and from ~10% to >90%, respectively [124]. Moreover, ML approaches have also been successfully used to investigate CNT genotoxicity prediction [125] and discover molecular recognition based on SWCNT corona phases [126].

In summary, ML techniques enhance CNT applications in the biomedical field by enabling precise and targeted functionalization of CNTs, improving their biocompatibility, and optimizing their interaction with biological systems. Additionally, ML-driven approaches provide powerful predictive tools to accelerate the discovery and development of CNT-based biomedical constructs, expanding the possibilities for innovative therapies and diagnostics.

## 5. Conclusions and Perspective

This review demonstrates the transformative impact of ML on CNT research, covering areas such as the optimization of growth conditions, automation of structure characterization, and application development. ML algorithms have proven highly effective in modeling the complex relationships between synthesis parameters and the resulting structures and properties of CNTs by identifying patterns and correlations within datasets. As a result, improvements in the height of CNT forests and the quality of CNT thin films have been achieved. Additionally, the integration of autonomous synthesis systems has further enhanced the CNT production, reducing reliance on manual trial-and-error processes.

ML has been widely adopted for materials characterization and has started to revolutionize the CNT field by automating traditionally manual tasks. Notable successful applications include determination of the chiral indices through HRTEM images and rapid identification of suspended CNTs using Raman spectroscopy, both of which utilized deep learning for automation. ML has demonstrated enhanced efficiency to extract information to automate the analysis process, markedly improving the speed and accuracy of data analysis, facilitating high-throughput studies and more consistent results.

Moreover, ML has been accelerating the innovation and deployment of CNT-based applications in electronics and sensors, through automating complex analyses, optimizing sensor designs, and providing rapid, data-driven insights. It has also provided valuable guidance to design composites with desired properties by optimizing fabrication processes and predicting composite characteristics. In the biomedical field, ML is playing a crucial role in optimizing CNTs' functionalization and predicting their interactions with biological molecules, furthering CNT applications in drug delivery, biosensors, and diagnostics.

Despite the significant progress, the integration of ML into CNT research is still in its early stages and holds immense potential for exponential growth in the near future. Up to now, application examples of ML in CNT characterizations are still very limited. A search using the key words "machine learning" and "carbon nanotube" in the "Web of Science Core Collection" yielded only 131 publications, with the number of publications increasing in a linear manner from 2 in 2015 to >30 in 2024.

A critical issue is the selection of an appropriate ML model for specific data types and problems. Many ML models, such as linear regression (LR), SVM, XGBoost, MLP, and Bayesian optimization, have been developed based on smart algorithms. LR is a basic model that assumes a linear relationship between input parameters and the target variable. SVM aims to find a hyperplane in a high-dimensional space that best fits the data and can model nonlinear relationships using kernel functions. XGBoost is an optimized gradient boosting algorithm that builds an ensemble of decision trees, with each tree focusing on correcting the errors of the previous ones. MLP is a type of neural network that learns nonlinear relationships through multiple layers. Bayesian optimization is a probabilistic model that optimizes by building a surrogate model (e.g., a Gaussian process) and uses it to guide exploration and exploitation. Up to now, model selection has been based primarily on performance metrics such as the coefficient of determination ($R^2$) for training and test datasets. It is expected that future models incorporating physical insights and expert knowledge of growth, characterization, and applications will be constructed to be more effective in addressing the problems in materials science and CNT research.

One of the main limitations is the lack of organized data, which are indispensable for training complex ML models. Currently, there is no centralized database for CNT research that aggregates experimental data from various groups. Most advances have been driven by individual groups accumulating years of experimental data. The hunger for data is a common challenge in data-driven science. Typically, a scale of tens of thousands of data points is needed for high-accuracy deep learning modeling. However, conducting experiments is time-consuming, as CNT growth can take hours, while capturing TEM images and obtaining Raman spectra typically take minutes. Collaboration within the CNT research community, among experimentalists, theorists, and data scientists, will be the key to constructing shared CNT databases and establishing protocols for data storage, sharing, usage, and importantly protecting the precious data. Another approach to data collection is through large-scale, high-throughput calculations, though computational power remains a limiting factor. Furthermore, there is a gap between theoretical and experimental studies regarding system size and timescales. For example, a typical simulation investigates a few hundred atoms by first-principles calculation over nanoseconds, while real-world reactions involve a large number of atoms on the order of $10^{23}$ over hours. Moreover, although the current main problem is the availability of datasets, attention should also be given to potential bias in datasets during the data collection process. For example, most catalysts used for CNT growth are transitional metals, and this imbalance could hinder the discovery of new catalyst systems. A possible solution is to integrate and collaborate with standardized databases when CNT-specific databases are lacking. If ML models are general enough, they could be applied to CNTs as a subset of materials science. Thus, more practical applications of ML will enable a deeper understanding of CNT data patterns.

As a subset of ML, generative AI has recently gained significant attention because it not only learns from data but also generates new content, such as text, audio, and images, based on the training data. In recent years, natural language processing (NLP) models like GPT-4 have demonstrated the ability to generate human-like text and engage in conversations. Although generative AI is starting to be applied to scientific research, including materials science, its use in CNT research remains limited. However, its potential is immense and should not be overlooked. Generative AI models, such as generative adversarial networks (GANs) and variational autoencoders (VAEs), could be highly useful in future CNT research for tasks such as optimizing synthesis pathways, predicting novel CNT structures, and enabling property-driven material discovery. The integration of generative AI into CNT research has the potential to further accelerate innovation by automating more creative aspects of material design and discovery. As this area continues to evolve, we anticipate that generative AI will play a pivotal role alongside ML, opening new frontiers in CNT research.

Looking ahead, another important direction will be the development of integrated AI systems that combine growth optimization, structural characterization, and property measurements into a single closed-loop framework. By leveraging ML techniques, these systems could continuously optimize the entire CNT research process, significantly accelerating discovery and innovation. As human researchers increasingly focus on designing AI systems and making critical scientific decisions, this integrated AI-driven approach is expected to drive revolutionary breakthroughs in CNT growth mechanisms, the discovery of novel structures, and the development of unprecedented applications.

**Author Contributions:** Conceptualization, G.C. and D.-M.T.; investigation, G.C. and D.-M.T.; writing—original draft preparation, G.C. and D.-M.T.; writing—review and editing, G.C. and D.-M.T.; visualization, G.C.; funding acquisition, G.C. and D.-M.T. All authors have read and agreed to the published version of the manuscript.

**Funding:** This work was supported by JSPS KAKENHI Grant Number JP23K04552. D.-M.T. discloses support from JSPS Kakenhi (grants JP25820336, JP20K05281, JP23H01796), JST-FOREST Program (grant JPMJFR223T, Japan), WPI-MANA 'Challenging Research Program (CRP)', NIMS 'Support system for curiosity-driven research', and "Advanced Research Infrastructure for Materials and Nanotechnology in Japan (ARIM)" of the Ministry of Education, Culture, Sports, Science and Technology (MEXT), proposal Number JPMXP1224NM5238.

**Data Availability Statement:** Not applicable.

**Conflicts of Interest:** The authors declare no conflicts of interest.

## References

1. Mjolsness, E.; DeCoste, D. Machine learning for science: State of the art and future prospects. *Science* **2001**, *293*, 2051–2055. [CrossRef] [PubMed]
2. Ramprasad, R.; Batra, R.; Pilania, G.; Mannodi-Kanakkithodi, A.; Kim, C. Machine learning in materials informatics: Recent applications and prospects. *NPJ Comput. Mater.* **2017**, *3*, 54. [CrossRef]
3. Jose, R.; Ramakrishna, S. Materials 4.0: Materials big data enabled materials discovery. *Appl. Mater. Today* **2018**, *10*, 127–132. [CrossRef]
4. Jordan, M.I.; Mitchell, T.M. Machine learning: Trends, perspectives, and prospects. *Science* **2015**, *349*, 255–260. [CrossRef]
5. Butler, K.T.; Davies, D.W.; Cartwright, H.; Isayev, O.; Walsh, A. Machine learning for molecular and materials science. *Nature* **2018**, *559*, 547–555. [CrossRef]
6. Huang, G.N.; Guo, Y.N.; Chen, Y.; Nie, Z.W. Application of Machine Learning in Material Synthesis and Property Prediction. *Materials* **2023**, *16*, 5977. [CrossRef]
7. Xu, P.C.; Ji, X.B.; Li, M.J.; Lu, W.C. Small data machine learning in materials science. *NPJ Comput. Mater.* **2023**, *9*, 42. [CrossRef]
8. Wei, J.; Chu, X.; Sun, X.-Y.; Xu, K.; Deng, H.-X.; Chen, J.; Wei, Z.; Lei, M. Machine learning in materials science. *InfoMat* **2019**, *1*, 338–358. [CrossRef]
9. Choi, R.Y.; Coyner, A.S.; Kalpathy-Cramer, J.; Chiang, M.F.; Campbell, J.P. Introduction to Machine Learning, Neural Networks, and Deep Learning. *Transl. Vis. Sci. Technol.* **2020**, *9*, 14. [CrossRef]
10. Sarker, I.H. Machine Learning: Algorithms, Real-World Applications and Research Directions. *SN Comput. Sci.* **2021**, *2*, 160. [CrossRef]
11. Rickman, J.M.; Lookman, T.; Kalinin, S.V. Materials informatics: From the atomic-level to the continuum. *Acta Mater.* **2019**, *168*, 473–510. [CrossRef]
12. Raccuglia, P.; Elbert, K.C.; Adler, P.D.F.; Falk, C.; Wenny, M.B.; Mollo, A.; Zeller, M.; Friedler, S.A.; Schrier, J.; Norquist, A.J. Machine-learning-assisted materials discovery using failed experiments. *Nature* **2016**, *533*, 73–76. [CrossRef]
13. Tshitoyan, V.; Dagdelen, J.; Weston, L.; Dunn, A.; Rong, Z.Q.; Kononova, O.; Persson, K.A.; Ceder, G.; Jain, A. Unsupervised word embeddings capture latent knowledge from materials science literature. *Nature* **2019**, *571*, 95–98. [CrossRef]
14. Rao, Z.Y.; Tung, P.Y.; Xie, R.W.; Wei, Y.; Zhang, H.B.; Ferrari, A.; Klaver, T.P.C.; Körmann, F.; Sukumar, P.T.; da Silva, A.K.; et al. Machine learning-enabled high-entropy alloy discovery. *Science* **2022**, *378*, 78–84. [CrossRef] [PubMed]
15. Rao, R.; Carpena-Nunez, J.; Nikolaev, P.; Susner, M.A.; Reyes, K.G.; Maruyama, B. Advanced machine learning decision policies for diameter control of carbon nanotubes. *NPJ Comput. Mater.* **2021**, *7*, 157. [CrossRef]
16. Lin, D.; Muroga, S.; Kimura, H.; Jintoku, H.; Tsuji, T.; Hata, K.; Chen, G.H.; Futaba, D.N. Addressing the Trade-Off between Crystallinity and Yield in Single-Walled Carbon Nanotube Forest Synthesis Using Machine Learning. *ACS Nano* **2023**, *17*, 22821–22829. [CrossRef]
17. Burnett, T.L.; Withers, P.J. Completing the picture through correlative characterization. *Nat. Mater.* **2019**, *18*, 1041–1049. [CrossRef] [PubMed]
18. Pagan, D.C.; Phan, T.Q.; Weaver, J.S.; Benson, A.R.; Beaudoin, A.J. Unsupervised learning of dislocation motion. *Acta Mater.* **2019**, *181*, 510–518. [CrossRef]
19. Qi, Y.P.; Hu, D.; Jiang, Y.C.; Wu, Z.P.; Zheng, M.; Chen, E.X.; Liang, Y.; Sadi, M.A.; Zhang, K.; Chen, Y.P. Recent Progresses in Machine Learning Assisted Raman Spectroscopy. *Adv. Opt. Mater.* **2023**, *11*, 2203104. [CrossRef]

20. Guo, S.X.; Popp, J.; Bocklitz, T. Chemometric analysis in Raman spectroscopy from experimental design to machine learning-based modeling. *Nat. Protoc.* **2021**, *16*, 5426–5459. [CrossRef]
21. Menevseoglu, A.; Entrenas, J.A.; Gunes, N.; Dogan, M.A.; Pérez-Marín, D. Machine learning-assisted near-infrared spectroscopy for rapid discrimination of apricot kernels in ground almond. *Food Control* **2024**, *159*, 110272. [CrossRef]
22. Windarsih, A.; Jatmiko, T.H.; Anggraeni, A.S.; Rahmawati, L. Machine learning-assisted FT-IR spectroscopy for identification of pork oil adulteration in tuna fish oil. *Vib. Spectrosc.* **2024**, *134*, 103715. [CrossRef]
23. Penfold, T.; Watson, L.; Middleton, C.; David, T.; Verma, S.; Pope, T.; Kaczmarek, J.; Rankine, C. Machine-learning strategies for the accurate and efficient analysis of x-ray spectroscopy. *Mach. Learn-Sci. Technol.* **2024**, *5*, 021001. [CrossRef]
24. Drera, G.; Kropf, C.M.; Sangaletti, L. Deep neural network for x-ray photoelectron spectroscopy data analysis. *Mach. Learn-Sci. Technol.* **2020**, *1*, 015008. [CrossRef]
25. de Curtò, J.; de Zarzà, I.; Roig, G.; Calafate, C.T. Large Language Model-Informed X-ray Photoelectron Spectroscopy Data Analysis. *Food Control* **2024**, *5*, 181–201. [CrossRef]
26. Pielsticker, L.; Nicholls, R.L.; DeBeer, S.; Greiner, M. Convolutional neural network framework for the automated analysis of transition metal X-ray photoelectron spectra. *Anal. Chim. Acta* **2023**, *1271*, 341433. [CrossRef] [PubMed]
27. Dee, N.T.; Schneider, M.; Zakharov, D.N.; Kidambi, P.R.; Hart, A. Automated processing of environmental transmission electron microscopy images for quantification of thin film dewetting and carbon nanotube nucleation dynamics. *Carbon* **2022**, *192*, 249–258. [CrossRef]
28. Lin, P.A.; Gomez-Ballesteros, J.L.; Burgos, J.C.; Balbuena, P.B.; Natarajan, B.; Sharma, R. Direct evidence of atomic-scale structural fluctuations in catalyst nanoparticles. *J. Catal.* **2017**, *349*, 149–155. [CrossRef]
29. Hajilounezhad, T.; Bao, R.; Palaniappan, K.; Bunyak, F.; Calyam, P.; Maschmann, M.R. Predicting carbon nanotube forest attributes and mechanical properties using simulated images and deep learning. *NPJ Comput. Mater.* **2021**, *7*, 134. [CrossRef]
30. Cao, B.; Adutwum, L.A.; Oliynyk, A.O.; Luber, E.J.; Olsen, B.C.; Mar, A.; Buriak, J.M. How To Optimize Materials and Devices via Design of Experiments and Machine Learning: Demonstration Using Organic Photovoltaics. *ACS Nano* **2018**, *12*, 7434–7444. [CrossRef]
31. Tang, D.-M.; Erohin, S.V.; Kvashnin, D.G.; Demin, V.A.; Cretu, O.; Jiang, S.; Zhang, L.; Hou, P.-X.; Chen, G.H.; Futaba, D.N.; et al. Semiconductor nanochannels in metallic carbon nanotubes by thermomechanical chirality alteration. *Science* **2021**, *374*, 1616–1620. [CrossRef] [PubMed]
32. LeMieux, M.C.; Roberts, M.; Barman, S.; Jin, Y.W.; Kim, J.M.; Bao, Z.N. Self-sorted, aligned nanotube networks for thin-film transistors. *Science* **2008**, *321*, 101–104. [CrossRef]
33. Feng, C.; Liu, K.; Wu, J.S.; Liu, L.; Cheng, J.S.; Zhang, Y.Y.; Sun, Y.H.; Li, Q.Q.; Fan, S.S.; Jiang, K.L. Flexible, Stretchable, Transparent Conducting Films Made from Superaligned Carbon Nanotubes. *Adv. Funct. Mater.* **2010**, *20*, 885–891. [CrossRef]
34. Choi, W.B.; Jin, Y.W.; Kim, H.Y.; Lee, S.J.; Yun, M.J.; Kang, J.H.; Choi, Y.S.; Park, N.S.; Lee, N.S.; Kim, J.M. Electrophoresis deposition of carbon nanotubes for triode-type field emission display. *Appl. Phys. Lett.* **2001**, *78*, 1547–1549. [CrossRef]
35. Chen, G.H.; Shin, D.H.; Iwasaki, T.; Kawarada, H.; Lee, C.J. Enhanced field emission properties of vertically aligned double-walled carbon nanotube arrays. *Nanotechnology* **2008**, *19*, 415703. [CrossRef]
36. Izadi-Najafabadi, A.; Yamada, T.; Futaba, D.N.; Yudasaka, M.; Takagi, H.; Hatori, H.; Iijima, S.; Hata, K. High-Power Supercapacitor Electrodes from Single-Walled Carbon Nanohorn/Nanotube Composite. *ACS Nano* **2011**, *5*, 811–819. [CrossRef]
37. Yalovega, G.E.; Brzhezinskaya, M.; Dmitriev, V.O.; Shmatko, V.A.; Ershov, I.V.; Ulyankina, A.A.; Chernysheva, D.V.; Smirnova, N.V. Interfacial Interaction in $MeO_x$/MWNTs (Me-Cu, Ni) Nanostructures as Efficient Electrode Materials for High-Performance Supercapacitors. *Nanomaterials* **2024**, *14*, 947. [CrossRef] [PubMed]
38. Züttel, A.; Sudan, P.; Mauron, P.; Kiyobayashi, T.; Emmenegger, C.; Schlapbach, L. Hydrogen storage in carbon nanostructures. *Int. J. Hydrog. Energy* **2002**, *27*, 203–212. [CrossRef]
39. Brzhezinskaya, M.; Belenkov, E.A.; Greshnyakov, V.A.; Yalovega, G.E.; Bashkin, I.O. New aspects in the study of carbon-hydrogen interaction in hydrogenated carbon nanotubes for energy storage applications. *J. Alloys Compd.* **2019**, *792*, 713–720. [CrossRef]
40. Landi, B.J.; Ganter, M.J.; Cress, C.D.; DiLeo, R.A.; Raffaelle, R.P. Carbon nanotubes for lithium ion batteries. *Energy Environ. Sci.* **2009**, *2*, 638–654. [CrossRef]
41. Lahiri, I.; Oh, S.W.; Hwang, J.Y.; Cho, S.; Sun, Y.K.; Banerjee, R.; Choi, W. High Capacity and Excellent Stability of Lithium Ion Battery Anode Using Interface-Controlled Binder-Free Multiwall Carbon Nanotubes Grown on Copper. *ACS Nano* **2010**, *4*, 3440–3446. [CrossRef] [PubMed]
42. Chen, G.H.; Dodson, B.; Hedges, D.M.; Steffensen, S.C.; Harb, J.N.; Puleo, C.; Galligan, C.; Ashe, J.; Vanfleet, R.R.; Davis, R.C. Fabrication of High Aspect Ratio Millimeter-Tall Free-Standing Carbon Nanotube-Based Microelectrode Arrays. *ACS Biomater. Sci. Eng.* **2018**, *4*, 1900–1907. [CrossRef] [PubMed]
43. Yang, N.; Chen, X.P.; Ren, T.L.; Zhang, P.; Yang, D.G. Carbon nanotube based biosensors. *Sensor Actuat. B-Chem.* **2015**, *207*, 690–715. [CrossRef]
44. Zhang, W.X.; Zhang, Z.Z.; Zhang, Y.G. The application of carbon nanotubes in target drug delivery systems for cancer therapies. *Nanoscale Res. Lett.* **2011**, *6*, 555. [CrossRef] [PubMed]
45. Chen, G.H.; Sundaram, R.; Sekiguchi, A.; Hata, K.; Futaba, D.N. Through-Silicon-Via Interposers with Cu-Level Electrical Conductivity and Si-Level Thermal Expansion Based on Carbon Nanotube-Cu Composites for Microelectronic Packaging Applications. *ACS Appl. Nano Mater.* **2021**, *4*, 869–876. [CrossRef]

46. Xu, M.; Du, F.; Ganguli, S.; Roy, A.; Dai, L.M. Carbon nanotube dry adhesives with temperature-enhanced adhesion over a large temperature range. *Nat. Commun.* **2016**, *7*, 13450. [CrossRef]
47. Qu, L.T.; Dai, L.M.; Stone, M.; Xia, Z.H.; Wang, Z.L. Carbon nanotube arrays with strong shear binding-on and easy normal lifting-off. *Science* **2008**, *322*, 238–242. [CrossRef]
48. Chen, G.H.; Futaba, D.N.; Kimura, H.; Sakurai, S.; Yumura, M.; Hata, K. Absence of an Ideal Single-Walled Carbon Nanotube Forest Structure for Thermal and Electrical Conductivities. *ACS Nano* **2013**, *7*, 10218–10224. [CrossRef]
49. Shibuya, A.; Chen, G.H.; Miyoshi, A.; Futaba, D.N. Improving the synthetic efficiency of single-wall carbon nanotube forests using a gas-analysis-designed mixed carbon feedstock. *Carbon* **2020**, *170*, 59–65. [CrossRef]
50. Chen, G.H.; Tsuji, T.; Yamada, M.; He, J.; Shimizu, Y.; Sakakita, H.; Hata, K.; Futaba, D.N.; Sakurai, S. Multi-step chemical vapor synthesis reactor based on a microplasma for structure-controlled synthesis of single-walled carbon nanotubes. *Chem. Eng. J.* **2022**, *444*, 136634. [CrossRef]
51. Sharma, P.; Pavelyev, V.; Kumar, S.; Mishra, P.; Islam, S.S.; Tripathi, N. Analysis on the synthesis of vertically aligned carbon nanotubes: Growth mechanism and techniques. *J. Mater. Sci.-Mater. Electron.* **2020**, *31*, 4399–4443. [CrossRef]
52. Chen, G.H.; Davis, R.C.; Futaba, D.N.; Sakurai, S.; Kobashi, K.; Yumura, M.; Hata, K. A sweet spot for highly efficient growth of vertically aligned single-walled carbon nanotube forests enabling their unique structures and properties. *Nanoscale* **2016**, *8*, 162–171. [CrossRef] [PubMed]
53. Sugime, H.; Sato, T.; Nakagawa, R.; Hayashi, T.; Inoue, Y.; Noda, S. Ultra-long carbon nanotube forest via in situ supplements of iron and aluminum vapor sources. *Carbon* **2021**, *172*, 772–780. [CrossRef]
54. Zhong, G.F.; Iwasaki, T.; Robertson, J.; Kawarada, H. Growth Kinetics of 0.5 cm Vertically Aligned Single-Walled Carbon Nanotubes. *J. Phys. Chem. B* **2007**, *111*, 1907–1910. [CrossRef]
55. Lee, S.H.; Park, J.; Park, J.H.; Lee, D.M.; Lee, A.; Moon, S.Y.; Lee, S.Y.; Jeong, H.S.; Kim, S.M. Deep-injection floating-catalyst chemical vapor deposition to continuously synthesize carbon nanotubes with high aspect ratio and high crystallinity. *Carbon* **2021**, *173*, 901–909. [CrossRef]
56. Kimura, H.; Futaba, D.N.; Yumura, M.; Hata, K. Mutual Exclusivity in the Synthesis of High Crystallinity and High Yield Single-Walled Carbon Nanotubes. *J. Am. Chem. Soc.* **2012**, *134*, 9219–9224. [CrossRef]
57. Chen, G.H.; Seki, Y.; Kimura, H.; Sakurai, S.; Yumura, M.; Hata, K.; Futaba, D.N. Diameter control of single-walled carbon nanotube forests from 1.3-3.0 nm by arc plasma deposition. *Sci. Rep.* **2014**, *4*, 3804. [CrossRef] [PubMed]
58. Tamiya, S.; Sato, T.; Kushida, M. Diameter control of vertically aligned carbon nanotubes using CoFe2O4 nanoparticle Langmuir-Blodgett films. *Jpn. J. Appl. Phys.* **2018**, *57*, 03EG15. [CrossRef]
59. Zhao, B.; Futaba, D.N.; Yasuda, S.; Akoshima, M.; Yamada, T.; Hata, K. Exploring Advantages of Diverse Carbon Nanotube Forests With Tailored Structures Synthesized by Supergrowth from Engineered Catalysts. *ACS Nano* **2009**, *3*, 108–114. [CrossRef]
60. Chen, G.H.; Futaba, D.N.; Sakurai, S.; Yumura, M.; Hata, K. Interplay of wall number and diameter on the electrical conductivity of carbon nanotube thin films. *Carbon* **2014**, *67*, 318–325. [CrossRef]
61. Xie, R.S.; Zhong, G.F.; Zhang, C.; Chen, B.A.; Esconjauregui, C.S.; Robertson, J. Diameter and wall number control of carbon nanotubes by chemical vapor deposition. *J. Appl. Phys.* **2013**, *114*, 244302. [CrossRef]
62. Hiraoka, T.; Izadi-Najafabadi, A.; Yamada, T.; Futaba, D.N.; Yasuda, S.; Tanaike, O.; Hatori, H.; Yumura, M.; Iijima, S.; Hata, K. Compact and Light Supercapacitor Electrodes from a Surface-Only Solid by Opened Carbon Nanotubes with 2200 $m^2$ $g^{-1}$ Surface Area. *Adv. Funct. Mater.* **2010**, *20*, 422–428. [CrossRef]
63. Futaba, D.N.; Goto, J.; Yamada, T.; Yasuda, S.; Yumura, M.; Hata, K. Outer-specific surface area as a gauge for absolute purity of single-walled carbon nanotube forests. *Carbon* **2010**, *48*, 4542–4546. [CrossRef]
64. Xu, M.; Futaba, D.N.; Yumura, M.; Hata, K. Alignment Control of Carbon Nanotube Forest from Random to Nearly Perfectly Aligned by Utilizing the Crowding Effect. *ACS Nano* **2012**, *6*, 5837–5844. [CrossRef]
65. Eres, G.; Rouleau, C.M.; Puretzky, A.A.; Geohegan, D.B.; Wang, H. Cooperative Behavior in the Evolution of Alignment and Structure in Vertically Aligned Carbon-Nanotube Arrays Grown using Chemical Vapor Deposition. *Phys. Rev. Appl.* **2018**, *10*, 024010. [CrossRef]
66. Ci, L.; Vajtai, R.; Ajayan, P.M. Vertically aligned large-diameter double-walled carbon nanotube arrays having ultralow density. *J. Phys. Chem. C* **2007**, *111*, 9077–9080. [CrossRef]
67. Zhong, G.F.; Warner, J.H.; Fouquet, M.; Robertson, A.W.; Chen, B.A.; Robertson, J. Growth of Ultrahigh Density Single-Walled Carbon Nanotube Forests by Improved Catalyst Design. *ACS Nano* **2012**, *6*, 2893–2903. [CrossRef] [PubMed]
68. Tang, B.J.; Lu, Y.H.; Zhou, J.D.; Chouhan, T.; Wang, H.; Golani, P.; Xu, M.Z.; Xu, Q.; Guan, C.T.; Liu, Z. Machine learning-guided synthesis of advanced inorganic materials. *Mater. Today* **2020**, *41*, 72–80. [CrossRef]
69. Yuan, R.H.; Liu, Z.; Balachandran, P.V.; Xue, D.Q.; Zhou, Y.M.; Ding, X.D.; Sun, J.; Xue, D.Z.; Lookman, T. Accelerated Discovery of Large Electrostrains in $BaTiO_3$-Based Piezoelectrics Using Active Learning. *Adv. Mater.* **2018**, *30*, 1702884. [CrossRef]
70. Lu, S.H.; Zhou, Q.H.; Ouyang, Y.X.; Guo, Y.L.; Li, Q.; Wang, J.L. Accelerated discovery of stable lead-free hybrid organic-inorganic perovskites via machine learning. *Nat. Commun.* **2018**, *9*, 3405. [CrossRef]
71. Xue, D.Z.; Balachandran, P.V.; Hogden, J.; Theiler, J.; Xue, D.Q.; Lookman, T. Accelerated search for materials with targeted properties by adaptive design. *Nat. Commun.* **2016**, *7*, 11241. [CrossRef] [PubMed]
72. Kaushal, A.; Alexander, R.; Rao, P.T.; Prakash, J.; Dasgupta, K. Artificial neural network, Pareto optimization, and Taguchi analysis for the synthesis of single-walled carbon nanotubes. *Carbon Trends* **2021**, *2*, 100016. [CrossRef]

73. Shields, B.J.; Stevens, J.; Li, J.; Parasram, M.; Damani, F.; Alvarado, J.I.M.; Janey, J.M.; Adams, R.P.; Doyle, A.G. Bayesian reaction optimization as a tool for chemical synthesis. *Nature* **2021**, *590*, 89–96. [CrossRef]
74. Chu, D.M.; Ji, Z.C.; Zhang, X.J.; Zhao, X.Y.; He, Y.; Bai, W.J. Machine learning for the regulation strategy and mechanism of the integrated growth of carbon nanotube arrays. *New J. Chem.* **2023**, *47*, 21883–21896. [CrossRef]
75. Bulmer, J.S.; Sloan, A.W.N.; Glerum, M.; Carpena-Nunez, J.; Waelder, R.; Humes, J.; Boies, A.M.; Pasquali, M.; Rao, R.; Maruyama, B. Forecasting carbon nanotube diameter in floating catalyst chemical vapor deposition. *Carbon* **2023**, *201*, 719–733. [CrossRef]
76. Ji, Z.H.; Zhang, L.L.; Tang, D.M.; Chen, C.M.; Nordling, T.E.M.; Zhang, Z.D.; Ren, C.L.; Da, B.; Li, X.; Guo, S.Y.; et al. High-throughput screening and machine learning for the efficient growth of high-quality single-wall carbon nanotubes. *Nano Res.* **2021**, *14*, 4610–4615. [CrossRef]
77. Krasnikov, D.V.; Khabushev, E.M.; Gaev, A.; Bogdanova, A.R.; Iakovlev, V.Y.; Lantsberg, A.; Kallio, T.; Nasibulin, A.G. Machine learning methods for aerosol synthesis of single-walled carbon nanotubes. *Carbon* **2023**, *202*, 76–82. [CrossRef]
78. Shin, S.; Song, H.; Shin, Y.S.; Lee, J.; Seo, T.H. Bayesian Optimization of Wet-Impregnated Co-Mo/$Al_2O_3$ Catalyst for Maximizing the Yield of Carbon Nanotube Synthesis. *Nanomaterials* **2024**, *14*, 75. [CrossRef]
79. Ding, R.; Chen, J.; Chen, Y.; Liu, J.; Bando, Y.; Wang, X. Unlocking the potential: Machine learning applications in electrocatalyst design for electrochemical hydrogen energy transformation. *Chem. Soc. Rev.* **2024**. [CrossRef]
80. Alwosheel, A.; van Cranenburgh, S.; Chorus, C.G. Is your dataset big enough? Sample size requirements when using artificial neural networks for discrete choice analysis. *J. Choice Model.* **2018**, *28*, 167–182. [CrossRef]
81. Nikolaev, P.; Hooper, D.; Perea-Lopez, N.; Terrones, M.; Maruyama, B. Discovery of Wall-Selective Carbon Nanotube Growth Conditions via Automated Experimentation. *ACS Nano* **2014**, *8*, 10214–10222. [CrossRef] [PubMed]
82. Nikolaev, P.; Hooper, D.; Webber, F.; Rao, R.; Decker, K.; Krein, M.; Poleski, J.; Barto, R.; Maruyama, B. Autonomy in materials research: A case study in carbon nanotube growth. *NPJ Comput. Mater.* **2016**, *2*, 16031. [CrossRef]
83. Chang, J.; Nikolaev, P.; Carpena-Nunez, J.; Rao, R.; Decker, K.; Islam, A.E.; Kim, J.; Pitt, M.A.; Myung, J.I.; Maruyama, B. Efficient Closed-loop Maximization of Carbon Nanotube Growth Rate using Bayesian Optimization. *Sci. Rep.* **2020**, *10*, 9040. [CrossRef]
84. Li, Y.; Wang, S.; Lv, Z.; Wang, Z.; Zhao, Y.; Xie, Y.; Xu, Y.; Qian, L.; Yang, Y.; Zhao, Z. Transforming the Synthesis of Carbon Nanotubes with Machine Learning Models and Automation. *arXiv* **2024**, arXiv:2404.01006. [CrossRef]
85. Iijima, S. Helical Microtubules of Graphitic Carbon. *Nature* **1991**, *354*, 56–58. [CrossRef]
86. Qin, L.C. Electron diffraction from carbon nanotubes. *Rep. Prog. Phys.* **2006**, *69*, 2761–2821. [CrossRef]
87. Jorio, A.; Saito, R.; Hafner, J.H.; Lieber, C.M.; Hunter, M.; McClure, T.; Dresselhaus, G.; Dresselhaus, M.S. Structural (n, m) determination of isolated single-wall carbon nanotubes by resonant Raman scattering. *Phys. Rev. Lett.* **2001**, *86*, 1118–1121. [CrossRef]
88. Dresselhaus, M.S.; Dresselhaus, G.; Saito, R.; Jorio, A. Raman spectroscopy of carbon nanotubes. *Phys. Rep.* **2005**, *409*, 47–99. [CrossRef]
89. Belianinov, A.; Vasudevan, R.; Strelcov, E.; Steed, C.; Yang, S.M.; Tselev, A.; Jesse, S.; Biegalski, M.; Shipman, G.; Symons, C.; et al. Big data and deep data in scanning and electron microscopies: Deriving functionality from multidimensional data sets. *Adv. Struct. Chem. Imaging* **2015**, *1*, 6. [CrossRef]
90. Aguiar, J.A.; Gong, M.L.; Unocic, R.R.; Tasdizen, T.; Miller, B.D. Decoding crystallography from high-resolution electron imaging and diffraction datasets with deep learning. *Sci. Adv.* **2019**, *5*, eaaw1949. [CrossRef]
91. Cui, A.Y.; Jiang, K.; Jiang, M.H.; Shang, L.Y.; Zhu, L.Q.; Hu, Z.G.; Xu, G.S.; Chu, J.H. Decoding Phases of Matter by Machine-Learning Raman Spectroscopy. *Phys. Rev. Appl.* **2019**, *12*, 054049. [CrossRef]
92. Trejo, O.; Dadlani, A.L.; De La Paz, F.; Acharya, S.; Kravec, R.; Nordlund, D.; Sarangi, R.; Prinz, F.B.; Torgersen, J.; Dasgupta, N.P. Elucidating the Evolving Atomic Structure in Atomic Layer Deposition Reactions with in Situ XANES and Machine Learning. *Chem. Mater.* **2019**, *31*, 8937–8947. [CrossRef]
93. Kaufmann, K.; Zhu, C.Y.; Rosengarten, A.S.; Maryanovsky, D.; Harrington, T.J.; Marin, E.; Vecchio, K.S. Crystal symmetry determination in electron diffraction using machine learning. *Science* **2020**, *367*, 564–568. [CrossRef] [PubMed]
94. Spurgeon, S.R.; Ophus, C.; Jones, L.; Petford-Long, A.; Kalinin, S.V.; Olszta, M.J.; Dunin-Borkowski, R.E.; Salmon, N.; Hattar, K.; Yang, W.C.D.; et al. Towards data-driven next-generation transmission electron microscopy. *Nat. Mater.* **2021**, *20*, 274–279. [CrossRef] [PubMed]
95. Uchida, T.; Tazawa, M.; Sakai, H.; Yamazaki, A.; Kobayashi, Y. Radial breathing modes of single-walled carbon nanotubes in resonance Raman spectra at high temperature and their chiral index assignment. *Appl. Surf. Sci.* **2008**, *254*, 7591–7595. [CrossRef]
96. Zhang, J.; Perrin, M.L.; Barba, L.; Overbeck, J.; Jung, S.; Grassy, B.; Agal, A.; Muff, R.; Brönnimann, R.; Haluska, M.; et al. High-speed identification of suspended carbon nanotubes using Raman spectroscopy and deep learning. *Microsyst. Nanoeng.* **2022**, *8*, 19. [CrossRef]
97. Kajendirarajah, U.; Avilés, M.O.; Lagugné-Labarthet, F. Deciphering tip-enhanced Raman imaging of carbon nanotubes with deep learning neural networks. *Phys. Chem. Chem. Phys.* **2020**, *22*, 17857–17866. [CrossRef]
98. Luo, Q.X.; Holm, E.A.; Wang, C. A transfer learning approach for improved classification of carbon nanomaterials from TEM images. *Nanoscale Adv.* **2021**, *3*, 206–213. [CrossRef]
99. Kalinin, S.V.; Ziatdinov, M.; Spurgeon, S.R.; Ophus, C.; Stach, E.A.; Susi, T.; Agar, J.; Randall, J. Deep learning for electron and scanning probe microscopy: From materials design to atomic fabrication. *MRS Bull.* **2022**, *47*, 931–939. [CrossRef]

100. Govind, K.; Oliveros, D.; Dlouhy, A.; Legros, M.; Sandfeld, S. Deep learning of crystalline defects from TEM images: A solution for the problem of 'never enough training data'. *Mach. Learn-Sci. Technol.* **2024**, *5*, 015006. [CrossRef]
101. Förster, G.D.; Castan, A.; Loiseau, A.; Nelayah, J.; Alloyeau, D.; Fossard, F.; Bichara, C.; Amara, H. A deep learning approach for determining the chiral indices of carbon nanotubes from high-resolution transmission electron microscopy images. *Carbon* **2020**, *169*, 465–474. [CrossRef]
102. Lee, B.; Yoon, S.; Lee, J.W.; Kim, Y.; Chang, J.; Yun, J.; Ro, J.C.; Lee, J.S.; Lee, J.H. Statistical Characterization of the Morphologies of Nanoparticles through Machine Learning Based Electron Microscopy Image Analysis. *ACS Nano* **2020**, *14*, 17125–17133. [CrossRef] [PubMed]
103. Chen, D.; Gao, J.; He, D.; He, J.; Li, Y.; Zhang, M.; Li, W.; Chen, X.; He, X.; Fu, T. Plasmonic Bridge Sensor Enabled by Carbon Nanotubes and Au-Ag Nano-Rambutan for Multifunctional Detection of Biomechanics and Bio/Chemical Molecules. *ACS Appl. Mater. Interfaces* **2023**, *15*, 8783–8793. [CrossRef]
104. Kelich, P.; Adams, J.; Jeong, S.; Navarro, N.; Landry, M.P.; Vukovic, L. Predicting Serotonin Detection with DNA-Carbon Nanotube Sensors across Multiple Spectral Wavelengths. *J. Chem. Inf. Model.* **2024**, *64*, 3992–4001. [CrossRef]
105. Tadokoro, Y.; Funayama, K.; Kawano, K.; Miura, A.; Hirotani, J.; Ohno, Y.; Tanaka, H. Artificial-intelligence-assisted mass fabrication of nanocantilevers from randomly positioned single carbon nanotubes. *Microsyst. Nanoeng.* **2023**, *9*, 32. [CrossRef]
106. Aliyana, A.K.; Kumar, S.K.N.; Marimuthu, P.; Baburaj, A.; Adetunji, M.; Frederick, T.; Sekhar, P.; Fernandez, R.E. Machine learning-assisted ammonium detection using zinc oxide/multi-walled carbon nanotube composite based impedance sensors. *Sci. Rep.* **2021**, *11*, 24321. [CrossRef]
107. Bian, L.; Wang, Z.H.; White, D.L.; Star, A. Machine learning-assisted calibration of $Hg^{2+}$ sensors based on carbon nanotube field-effect transistors. *Biosens. Bioelectron.* **2021**, *180*, 113085. [CrossRef] [PubMed]
108. Fan, G.; Low, K.L. Physics-Integrated Machine Learning for Efficient Design and Optimization of a Nanoscale Carbon Nanotube Field-Effect Transistor. *ECS J. Solid. State Sci. Technol.* **2023**, *12*, 091005. [CrossRef]
109. Kelich, P.; Jeong, S.; Navarro, N.; Adams, J.; Sun, X.Q.; Zhao, H.H.; Landry, M.P.; Vukovic, L. Discovery of DNA-Carbon Nanotube Sensors for Serotonin with Machine Learning and Near-infrared Fluorescence Spectroscopy. *ACS Nano* **2022**, *16*, 736–745. [CrossRef]
110. Yu, T.Y.; Mo, X.D.; Chen, M.J.; Yao, C.F. Machine-learning-assisted microstructure-property linkages of carbon nanotube-reinforced aluminum matrix nanocomposites produced by laser powder bed fusion. *Nanotechnol. Rev.* **2021**, *10*, 1410–1424. [CrossRef]
111. Ranaiefar, M.; Singh, M.; Halbig, M.C. Additively Manufactured Carbon-Reinforced ABS Honeycomb Composite Structures and Property Prediction by Machine Learning. *Molecules* **2024**, *29*, 2736. [CrossRef] [PubMed]
112. Jalal, M.; Moradi-Dastjerdi, R.; Bidram, M. Big data in nanocomposites: ONN approach and mesh-free method for functionally graded carbon nanotube-reinforced composites. *J. Comput. Des. Eng.* **2019**, *6*, 209–223. [CrossRef]
113. Huang, J.S.; Liew, J.X.; Liew, K.M. Data-driven machine learning approach for exploring and assessing mechanical properties of carbon nanotube-reinforced cement composites. *Compos. Struct.* **2021**, *267*, 113917. [CrossRef]
114. Bagherzadeh, F.; Shafighfard, T. Ensemble Machine Learning approach for evaluating the material characterization of carbon nanotube-reinforced cementitious composites. *Case Stud. Constr. Mat.* **2022**, *17*, e01537. [CrossRef]
115. Okasha, N.M.; Mirrashid, M.; Naderpour, H.; Ciftcioglu, A.O.; Meddage, D.P.P.; Ezami, N. Machine learning approach to predict the mechanical properties of cementitious materials containing carbon nanotubes. *Dev. Built Environ.* **2024**, *19*, 100494. [CrossRef]
116. Le, T.-T. Prediction of tensile strength of polymer carbon nanotube composites using practical machine learning method. *J. Compos. Mater.* **2021**, *55*, 787–811. [CrossRef]
117. Liu, B.K.; Vu-Bac, N.; Zhuang, X.Y.; Fu, X.L.; Rabczuk, T. Stochastic integrated machine learning based multiscale approach for the prediction of the thermal conductivity in carbon nanotube reinforced polymeric composites. *Compos. Sci. Technol.* **2022**, *224*, 109425. [CrossRef]
118. Liu, B.K.; Vu-Bac, N.; Zhuang, X.Y.; Fu, X.L.; Rabczuk, T. Stochastic full-range multiscale modeling of thermal conductivity of Polymeric carbon nanotubes composites: A machine learning approach. *Compos. Struct.* **2022**, *289*, 115393. [CrossRef]
119. Champa-Bujaico, E.; Díez-Pascual, A.M.; Garcia-Diaz, P.; Sessini, V.; Mosquera, M.E.G. Machine learning algorithms to optimize the properties of bio-based poly (butylene succinate-co- butylene adipate) nanocomposites with carbon nanotubes. *Ind. Crop Prod.* **2024**, *219*, 119018. [CrossRef]
120. Zarei, A.; Farazin, A. Machine Learning Insights into the Influence of Carbon Nanotube Dimensions on Nanocomposite Properties: A Comprehensive Exploration. *J. Comput. Appl. Mech.* **2024**, *55*, 462–472. [CrossRef]
121. Chen, G.H.; Dodson, B.; Johnson, F.; Hancu, I.; Fiveland, E.; Zhang, W.; Galligan, C.; Puleo, C.; Davis, R.C.; Ashe, J.; et al. Tissue-susceptibility matched carbon nanotube electrodes for magnetic resonance imaging. *J. Magn. Reson.* **2018**, *295*, 72–79. [CrossRef] [PubMed]
122. Ouassil, N.; Pinals, R.L.; Del Bonis-O'Donnell, J.T.; Wang, J.W.; Landry, M.P. Supervised learning model predicts protein adsorption to carbon nanotubes. *Sci. Adv.* **2022**, *8*, eabm0898. [CrossRef] [PubMed]
123. Lee, D.; Lee, J.; Kim, W.; Suh, Y.; Park, J.; Kim, S.; Kim, Y.; Kwon, S.; Jeong, S. Systematic Selection of High-Affinity ssDNA Sequences to Carbon Nanotubes. *Adv. Sci.* **2024**, *11*, 2308915. [CrossRef] [PubMed]
124. Lin, Z.W.; Yang, Y.; Jagota, A.; Zheng, M. Machine Learning-Guided Systematic Search of DNA Sequences for Sorting Carbon Nanotubes. *ACS Nano* **2022**, *16*, 4705–4713. [CrossRef]

125. Kotzabasaki, M.; Sotiropoulos, I.; Charitidis, C.; Sarimveis, H. Machine learning methods for multi-walled carbon nanotubes (MWCNT) genotoxicity prediction. *Nanoscale Adv.* **2021**, *3*, 3167–3176. [CrossRef]
126. Gong, X.; Renegar, N.; Levi, R.; Strano, M.S. Machine learning for the discovery of molecular recognition based on single-walled carbon nanotube corona-phases. *NPJ Comput. Mater.* **2022**, *8*, 135. [CrossRef]

MDPI AG
Grosspeteranlage 5
4052 Basel
Switzerland
Tel.: +41 61 683 77 34

*Nanomaterials* Editorial Office
E-mail: nanomaterials@mdpi.com
www.mdpi.com/journal/nanomaterials

www.ingramcontent.com/pod-product-compliance
Lightning Source LLC
Chambersburg PA
CBHW040835250926
51850CB00007BA/1321
* 9 7 8 3 7 2 5 8 7 7 0 7 2 *